NOVEL APPLICATIONS IN POLYMERS AND WASTE MANAGEMENT

NOVEL APPLICATIONS IN POLYMERS AND WASTE MANAGEMENT

Edited by

Badal Jageshwar Prasad Dewangan, PhD
Maheshkumar Narsingrao Yenkie, PhD

Apple Academic Press Inc.
3333 Mistwell Crescent
Oakville, ON L6L 0A2 Canada

Apple Academic Press Inc.
9 Spinnaker Way
Waretown, NJ 08758 USA

© 2018 by Apple Academic Press, Inc.
No claim to original U.S. Government works
Printed in the United States of America on acid-free paper
International Standard Book Number-13: 978-1-77188-475-4 (Hardcover)
International Standard Book Number-13: 978-1-315-36584-8 (eBook)

All rights reserved. No part of this work may be reprinted or reproduced or utilized in any form or by any electronic, mechanical or other means, now known or hereafter invented, including photocopying and recording, or in any information storage or retrieval system, without permission in writing from the publisher or its distributor, except in the case of brief excerpts or quotations for use in reviews or critical articles.

This book contains information obtained from authentic and highly regarded sources. Reprinted material is quoted with permission and sources are indicated. Copyright for individual articles remains with the authors as indicated. A wide variety of references are listed. Reasonable efforts have been made to publish reliable data and information, but the authors, editors, and the publisher cannot assume responsibility for the validity of all materials or the consequences of their use. The authors, editors, and the publisher have attempted to trace the copyright holders of all material reproduced in this publication and apologize to copyright holders if permission to publish in this form has not been obtained. If any copyright material has not been acknowledged, please write and let us know so we may rectify in any future reprint.

Trademark Notice: Registered trademark of products or corporate names are used only for explanation and identification without intent to infringe.

Library and Archives Canada Cataloguing in Publication

Novel applications in polymers and waste management / edited by Badal Jageshwar Prasad Dewangan, PhD, Maheshkumar Narsingrao Yenkie, PhD.

Includes bibliographical references and index.
Issued in print and electronic formats.
ISBN 978-1-77188-475-4 (hardcover).--ISBN 978-1-315-36584-8 (PDF)
1. Polymers. 2. Factory and trade waste. I. Dewangan, Badal Jageshwar Prasad, editor II. Yenkie, Maheshkumar Narsingrao, editor

TP156.P6N68 2017   668.9   C2017-906368-5   C2017-906369-3

Library of Congress Cataloging-in-Publication Data

Names: Dewangan, Badal Jageshwar Prasad, editor. | Yenkie, Maheshkumar Narsingrao, editor.
Title: Novel applications in polymers and waste management / editors: Badal Jageshwar Prasad Dewangan, Maheshkumar Narsingrao Yenkie.
Description: Toronto ; New Jersey : Apple Academic Press, 2018. | Includes bibliographical references and index.
Identifiers: LCCN 2017044764 (print) | LCCN 2017045813 (ebook) | ISBN 9781315365848 (eBook) | ISBN 9781771884754 (hardcover : alk. paper)
Subjects: | MESH: Polymers | Nanocomposites | Waste Management | Industrial Waste
Classification: LCC QP801.P64 (ebook) | LCC QP801.P64 (print) | NLM QT 37.5.P7 | DDC 572/.33--dc23
LC record available at https://lccn.loc.gov/2017044764

Apple Academic Press also publishes its books in a variety of electronic formats. Some content that appears in print may not be available in electronic format. For information about Apple Academic Press products, visit our website at **www.appleacademicpress.com** and the CRC Press website at **www.crcpress.com**

ABOUT THE EDITORS

Badal Jageshwar Prasad Dewangan, PhD

Badal Jageshwar Prasad Dewangan, PhD, is presently working as Head and Assistant Professor in Plastics and Polymer Technology Department at the Laxminarayan Institute of Technology, R.T.M. Nagpur University, Nagpur, India. He has almost five years of teaching experience, including PhD work, and while teaching, he has 10 years of research experience. He has participated and presented his work at several national and international conferences and has five national and international publications and one book chapter to his credit. As a convenor, he has organized a three-day national workshop and a two-day national conference funded by government agencies. He has also worked as a member of the organizing committees of national conferences and workshops. He is a member of several professional organizations, such as the Society of Plastics Engineers (SPE), the Indian Plastics Institute (IPI), the All India Plastics Manufactures Association (AIPMA), the Colour Society, and the Indian Institute of Chemical Engineers (IIChE). His areas of research interest are polymer blends, polymer nanocomposites, and synthesis of tailor-made functional copolymers via controlled living radical polymerization (ATRP, RAFT, and NMP) and their application as polymer/coating additives mainly surfactants (emulsifier, wetting, and dispersing agents), compatibilizers, and corrosion inhibitors.

Maheshkumar Narsingrao Yenkie, PhD

Maheshkumar Narsingrao Yenkie, PhD (yenkiemskm@gmail.com) is a PhD in chemistry and is presently working as Professor of Chemistry at the Laxminarayan Institute of Technology, R.T.M. Nagpur University, Nagpur. He is also the former director-in-charge of the Laxminarayan Institute of Technology and former registrar (Oct. 2010–Sept. 2012) and former Pro-Vice Chancellor (Sept. 2012–March 2014) of R.T.M. Nagpur University, Nagpur. He has 35 years of undergraduate and postgraduate teaching experience and is a former German Academic Exchange Fellowship holder, having worked at the Institut fuer Thermische Verfahrenstechnik, Technical University of Clausthal, Clausthal-Zellerfeld, F.R. Germany (1988–1991).

His areas of research interest are kinetics and equilibrium studies on adsorption of hazardous pollutants from synthetic and industrial wastewater onto powdered and granular activated carbon and synthetic adsorbents in batch and flow systems, air and water pollution monitoring and control, advanced oxidation processes using strong oxidizing agents like ozone and hydrogen peroxide catalyzed by UV radiations for their removal from hazardous wastes, biological treatment (activated sludge process) of industrial and domestic wastewater, landfill leachates, etc. in high performance loop reactors (compact reactor) and biofilm reactors for eliminating BOD, COD, and nutrients and fluoride removal from water and wastewater. He has coordinated several research and development projects as a principal investigator, which includes research work related to photocatalytic degradation of dyes in wastewater, detoxification of wastewater, and defluoridation of water.

He has guided 21 students for their PhD degrees to date and has more than 55 national and international publications to his credit. He has participated in several national and international conferences and has presented his work. He is also a guest faculty at the National Civil Defence College, Ministry of Home Affairs, Govt. of India, and at the Department of Linguistics and Foreign Languages for teaching the German language to certificate and diploma students. He is a life member of six professional organizations, namely the Indian Science Congress Association, Oil Technologist's Association, Indian Society of Analytical Scientists, Chemical Research Society of India (I.I.S. Bangalore), Nagpur University Teachers' Association, and Indian Society for Technical Education.

CONTENTS

List of Contributors ix

List of Abbreviations xiii

Preface xvii

PART I: NOVEL APPLICATION OF POLYMERS 1

1. **Novel Comparative Studies: Synthesis, Characterization of Copolymers and Their Composites for the Removal of Heavy Metals** 3
   K. A. Nandekar and W. B. Gurnule

2. **Electrical Transport Properties of Poly (Aniline-*co*-*N*-Phenylaniline) Copolymers** 29
   A. D. Borkar

3. **Surface-Altered Poly(Methyl Methacrylate) Resin for Antifungal Dentures** 41
   Abhay Narayane, Akshay Mohan, and Siddharth Meshram

4. **Polybenzimidazole as Membranes in Direct Methanol Fuel Cells** 55
   P. P. Kundu and S. Mondal

5. **Applications of Smart Polymers in Emerging Areas** 75
   U. V. Gaikwad, A. R. Chaudhari, and S. V. Gaikwad

PART II: POLYMER NANOCOMPOSITES AND THEIR APPLICATIONS 91

6. **Advancements in Polymer Nanotechnology** 93
   Satyendra Mishra and Dharmesh Hansora

7. **Synthesis and Characterization of SnO_2/Polyaniline and Al-doped SnO_2/Polyaniline Composite Nanofiber-based Sensors for Hydrogen Gas Sensing** 123
   Hemlata J. Sharma and Subhash B. Kondawar

8. **Nanocomposites for Food Packaging Applications** 137
   Badal Dewangan and Umesh Marathe

9. Nanotechnology in Wastewater Treatment: A Review 173

   Bais Madhuri, S. P. Singh, and R. D. Batra

10. A Review on Preparation of Conductive Paints with
    CNTs As Fillers 183

    Sahithi Ravuluri, Mansi Khandelwal, Harshit Bajpai, G. S. Bajad, and R. P. Vijayakumar

PART III: BIOPOLYMERS AND THEIR APPLICATIONS 205

11. Removing Heavy Metals From Industrial Wastewater Using
    Economically Modified Biopolymers and Hydrogel Adsorbents 207

    Vedant Manojkumar Danak and Yashawant P. Bhalerao

12. Development of Biopolymers for Detergent 223

    Dhakite Pravin, Burande Bharati, and Gogte Bhalchandra

13. The Therapeutic Role of the Components of *Aloe vera* in Activating
    the Factors that Induce Osteoarthritic Joint Remodeling 237

    Abhipriya Chatterjee and Patit Paban Kundu

PART IV: INDUSTRIAL WASTE MITIGATION 267

14. Adsorption of Reactive Dye 21 on Fly Ash and MnO_2-Coated
    Fly Ash Adsorbent: Batch and Continuous Studies 269

    Deepika Brijpuriya, Manoj Jamdarkar, Pratibha Agrawal, and Tapas Nandy

15. Turning Waste Into Zeolite 4A Resin and Delineation of
    Their Environmental Applications: A Review 291

    S. U. Meshram, B. R. Gawhane, and P. B. Suhagpure

16. Study of Eco-Friendly Additives for Wood–Plastics Composites:
    A Step Toward a Better Environment 309

    S. A. Puranik, Dinesh Desai, and Kintu Jain

17. Synthesis of Crude Oil by Catalytic Pyrolysis of Waste Plastics 321

    Siddharth Avnesh Mehta

PART V: MODIFICATION OF INORGANIC MATERIALS 331

18. Growth of $KNbO_3$ Crystals and Their Appearance 333

    Naresh M. Patil, Vivek B. Korde, and Sanjay H. Shamkuwar

Index 339

LIST OF CONTRIBUTORS

Pratibha Agrawal
Laxminarayan Institute of Technology, Nagpur, India

G. S. Bajad
Visvesvaraya National Institute of Technology, Nagpur 440010, Maharashtra, India

Harshit Bajpai
Visvesvaraya National Institute of Technology, Nagpur 440010, Maharashtra, India

R. D. Batra

Gogte Bhalchandra
Department of Oil Technology, Laxminarayan Institute of Technology, R.T.M. Nagpur University, Nagpur, India

Yashawant P. Bhalerao
Department of Chemical Engineering, Shroff S. R. Rotary Institute of Technology (SRICT), Ankleshwar 393002, Gujarat, India

Burande Bharati
Department of Applied Chemistry, Indira Gandhi Priyadarshini College of Engineering, Nagpur, India

A. D. Borkar
Department of Chemistry, Nabira Mahavidyalaya, Katol District, Nagpur 441302, India. E-mail: arun.borkar@rediffmail.com

Deepika Brijpuriya
NEERI, Nehru Marg, Nagpur, India. E-mail: dipikagupta.nagpur@gmail.com

Abhipriya Chatterjee
Advanced Polymer Laboratory, Department of Polymer Science and Technology, University of Calcutta, 92 APC Road, Kolkata 700009, India

A. R. Chaudhari
Department of Chemistry, Priyadarshini Bhagwati College of Engineering, Nagpur, India

Vedant Manojkumar Danak
Department of Chemical Engineering, Shroff S. R. Rotary Institute of Technology (SRICT), Ankleshwar 393002, Gujarat, India

Dinesh Desai
Plastics Technology Department, Lalbhai Dalpatbhai College of Engineering, Ahmedabad, Gujarat, India

Badal Dewangan
Department of Polymer Technology, Laxminarayan Institute of Technology, Nagpur 440001, Maharashtra, India

S. V. Gaikwad
Department of Chemistry, Dr. Babasaheb Ambedkar College of Engineering & Research, Nagpur, India

U. V. Gaikwad
Department of Physics, Priyadarshini Bhagwati College of Engineering, Nagpur, India. E-mail: umagaikwad353@gmail.com

B. R. Gawhane
Research Assistant, Department of Oil, Fats and Surfactants Technology, Laxminarayan Institute of Technology, RTM Nagpur University, Amravati Road, Nagpur-440033, India

W. B. Gurnule
Department of Chemistry, Kamla Nehru Mahavidyalaya, Sakkardara Square, Nagpur 440024, India. E-mail: wbgurnule@yahoo.co.in

Dharmesh Hansora
University Institute of Chemical Technology, North Maharashtra University, Jalgaon 425001, Maharashtra, India. E-mail: profsm@rediffmail.com

Kintu Jain
Plastics Technology Department, Lalbhai Dalpatbhai College of Engineering, Ahmedabad, Gujarat, India. E-mail: dineshdesai2002@gmail.com

Manoj Jamdarkar
Shri M. Mohota College of Science, Nagpur, India

Mansi Khandelwal
Visvesvaraya National Institute of Technology, Nagpur 440010, Maharashtra, India

Subhash B. Kondawar
Department of Physics, Polymer Nanotech Laboratory, Rashtrasant Tukadoji Maharaj Nagpur University, Nagpur 440033, India

Vivek B. Korde
Department of Applied Physics, Laxminarayan Institute of Technology, RTM Nagpur, Nagpur 440033, India

Patit Paban Kundu
Advanced Polymer Laboratory, Department of Polymer Science and Technology, University of Calcutta, 92 APC Road, Kolkata 700009, India. E-mail: ppk923@yahoo.com

Bais Madhuri
E-mail: madhuribais8@gmail.com

Jekin Bharat Makwana
Department of Plastics and Polymer Engineering, Maharashtra Institute of Technology, Aurangabad, Maharashtra, India

Umesh Marathe
Center for Polymer Science and Engineering, Indian Institute of Technology, New Delhi 110016, India

Siddharth Avnesh Mehta
Department of Plastics and Polymer Engineering, Maharashtra Institute of Technology, Aurangabad, Maharashtra, India

List of Contributors

Siddharth Meshram
Assistant Professor, Department of Applied Chemistry, Laxminarayan Institute of Technology, RTM Nagpur University, Nagpur 440033, India. E-mail: sidmesh2@gmail.com

Satyendra Mishra
University Institute of Chemical Technology, North Maharashtra University, Jalgaon 425001, Maharashtra, India

Akshay Mohan
Research Assistant, Department of Chemical Engineering, Laxminarayan Institute of Technology, RTM Nagpur University, Nagpur 440033, India

S. Mondal
Advanced Polymer Laboratory, Department of Polymer Science and Technology, University of Calcutta, 92 APC Road, Kolkata 700009, West Bengal, India

K. A. Nandekar
Department of Chemistry, Priyadarshani College of Engineering and Technology, Nagpur 440019, India

Tapas Nandy
NEERI, Nehru Marg, Nagpur, India

Abhay Narayane
Post Graduate Research Student, Department of Prosthodontics, SDKS Dental College and Hospital, Nagpur, India

Naresh M. Patil
Department of Applied Physics, Laxminarayan Institute of Technology, RTM Nagpur, Nagpur 440033, India

Dhakite Pravin
Department of Chemistry, S.N. Mor Arts, Commerce and Smt. G.D. Saraf Science College, Tumsar, Bhandara, India. E-mail: pravinchemkb@gmail.com

S. A. Puranik
Atmiya Institute of Technology & Science, Rajkot, Gujarat, India

Sahithi Ravuluri
Visvesvaraya National Institute of Technology, Nagpur 440010, Maharashtra, India

Sanjay H. Shamkuwar
Arts, Commerce & Science College, Kiran Nagar, Amravati 444 606, India

Hemlata J. Sharma
Department of Physics, Polymer Nanotech Laboratory, Rashtrasant Tukadoji Maharaj Nagpur University, Nagpur 440033, India

P. B. Suhagpure
Research Assistant, Department of Oil, Fats and Surfactants Technology, Laxminarayan Institute of Technology, RTM Nagpur University, Amravati Road, Nagpur 440033, India

R. P. Vijayakumar
Visvesvaraya National Institute of Technology, Nagpur 440010, Maharashtra, India

LIST OF ABBREVIATIONS

1D	one-dimensional
3D GBM	three-dimensional graphene-based macrostructures
BET	Brunauer–Emmett–Teller
BMP	bone morphogenetic protein
CFA	coal fly ash
CNTs	carbon nanotubes
CPCB	Central Pollution Control Board of India
CSA	camphor sulfonic acid
CVD	chemical vapor deposition
DMF	dimethyl formamide
DMFCs	direct methanol fuel cells
DMSO	dimethyl sulphoxide
EPDM	ethylene propylene diene monomer
ER	epoxy resin
EVA	ethylene-co-vinyl acetate
HLB	hydrophyle–lyphophyle balance
IL-1β	interleukin-1β
LCST	lower critical solution temperature
LDPE	low-density polyethylene
LLDPE	linear low density polyethylene
LOI	loss on ignition
MA	maleic anhydride
MAC	membrane attack protein
MBTS	2,2-dibenzothiazyl disulfide
MCL	maximum contaminant level
MFI	melt flow index
MMPs	matrix metalloproteinases
MWCNTs	multiwall carbon nanotubes
NaDDBS	dodecyl-benzene sodium sulfonate
NMP	n-methyl-2-pyrrolidone
nPS	polystyrene nanoparticles
OP-1	osteogenic protein-1
PAMAM	poly(amidoamine)

PANi	polyaniline
PANI-co-PNPANI	poly(aniline-co-N-phenylaniline)
PBIs	polybenzimidazoles
PCMs	phase change materials
PEMFCs	polymer electrolyte membrane fuel cells
PEMs	polymer electrolyte membranes
PEO	polyethylene oxide
PHAs	poly-hydroxyl alkanoates
PKNS	palm kernel nut shell
PMMA	polymethyl methacrylate
PNCs	polymer nanocomposites
PNIPAAm	poly(N-isopropylacrylamide)
PNPANI	poly(N-phenylaniline)
PP	polymer matrix
PPA	polyphosphoric acid
PPL	polypyrrolone
PPMA	methane sulfonic acid/P2O5
PS	polystyrene
PU	polyurathene rubber
PVC	poly-vinyl chloride
PVP	polyvinyl pyrrolidone
QACs	quaternary ammonium compounds
QAMS	quaternary ammonium silane-functionalized methacrylate
QASs	quaternary ammonium salts
RB 21	Reactive Blue 21
SBR	styrene–butadiene rubber
SEM	scanning electron microscopy
SMPs	shape memory polymers
SPs	smart polymers
SR	silicon rubber
SWCNTs	single-wall carbon nanotubes
TEM	transmittance electron microscopy
TGF-β	transforming growth factor β
TNF-α	tumor necrosis factor-α
TPKSP	treated palm kernel nut shell powder
TR-PBI	thermally rearranged polybenzimidazole
UPKSP	untreated palm kernel nut shell powder

US EPA	United States Environmental Protection Agency
UV–vis	UV–visible
VRH	variable range hopping
WPCs	wood–plastics composites
XRD	X-ray diffraction
XRF	X-ray fluorescence
ZnO	zinc oxide

PREFACE

Polymers have been playing a major role from the very beginning of the human civilization. Their development is considered as one of the important aspects in the overall economic growth of the country.

At the end of 19th century, the scarcity of some natural polymers led to the need for the development of the artificial polymer. Bakelite was the first commercialized polymer, and since then significant innovation in this field had started, and there is no turning back. From rocket science to artificial body parts, polymers have served society in every possible way and is simplifying lives.

The first section of the book, "Novel Application of Polymers," as the name suggests, deals with the various advancements in various attributes of technology with polymers playing the major role. The polymers that would be responsible for the progress in the field of energy, electronics, and medical sciences are discussed in this section. Polymer nanocomposites and nanomaterials are the major emerging fields in the materials sciences, promising a dominating future ahead. Composites are becoming more important because they can help to improve quality of life.

The second section of the book highlights this aspect of the macromolecules, while the third section emphasizes the concept of the biopolymers, their development, and applications.

In the discussion of any chemical industry, the major concern is the proper disposal and management of chemical wastes for the harmony and balance of our mother earth. The fourth section of this composition discusses industrial waste mitigation and extracting the potential of waste for other processes.

This book will provide important information to enthusiasts of this field and will encourage them to think about other permutations and possibilities in order to encourage more experimental and technical approaches to the uses of polymers and waste management.

PART I
Novel Application of Polymers

CHAPTER 1

NOVEL COMPARATIVE STUDIES: SYNTHESIS, CHARACTERIZATION OF COPOLYMERS AND THEIR COMPOSITES FOR THE REMOVAL OF HEAVY METALS

K. A. NANDEKAR[1] and W. B. GURNULE[2*]

[1]Department of Chemistry, Priyadarshani College of Engineering and Technology, Nagpur 440019, India

[2]Department of Chemistry, Kamla Nehru Mahavidyalaya, Sakkardara Square, Nagpur 440024, India

*Corresponding author. E-mail: wbgurnule@yahoo.co.in

CONTENTS

Abstract	4
1.1 Introduction	4
1.2 Experimental	6
1.3 Results and Discussion	8
1.4 Conclusion	25
Keywords	26
References	26

ABSTRACT

A novel comparative account for the ion-exchange and copolymer/activated charcoal composite was performed. The copolymer resin was synthesized involving *p*-hydroxybenzoic acid and semicarbazide with formaldehyde and the novel composite was prepared using copolymer and activated charcoal. The structure and properties of copolymer and copolymer/activated charcoal composite were observed by various characterizations such as elemental analysis, FTIR, NMR (^{13}C and ^{1}H), and SEM. The heavy metal ion removal by the copolymer and the composite was performed by batch separation technique for the selected divalent metal ions like Cu(II), Zn(II), Co(II), Pb(II), and Cd(II). The study was extended to various concentrations in different electrolyte, wide pH ranges, and at different rate. The selectivity of the order of removal of metal ion by the copolymer is Zn(II) > Cu(II) > Co(II) > Pb(II) > Cd(II) and by the composite is Pb(II) > Cd(II) > Cu(II) > Co(II) > Zn(II). The difference in the selectivity of order of metal ions may be due to the particle size, high porosity nature, large surface area, nature of the material, and the metal ions. Moreover, the ion-exchange results of the copolymer and its composite were compared with the commercially available resin.

1.1 INTRODUCTION

The removal of traces of heavy toxic metal ions present in industrial wastewater, domestic, and nuclear wastes has been given much attention in the last decades, because of their tendency to accumulate in living organisms and induces harmful effects to natural resources. Therefore, an attempt has been made to synthesize a novel chelating copolymer and to assess the ion-exchange characteristics of the copolymer. Copolymers were applied in various fields of research as ion-exchangers, high thermal resistance materials, and electrical appliances.[1] Among all water contaminations, heavy metal ions such as Pb^{2+}, Cd^{2+}, Zn^{2+}, Ni^{2+}, and Hg^{2+} have high-toxic properties and can cause severe health problems in animals and human beings.

It is well known that chronic cadmium toxicity is the inducement of Japan Itai-Itai disease. The harmful effects of Cd also lead a number of acute and chronic disorders, such as renal damage, emphysema, hypertension, testicular atrophy, and skeletal malformation, anemia, liver, kidney, and brain damages. Slow inhalation of Cu(II) causes lung cancer.

Several techniques have been developed for the qualitative and quantitative determination of such metal ions by ion exchange, solvent extraction, reverse osmosis, adsorption, complexation, and precipitation. Ion-exchange process has been used in the recent years for the application in wastewater treatment, identification and removal of heavy metal ions.

The copolymer resin synthesized by 4-hydroxy benzoic acid and thiourea with formaldehyde was involved to ion-exchange process to remove specific metal ions. The order of metal ion uptake by the resin is $Fe^{3+} > Cu^{2+} > V^{2+} > Cd^{2+} > Mg^{2+}$ ions which may be due to the presence of S and N atom. The N atom showed a great contribution for the formation of chelation. Moreover, the 4-HBTUF copolymer has highly amorphous nature which evidences the better removal of heavy metal ions.[1] A copolymer resin was synthesized using *p*-cresol and oxamide with formaldehyde. The chelating ion-exchange properties of the copolymer were studied for Fe(III), Cu(II), Ni(II), Co(II), Zn(II), Cd(II), and Pb(II) ions. Batch equilibrium method was employed to remove the selective metal ions. The copolymer reported a good metal ion uptake for Fe(III), Cu(II), and Ni(II) ions than for Co(II), Zn(II), Cd(II), and Pb(II) ions.[2,3]

A comparative study was done between the commercial activated carbon, chitosan, and chitosan/activated carbon composite for the removal of cadmium ions from aqueous solution. The adsorbent of pH solution is the most important factor for the removal of heavy metal ions.[4] Chitosan–charcoal composite was prepared by chitosan and charcoal by a simple solution–evaporation method. The prepared composite can effectively be used for the treatment of chromium from wastewater. The composite was employed for the adsorption of chromium by various rate, pH, and dose of adsorbent. The adsorption capacity is dependent on pH maximum.[5] Magnesium and coconut shell-activated carbon composite were involved to selectively remove the heavy metal ions in aqueous solution. The prepared magnesium and coconuts shell-activated carbon composite effectively removed the Zn(II) and Cd(II) from waste aqueous solution by ion-exchange method.[6]

Based on the literature, we have concluded that there were no earlier reports for the copolymer composites. Hence, a novel copolymer was synthesized from *p*-hydroxybenzoic acid and semicarbazide with formaldehyde and a novel copolymer composite was prepared using activated charcoal and the synthesized copolymer. Batch separation method was adopted for the removal of heavy metal ions using the copolymer and its composite and reported. The thermal stability of the both was determined by TGA. Based on the TG data, various kinetics of thermodynamic parameter have been evolved. Then, a comparative account was made between the copolymer and its composite and commercial resins for their ion-exchange and thermal studies and reported.

1.2 EXPERIMENTAL

1.2.1 MATERIALS

p-Hydroxybenzoic acid, semicarbazide, and formaldehyde were used as purchased from Merck, India. Metal ion solutions such as Cu^{2+}, Zn^{2+}, Co^{2+}, Pb^{2+}, and Cd^{2+} were prepared by its nitrate salts dissolving in deionized water. The other chemicals and solvents procured from Merck were used as received without further purification.

1.2.2 SYNTHESIS OF COPOLYMER AND COMPOSITE

The copolymer resin was synthesized involving *p*-hydroxybenzoic acid (0.1 mol) and semicarbazide (0.1 mol) with formaldehyde (0.2 mol) in 1:1:2 ratio by polycondensation technique using DMF medium for 6 h at 126 ± 2°C in oil bath, followed by the procedure based on earlier literature.[9] The mechanism of the synthesis of copolymer is shown in Scheme 1.1. The novel copolymer/activated charcoal composite was prepared in 1:2 ratio. The copolymer was dissolved in 25 ml of DMF and the activated charcoal was added into it and subjected to ultrasonication for 3 h with constant stirring for 24 h. Finally, the obtained black colored composite was dried in an air oven at 70°C for 24 h.

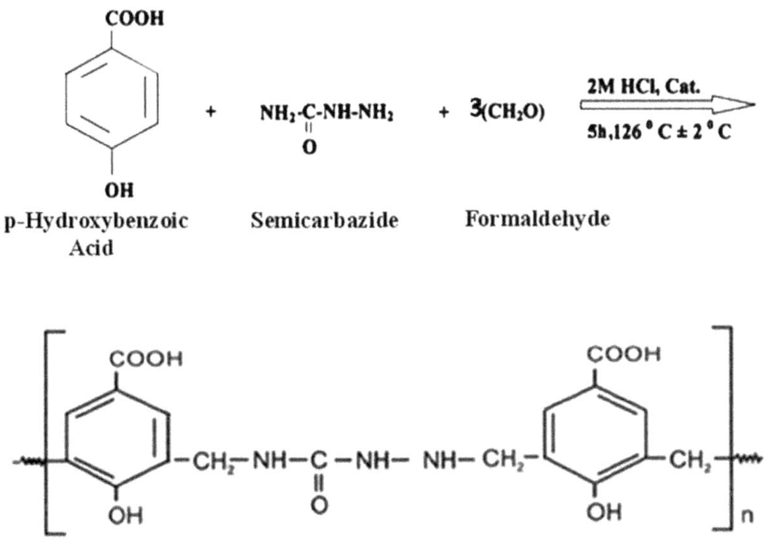

Novel Comparative Studies: Synthesis, Characterization

SCHEME 1.1 Synthetic route of *p*-HBSF-II copolymer.

1.2.3 ION-EXCHANGE STUDIES

1.2.3.1 ANALYSIS OF METAL ION UPTAKE AT DIFFERENT ELECTROLYTES

Batch equilibrium method was involved to determine the metal ion uptake of selective metal ions like Cu^{2+}, Zn^{2+}, Co^{2+}, Pb^{2+}, and Cd^{2+}. The pH of the solution was adjusted to the required value using either 0.1 M HCl or 0.1 M NaOH. The copolymer and copolymer/activated composite samples (25 mg) were taken in a pre-cleaned glass bottles and 25 ml of each of the different electrolytes, such as NaCl, $NaClO_4$, and $NaNO_3$ at various concentrations, namely, 0.1, 0.5, and 1.0 N, are added. These solutions were mechanically vigorously stirred for 24 h to allow the copolymer and its composite to swell. Exactly 2 ml of 0.1 mol of a specific metal ion solution was added to each glass bottle and again stirred for 24 h. The copolymer and its composite were then filtered off from each bottle and washed with deionized water. The filtrate solution was collected and then titrated against standard Na_2EDTA solution to estimate the metal ions. The same procedure was followed for the blank experiment without the addition of the copolymer and its composite sample.

1.2.3.2 ANALYSIS OF METAL IONS AT DIFFERENT pH

The removal of metal ions at various pH range between 1.5 and 6.0 of the copolymer and copolymer/activated charcoal composite phase and the aqueous phase was determined in the presence of 0.1 M NaCl at 25°C.

$$Q = V(C_0 - C_e)/M$$

where V is the volume, C_0 is the initial concentration, C_e is the final concentration, and M is the weight of the polymer/composite.

1.2.3.3 ANALYSIS OF RATE OF METAL ION UPTAKE

To evaluate the time required to achieve the state of equilibrium under the experimental conditions, a series of experiments was carried out to estimate the amount of metal ion adsorbed by the copolymer and its composite at specific time intervals. The rate of metal ion uptake is expressed as the percentage of the metal ion uptake after a specific time related to that in the state of equilibrium. It is given as follows:

$$\text{Metal ion taken up at different times (\%)} = \frac{\text{Metal ion adsorbed}}{\text{Metal ion adsorbed at equilibrium}} \times 10.$$

1.3 RESULTS AND DISCUSSION

1.3.1 ELEMENTAL ANALYSIS

From the elemental data, the empirical formula of the copolymer is found to be $C_{18}H_{17}O_7N_3$.

The percentage of the elements present in the copolymer that was concluded by the calculated and experimental values found for carbon is 55.90 (55.81), hydrogen is 4.74 (4.71), nitrogen is 10.92 (10.85), and oxygen is 28.94 (28.63) which are in good agreement with each other. The values of elemental analysis confirm the proposed structure of the copolymer presented in Scheme 1.1.

1.3.2 FTIR SPECTRAL ANALYSIS

The FTIR spectrum of the synthesized copolymer has a whole structure and contains various functional groups present, which is clearly exhibited in Figure 1.1. From the spectrum, a characteristic strong band at 1272.1 cm^{-1} is assigned to C–N stretching of Ar–NH$_2$.[7] The absorption peak appeared at 2852.90 cm^{-1} belongs to the aromatic ring stretching vibration modes. It is evident that the 1,2,3,5-tetra substitution of aromatic benzene ring by sharp, medium/weak absorption bands appeared between 1030.37 and 813.86 cm^{-1}. The absorption band at 1676.52 cm^{-1} was caused by stretching vibrations of the –C=O (amide moiety). The absorption band at 3380.10 cm^{-1} corresponds to –NH bridge in the copolymer. The band located at 2916.7 cm^{-1} was assigned to the –CH$_2$ linkage present in the copolymer. A weak band in the region appeared at 1370.60 cm^{-1} may be assigned to –CH$_3$ groups present in the copolymer.

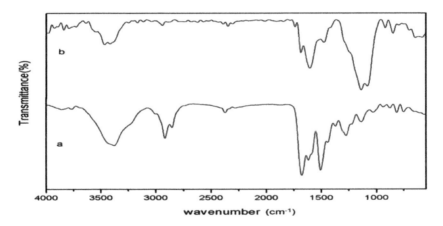

FIGURE 1.1 FTIR spectra of (a) *p*-HBSF-II copolymer and (b) composite.

The FTIR spectrum of composite is also showed in Figure 1.1. From the figure, the specific adsorption band at 3410.60 was caused by –NH bridge from copolymer.[8] The adsorption of band around 2851.95 can be attributed to aromatic ring stretching (C–H) present in the copolymer. A band at 2925.26 indicates –CH$_2$ stretching involved in the copolymer.

Based on the above data, it is confirmed that the copolymer has got adsorbed on the charcoal to make composite.

1.3.3 NMR SPECTRAL ANALYSIS

The ^{1}H NMR spectrum of the *p*-HBSF-II copolymer is exhibited in Figure 1.2. Based on the obtained results, multiple signals presented at 6.350–7.062 ppm indicate the aromatic protons. The aromatic methylene protons of Ar–CH$_2$ bridge are appeared at 1.9–2.2 ppm.[7] A singlet observed in the region 4.150 ppm is due to the 2H methylene proton of Ar–CH$_2$–N moiety of the copolymer. A weak signal may be attributed to the –NH bridge proton in the region at 8.055 ppm. The signal at 9.504 ppm is assigned to the Ar–NH$_2$ groups present in the copolymer. The ^{13}C NMR spectrum of *p*-HBSF-II copolymer resin is showed in Figure 1.2. The ^{13}C NMR spectrum shows the carbon present in the aromatic ring with the corresponding peaks at 127.1, 129.1, 130.5, 131.8, 134.2, and 135.8 ppm. A peak appeared at 160 ppm is attributed to the >C=O group of amide moiety present in the copolymer.[9] The peak appeared at 55 ppm may be assigned to the –CH$_2$ bridge in the copolymer. Followed on the spectral data acquired from FTIR, ^{1}H NMR, and ^{13}C NMR, the structure of the *p*-HBSF-II copolymer is proposed.

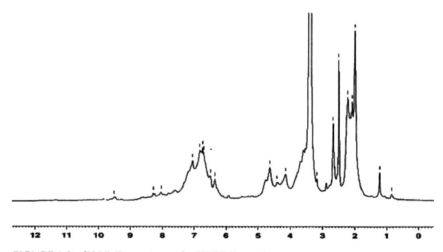

FIGURE 1.2 ^{1}H NMR spectrum of *p*-HBSF-II copolymer.

1.3.4 SURFACE ANALYSIS

Scanning electron microscopy was used to analyze the surface nature of the copolymer and its composite. The SEM images of activated charcoal, copolymer and its composite are shown in Figure 1.3(a)–(c), respectively. From the image, the more pores' structure are found to be present in the copolymer and its composite which can act as the more active site for the better adsorption of heavy metal ions. Figure 1.3(b) suggested that the structure of the copolymer was observed with large void volume on the surface which indicates the metals can be easily adsorbed on the surface. The large dimple was associated with the reinforcing particles combine to form the copolymer structure. The copolymer surface image possesses irregular honey comb structure with amorphous, high porosity, and permeability surface.[10,11] The SEM photograph of the copolymer composite is showed in Figure 1.3(c). It has wide varieties of pores present in the composite with more amorphous structure. The image shows new active site adhesive on the surface in sputter cluster which confirms the copolymer is composited with the activated charcoal. The increase in the surface area of the composite than the copolymer, the metal ion uptake, is found to be high.

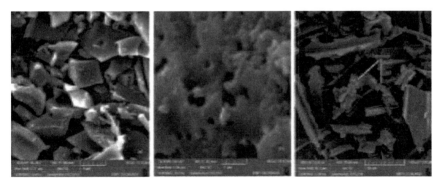

FIGURE 1.3 SEM photographs of (a) charcoal, (b) p-HBSF-II copolymer, and (c) composite.

1.3.5 ION-EXCHANGE STUDIES

1.3.5.1 EFFECT OF METAL ION UPTAKE IN DIFFERENT ELECTROLYTES WITH VARIATION IN CONCENTRATIONS

The comparative study of the effect of amount of metal ion uptake of p-HBSF-II copolymer and copolymer/activated charcoal composite was examined by ion-exchange. Influence of different electrolytes solution like Cl^-, NO_3^-, and ClO_4^- for interaction of different metal ion and adsorbent using different concentration 0.1, 0.5, and 1.0 M was studied. The outcome of the metal ion uptake results is exhibited in Table 1.1 and Figures 1.4–1.6. The sorption capacity was enhanced with increasing the concentration of the electrolytes for the both the adsorbents. At lower concentration, the number of metal ions is less compared to the available surface cavities.[12]

TABLE 1.1 Effect of Metal Ion Uptake by p-HBSF-II Copolymer and its Composite.

Metal ions	Concentration of electrolytes (mol L^{-1})	Metal ion uptake in the presence of electrolytes (m mol g^{-1})					
		NaCl		NaNO$_3$		NaClO$_4$	
		p-HBSF-II	Composite	p-HBSF-II	Composite	p-HBSF-II	Composite
Pb^{2+}	0.1	0.63	2.6	1.08	1.32	0.52	1.64
	0.5	0.86	2.7	1.08	2.05	0.75	1.72
	1	1.39	2.7	1.26	2.16	1.21	2.20
Cd^{2+}	0.1	0.64	1.12	0.86	0.86	0.75	0.75
	0.5	1.08	1.24	1.08	1.08	1.08	0.97
	1	1.08	1.61	1.08	1.42	1.08	1.72
Cu^{2+}	0.1	0.75	0.75	1.08	0.43	0.97	0.54
	0.5	2.46	0.98	1.25	0.86	1.51	0.75
	1	2.71	1.10	1.48	1.08	2.24	1.40
Co^{2+}	0.1	0.64	0.45	0.54	0.54	0.54	0.75
	0.5	0.95	0.78	1.08	0.75	1.40	0.86
	1	1.61	1.08	1.16	0.86	1.83	0.97
Zn^{2+}	0.1	1.42	0.52	1.40	0.43	2.05	0.44
	0.5	2.59	0.77	1.62	0.55	2.16	0.46
	1	3.00	0.88	2.05	0.64	2.59	0.64

The heavy metal adsorbed by p-HBSF-II copolymer and its composite depends upon the concentration and nature of the electrolytes, physical structure of pore size, various physical properties, and diffusion of counter ions.

The influence of the above factors is affected for the p-HBSF-II copolymer and copolymer/charcoal composite. The amount of metal ion uptake (Cu^{2+}, Zn^{2+}, Co^{2+}, Pb^{2+}, and Cd^{2+}) ions increases with increasing the concentration of the electrolytes. From the ion exchange results of the copolymer, it indicates that the uptake of Zn^{2+} and Cu^{2+} ions was found to be higher as compared to Co^{2+}, Pb^{2+}, and Cd^{2+}. This may be explained on the basis of Kitchner and Gregor et al. They found that the rate of diffusion, large size of metal ion, very slowly penetrates to the cation exchange in the copolymer surface cavities.[13] The small size of the Zn^{2+} easily diffuses.

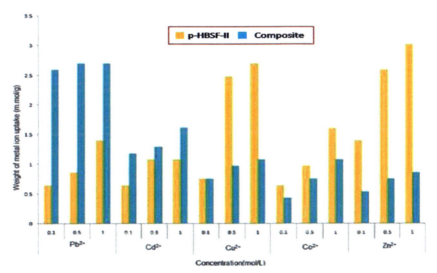

FIGURE 1.4 Effect of NaCl electrolyte on metal ion uptake by p-HBSF-II copolymer and its composite.

Through the available cavity, the heavy metal adsorbed by p-HBSF-II copolymer depends upon the distribution of pore size and diffusion of counter ions. The cation exchange capacity value depends upon the interionic force of attraction.

The higher selectivity of metal ion uptake for copolymer composite is Pb^{2+} and Cd^{2+}. Because the reason is that the adsorption ability was closely based on the ionic charges, hydrated ionic radius, and outermost electronic configuration of the metal ion. The hydrated ionic radius and the outermost electronic configuration of Pb^{2+} were different from those metal Cd^{2+}, Cu^{2+}, Co^{2+}, and Zn^{2+}. In addition, the hydrated ionic radii of Pb^{2+} was larger compared to that of other metal ions. From these, it can be concluded that the removal of Pb^{2+} metal ions is higher onto the composites comparatively to the p-HBSF-II copolymer.[14]

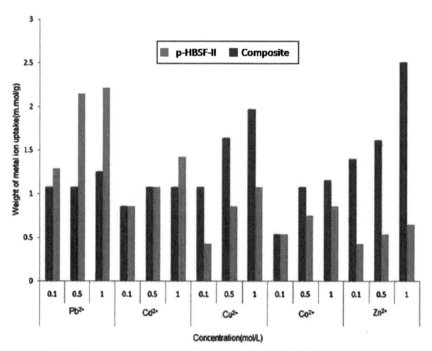

FIGURE 1.5 Effect of $NaNO_3$ electrolyte on metal ion uptake by *p*-HBSF-II copolymer and its composite.

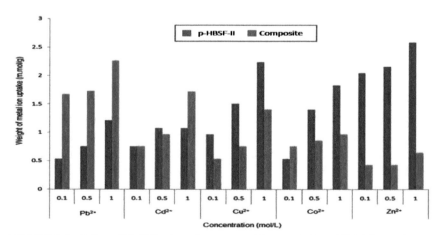

FIGURE 1.6 Effect of $NaClO_4$ electrolyte on metal ion uptake by *p*-HBSF-II copolymer and its composite.

On the other hand, Cu^{2+} metal ion uptake is higher than Zn^{2+} and Co^{2+} due to the ionic radius of Cu^{2+} (ionic radius 0.79 Å) which is higher than the Co^{2+} and Zn^{2+} due to intermediate behavior of high electro-negativity. From the data observed, the amount of Cu^{2+} and Co^{2+} was adsorbed lower in initial concentration because of the loading capacity of surface adsorbent. The copolymer/activated charcoal have more amorphous and pores on the surface. Hence, they can accommodate the metal ions easily into the cavities that depends on the specific particle size, that is, if the particle size increases, then the surface area also increases.

Hence, comparing the copolymer and its composite, the composite has higher metal ion uptake capacity than the copolymer. The order of selectivity of metal ion uptake for copolymer is $Zn^{2+} > Cu^{2+} > Co^{2+} > Pb^{2+} > Cd^{2+}$ and its copolymer composite is $Pb^{2+} > Cd^{2+} > Cu^{2+} > Co^{2+} > Zn^{2+}$. The metal-binding property of copolymer/activated charcoal composite is also found to be higher than that of the other reported earlier.[15,16]

1.3.5.2 ANALYSIS OF METAL IONS AT DIFFERENT PH

The effect of pH on the amount of metal ions uptake between copolymer and its composite phases and liquid phases can be explained by the results given in Figures 1.7 and 1.8. The obtained result indicates the relative amount of metal ion uptake by the copolymer and its composite increases with increasing pH of the medium. From the graph, the minimum adsorption observed at low pH could be related to the higher concentration and mobility of H^+ ions. At low pH, the surface of the adsorbent would be closely associated with hydronium ions (H_3O^+) and reduced the active site. In contrast, as the pH increases, the adsorption of metal ions increased. In the case of copolymer, metal ion uptake of Zn^{2+} and Cu^{2+} is very high compared to other metal ions, because of the stability of the complex formed during the adsorption process. The Zn^{2+} and Cu^{2+} have higher metal ion uptake which may be due to higher stability constant of the copolymer and metal complexes (Tables 1.2 and 1.3). The highest metal ion uptake for copolymer composite is Pb^{2+} and Cd^{2+}. It is due to the electrostatic interaction between the composite with negative surface charges and cations in the broad pH ranges.[17–19]

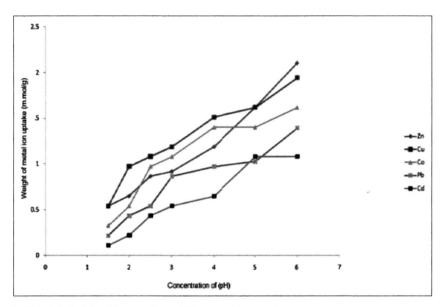

FIGURE 1.7 Metal ion uptake at different pH by copolymer.

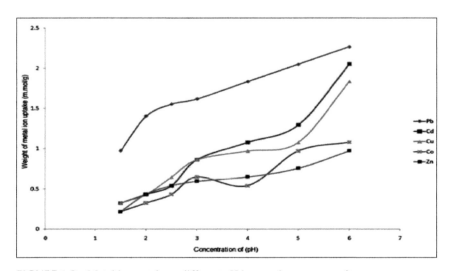

FIGURE 1.8 Metal ion uptake at different pH by copolymer composite.

TABLE 1.2 Metal Ion Uptake at Different pHs by *p*-HBSF-II Copolymer.

Metal ions	pH of the medium						
	1.5	2	2.5	3	4	5	6
Pb^{2+}	0.20	0.44	0.58	0.86	0.95	1.02	1.30
Cd^{2+}	0.10	0.20	0.44	0.54	0.62	1.10	1.12
Cu^{2+}	0.53	0.96	1.08	1.20	1.54	1.60	1.90
Co^{2+}	0.34	0.58	0.92	1.08	1.46	1.50	1.68
Zn^{2+}	0.55	0.64	0.86	0.90	1.12	1.62	2.10

TABLE 1.3 Metal Ion Uptake at Different pHs by *p*-HBSF-II Composite.

Metal ions	pH of the medium						
	1.5	2	2.5	3	4	5	6
Pb^{2+}	0.90	1.38	1.52	1.61	1.82	2.05	2.25
Cd^{2+}	0.32	0.42	0.55	0.88	1.08	1.29	2.05
Cu^{2+}	0.32	0.43	0.64	0.86	0.97	1.08	1.83
Co^{2+}	0.22	0.34	0.44	0.65	0.55	0.98	1.08
Zn^{2+}	0.21	0.43	0.54	0.59	0.64	0.75	0.97

$M^2 + (NO_3) = 0.1$ M; volume = 2 mL; NaCl = 1 M; volume = 25 mL; weight of resin = 25 mg; time = 24 h; room temperature.

1.3.5.3 ANALYSIS OF METAL ION UPTAKE IN DIFFERENT ELECTROLYTES WITH THE VARIATION IN RATE

The comparative study of the copolymer and copolymer/charcoal composite is extended to the important parameter, that is, the effect of rate for different metal ions was Cu^{2+}, Zn^{2+}, Co^{2+}, Pb^{2+}, and Cd^{2+}. The results are presented in Tables 1.4 and 1.5 and Figures 1.9 and 1.10. The rate of metal ion adsorption was determined to which the metal ion uptake has the shortest period of time to attain the close to equilibrium condition during the ion-exchange process. From the observed results, the time taken for uptake of Zn^{2+} ion is the shortest period of time to attain equilibrium at 4 h.[20] The rate of adsorption of heavy metal ion has been generally described to difference in ionic radii of metal ions, difference in the affinity of the metal ions for active groups on the *p*-HBSF-II adsorbent, and also nature of the anions of the salt of the metal ions. Kitchener and Greger et al. found that the rate of diffusion, large size of metal ion penetrates slowly in to the polymer lattices. In the copolymer composite, the time taken for Pb^{2+} and Cd^{2+} metal ion to reach close to equilibrium is 4 h by the influence of different electrolytes and adsorbent. Furthermore, the metal ion with smaller size can be easily adsorbed. In the case of larger size and bulkier, metal ion makes less hydrated like Pb^{2+} ion, that is, the heavily hydrated ions migrated slowly through into aqueous solutions. The fast rate of exchange can be explained on the basis of low of mass action.[21] From the above data observed, the order of rate of metal ion uptake by the p-HBSF-II copolymer is $Zn^{2+} > Cu^{2+} > Co^{2+} > Pb^{2+} > Cd^{2+}$ and its composite is $Pb^{2+} > Cd^{2+} > Cu^{2+} > Co^{2+} > Zn^{2+}$.

TABLE 1.4 Rate of Metal Ion Uptake by p-HBSF-II Copolymer.

Metal ions	Equilibrium attainment (%)						
	Time (h)						
	1	2	3	4	5	6	7
Pb^{2+}	16	31	54	63	67	72	83
Cd^{2+}	0	16	31	56	80	–	–
Cu^{2+}	27	55	68	84	92	–	–
Co^{2+}	28	41	50	73	80	85	–
Zn^{2+}	52	64	78	96	–	–	–

TABLE 1.5 Rate of Metal Ion Uptake by p-HBSF-II Composite.

Metal ions	Equilibrium attainment (%>)						
	Time (h)						
	1	2	3	4	5	6	7
Pb^{2+}	60	74	90	96	97	–	–
Cd^{2+}	64	81	88	95	95	–	–
Cu^{2+}	10	30	50	60	72	84	87
Co^{2+}	14	42	57	61	71	76	84
Zn^{2+}	19	40	60	70	80	84	82

$M^2 + (NO_3) = 0.1$ M; volume = 2 mL; NaCl = 1 M; volume = 25 mL; weight of resin = 25 mg; time = 24 h; room temperature.

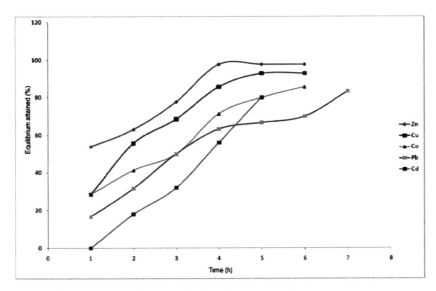

FIGURE 1.9 Rate of metal ion uptake by *p*-HBSF-II copolymer.

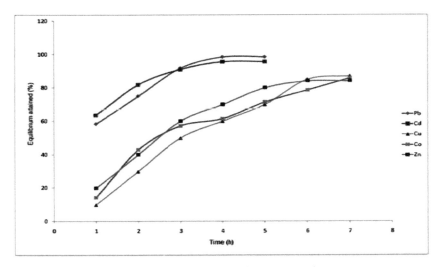

FIGURE 1.10 Rate of metal ion uptake by copolymer composite.

1.3.6 THERMAL STUDIES

Thermogravimetric analysis is one of the commonly used techniques for rapid evaluation of the thermal stability of different materials and also indicates the decomposition of polymers at various temperatures. The decomposition pattern of the *p*-HBSF-II copolymer has been shown in Figure 1.12. There are three steps of weight loss, as illustrated in the decomposition patterns in the temperature range 50–580°C (Table 1.6). The first stage corresponds to 250°C, until at 330°C with a decrease in the weight loss of 13.22% (calc. 15.42%) and this is due to the elimination of $-NH_2$ and CH_3 group. The second stage begins at 330°C and it finishes in 520°C with a decrease in the weight loss of 34.74% (calc. 36.68%) from the total weight. This weight loss may be attributed to the removal of benzene ring. At the temperature intervals of 520–590°C which corresponds to the degradation of other organic compounds present in the copolymer ligand with a weight loss of 48.5% (calc. 49.1%).[9]

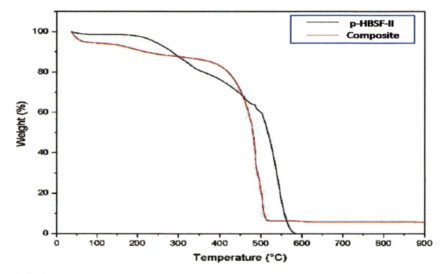

FIGURE 1.11 Thermograms of *p*-HBSF-II copolymer and its composite.

Thermogram of the copolymer composite also indicates a three step decomposition temperature which was observed in the range of 50–700°C. The first stage degradation of copolymer composite can be attributed to the decomposition of $-NH_2$ groups start at 110°C, and it finish in 220°C. The second stage begins at 220°C and ends at 450°C may be attributed to the removal of $-CH_3$. The third stage of degradation observed between

450 and 700°C, which corresponds to the complete degradation of whole aromatic rings, cross linking present in the copolymer composite. On comparing the thermograms of the both, the copolymer composite has more thermal stability than the copolymer which may be more composite residue left compare to copolymer.

TABLE 1.6 TGA Data of p-HBSF-II Copolymer and its Composite.

Compound	Weight loss (%) at various temperatures (°C)							Decomposition temperature (°C)	T_{max}*	T_{50}**
	100°C	200°C	300°C	400°C	500°C	600°C	700°C			
p-HBSF-II	1.4	2.24	12.80	24.0	42.72	99.6	–	250–600	600	540
p-HBSF-II composite	5.72	9.12	12.38	16.9	79.50	93.9	94.3	110–700	700	480

1.3.7 KINETICS OF THERMAL DEGRADATION

The kinetic and thermodynamic parameters' data for the both copolymer and its composite calculated by SW and FC methods (Table 1.7) and its results are summarized in Table 1.8. Increasing molecular weight provided an effective way to improve the thermal stability of copolymer and its composite. The p-HBSF-II copolymer ligand and its composite thermodynamic parameters that were calculated by SW and FC methods were in good agreement with each other. The activation energy of the copolymer composite was found to be higher than copolymer which reflects that the thermal stability of the copolymer composite was higher than copolymer. The negative values of entropy change (ΔS) for the decomposition step indicate that the composite and copolymer are more activated and involve slow decomposition reaction. The copolymer composite has observed low value of frequency factor (Z) than copolymer which reveals that the copolymer composite has slow decomposition than copolymer. The positive values of free energy change confirm that the endothermic effect was involved in the degradation reaction. The order of reaction for p-HBSF-II copolymer is nearly two but the composite is high, from this data may be composite have more thermal stability.[22,23] The E_a and n plots are shown in Figures 1.12–1.17.

TABLE 1.7 Formulae for Calculating Kinetic Parameters.

(1) Entopy change:

$$\text{Intercept} = [\log KR/h\phi E] + \Delta S/2.303R \quad (1.1)$$

where $K = 1.3806 \times 10^{-16}$ erg/deg/mol

$R = 1.987$ cal/deg/mol

$h = 6.625 \times 10^{-27}$ erg s

$\phi = 0.166$

ΔS = change in entropy

E = activation energy from graph.

(2) Free energy change:

$$\Delta F = \Delta H - T\Delta S \quad (1.2)$$

ΔH = Enthalpy change = activation energy

T = Temperature in K

ΔS = Entropy change from (1.1) used.

(3) Frequency factor:

$$B_{2/3} = \log ZE_a / \phi R \quad (1.3)$$

$$B_{2/3} = \log 3 + \log[1 - 3\sqrt{1-\alpha}] - \log P(x) \quad (1.4)$$

Z = Frequency factor

B = Calculated from equation (1.4)

$\log P(x)$ = calculated from Doyle's table corresponding to activation energy.

(4) Apparent entropy change:

$$S^* = 2.303 \log Zh/KT^* \quad (1.5)$$

Z = From relation (1.3)

T^* = Temperature at which half of the compound is decomposed from it total loss.

TABLE 1.8 Kinetic and Thermodynamic Parameters of p-HBSF-II and its Composite.

Compound	Activation energy (kJ/mol)		Entropy change $-\Delta S$ (J)	Free energy change ΔF (kJ)	Frequency factor Z (S^{-1})	Apparent entropy change (S^*)	Order of reaction (n)
	SW	FC					
p-HBSF-II	24.11	23.92	−62.52	115.54	700	23.74	0.98
composite	24.49	24.30	−62.01	115.57	564	24.01	0.92

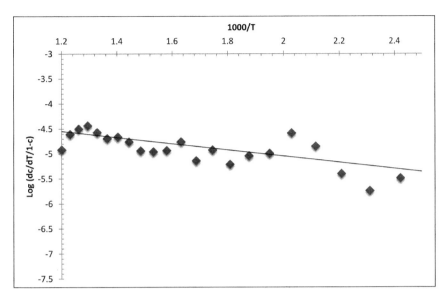

FIGURE 1.12 Sharp–Wentworth plot (E_a) for *p*-HBSF-II copolymer.

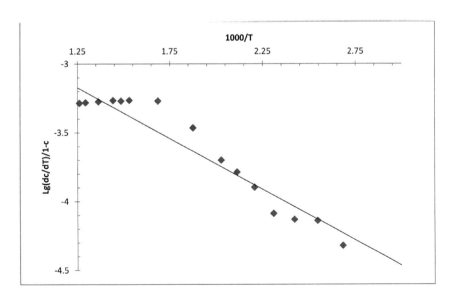

FIGURE 1.13 Sharp–Wentworth plot (E_a) for *p*-HBSF-II composite.

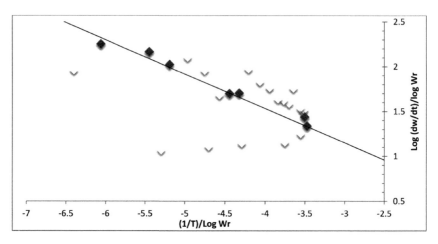

FIGURE 1.14 Freeman–Carroll plot (*n*) for *p*-HBSF-II copolymer.

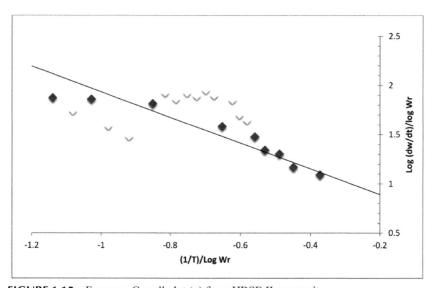

FIGURE 1.15 Freeman–Carroll plot (*n*) for *p*-HBSF-II composite.

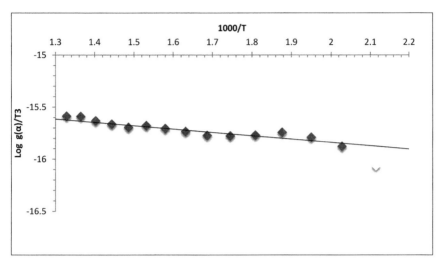

FIGURE 1.16 Freeman–Carroll plot (E_a) for p-HBSF-II copolymer.

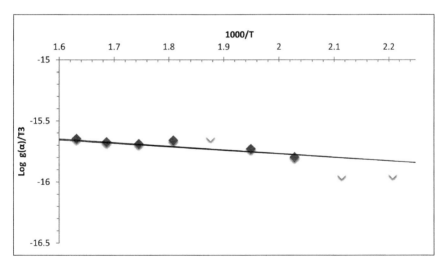

FIGURE 1.17 Freeman–Carroll plot (E_a) for p-HBSF-II composite.

1.4 CONCLUSION

A novel comparative account was made between the copolymer and its composite for the studies of ion-exchange and thermal decomposition kinetics. The structure and properties of the copolymer and its composite were confirmed by the spectral, morphology, and elemental analysis. From the ion-exchange results, the copolymer and its composite can act as an effective metal ion adsorbent due to the more porous and amorphous in nature. The copolymer composite is found to be an excellent metal ion exchanger than the copolymer, due to the increased surface area to form new active sites. This is also confirmed by SEM images. The results showed that the copolymer has higher selectivity metal ion uptake for Zn^{2+} and Cu^{2+} and for the composite, metal ion is Pb^{2+} and Cd^{2+} because of the different particle size. TGA results show that the copolymer composite exhibits a higher thermal stability and decomposition temperature compared to the copolymer, because composite has high activation energy and more residue left out at the end of decomposition process. From the above reasons, it is concluded that the copolymer composite can act as a better ion-exchanger and possesses higher thermal stability than the copolymer.

KEYWORDS

- **copolymer**
- **composite**
- **adsorption**
- **morphology**
- **ion-exchange**
- **thermal degradation**

REFERENCES

1. Gurnule, W. B.; Charulata, M. S.; Mudrika, A. Synthesis, Characterization, Morphology, Thermal, Electrical and Chelation Ion-exchange Properties of a Copolymer Resin. *J. Environ. Res. Dev.* **2013**, *7*, 1183–1192.
2. Singru, R. N.; Gurnule, W. B. Chelating Ion-exchange Properties of Copolymer Resins Derived from *p*-Cresol, Oxamide and Formaldehyde. *Iran. Polym. J.* **2010**, *19*, 169–183.
3. Riswan, A. M.; Azarudeen, R.; Karunakaran, M.; Karikalan, T.; Manikandan, R.; Burkanudeen, A. R. Cation Exchange Properties of a Copolymer: Synthesis and Characterization. *Int. J. Chem. Environ. Eng.* **2010**, *1*, 7–12.

4. Azarudeen, R. S.; Ahamed, M. A.; Burkanudeen, A. R. *Desalination* **2011**, *268*, 90–96.
5. Shah, B. A.; Shah, A. V.; Bhatt, R. R. Studies of Chelation Ion-exchange Properties of Copolymer Resin Derived from Salicylic Acid and its Analytical Applications. *Iran. Polym. J.* **2007**, *16*, 173–184.
6. Azarudeen, R. S.; Burkanudeen, A. R. Synthesis, Spectral, Morphology, Thermal Degradation Kinetics and Antibacterial Studies of Copolymer Metal Complexes. *J. Inorg. Organomet. Polym.* **2012**, *22*, 791–806.
7. Hydari, S.; Sharififard, H.; Nabavinia, M.; Parvizi, M. R. A Comparative Investigation on Removal Performances of Commercial Activated Carbon, Chitosan Biosorbent and Chitosan/Activated Carbon Composite for Cadmium. *Chem. Eng. J.* **2012**, *193*, 276–282.
8. Azarudeen, R. S.; Burkanudeen, A. R. Synthesis, Spectral, Morphology, Thermal Degradation Kinetics and Antibacterial Studies of Copolymer Metal Complexes. *J. Inorg. Organomet. Polym.* **2012**, *22*, 791–806.
9. Azarudeen, R. S.; Subha, R.; Jeyakumar, D.; Burkanudeen, A. R. Batch Separation Studies for the Removal of Heavy Metal Ions Using a Chelating Copolymer: Synthesis, Characterization and Isotherm Models. *Sep. Purif. Technol.* **2013**, *116*, 366–377.
10. Kalkan. E.; Nadaroglu, H.; Dikbas, N.; Tasgin, E.; Celebi, N. Bacteria-modified Red Mud for Adsorption Cadmium Ions from Aqueous Solutions. *Pol. J. Environ. Stud.* **2013**, *22*, 417–429.
11. Kalalagh, S. H. S.; Babazadeh, H.; Nazemi, A. H.; Manshouri, M. Isotherm and Kinetic Studies on Adsorption of Pb, Zn and Cu by Kaolinite. *Caspian J. Environ. Sci.* **2011**, *9*, 243–255.
12. Girija, A.; Seenivasan, R. K. Synergism Modified Ion-exchange Process of Selective Metal Ions and its Binary Mixtures by Using Low Cost Ion Exchange Resin and its Binary Mixture. *Rasayan J. Chem.* **2013**, *6*, 212–222.
13. Zhu, L.; Zhang, L.; Tang, Y. Synthesis of Montmorillonite/Poly(Acrylic Acid-*co*-2-Acrylamido-2-Methyl-1-Propane Sulfonic Acid) Superabsorbent Composite and the Study of its Adsorption. *Bull. Korean Chem. Soc.* **2012**, *33*(5), 1669.
14. Gurnule, W. B.; Patle, D. B. Preparation, Characterization and Chelating Ion-exchange Properties of Copolymer Resins Derived from *o*-Aminophenol, Urea and Formaldehyde. *Elixir Appl. Chem.* **2012**, *50*, 10338–10345.
15. Shah, B. A. Selective Sorption of Heavy Metal Ions from Aqueous Solutions Using *m*-Cresol Based Chelating Resin and its Analytical Applications. *Iran. J. Chem. Chem. Eng.* **2010**, *29*, 49–58.
16. Azarudeen, R.; Riswan, A. M.; Arunkumar, P.; Prabu, N.; Jeyakumar, D.; Burkanudeen, A. Metal Sorption Studies of a Novel Copolymer Resin. *Int. J. Chem. Environ. Eng.* **2010**, *1*, 23–28.
17. Naushad, M. U.; Al-Othman, Z. A.; Islam, M. Adsorption of Cadmium Ion Using a New Composite Cation Exchanger Polyaniline Sn(IV) Silicate: Kinetics, Thermodynamic and Isotherm Studies. *Int. J. Environ. Sci. Technol.* **2013**, *10*, 567–578.
18. Riswan, A. M. A.; Jeyakumar, D.; Burkanudeen, A. R. Removal of Cations Using Ion-binding Copolymer Involving 2-Amino-6-nitro-benzothiazole and Thiosemicarbazide with Formaldehyde by Batch Equilibrium Technique. *J. Hazard. Mater.* **2013**, *248–249*, 59–68.

19. Tehrani, M. S.; Azar, P. A.; Namin, P. E.; Dehaghi, S. M. Removal of Lead Ions from Wastewater Using Functionalized Multiwalled Carbon Nanotubes with Tris(2-aminoethyl)amine. *J. Environ. Prot.* **2013**, *4*, 529–536.
20. Bhatt, R. R.; Shah, B. A.; Shah, A. V. Uptake of Heavy Metal Ions by Chelating Ion-exchange Resin Derived from *p*-Hydroxybenzoic Acid–Formaldehyde–Resorcinol: Synthesis, Characterization and Sorption Dynamics. *Malaysian J. Anal. Sci.* **2012**, *16*, 117–133.
21. Burkanudeen, A. R.; Azarudeen, R. S.; Riswan, A. M.; Gurnule, W. B. Kinetics of Thermal Decomposition and Antimicrobial Screening of Copolymer Resins. *Polym. Bull.* **2010**, *67*, 1553–1568.
22. Abd El-halim, H. F.; Omar, M. M.; Mohamed, G. G. Synthesis, Structural, Thermal Studies and Biological Activity of a Tridentate Schiff Base Ligand and their Transition Metal Complexes. *Spectrochim. Acta* **2011**, *78*, 36–44.
23. Michael, P. E. P.; Barbe, J. M.; Juneja, H. D.; Paliwal, L. J. Synthesis, Characterization and Thermal Degradation of 8-Hydroxyquinoline–Guanidine–Formaldehyde Copolymer. *Eur. Polym J.* **2007**, *43*, 4995–5000.

CHAPTER 2

ELECTRICAL TRANSPORT PROPERTIES OF POLY(ANILINE-*CO*-*N*-PHENYLANILINE) COPOLYMERS

A. D. BORKAR[*]

Department of Chemistry, Nabira Mahavidyalaya, Katol District, Nagpur 441302, India

[*]*Corresponding author. E-mail: arun.borkar@rediffmail.com*

CONTENTS

Abstract 30
2.1 Introduction 30
2.2 Experimental Synthesis of Homopolymers 31
2.3 Synthesis of Copolymers 31
2.4 Characterization 32
2.5 Results and Discussion 32
2.6 Conclusions 37
Acknowledgments 38
Keywords 38
References 38

ABSTRACT

Poly(aniline-*co*-*N*-phenylaniline) copolymers have been synthesized by the chemically oxidative copolymerization of aniline and *N*-phenylaniline in aqueous hydrochloric acid medium under nitrogen atmosphere at 0–4°C. The molar feed ratio of monomers is varied to prepare copolymers of different compositions. The copolymers are characterized by FTIR, UV–visible spectroscopy. The solubility and spectroscopic analysis suggest that the product is a copolymer of aniline and *N*-phenylaniline. The electrical conductivity of the compressed pellets was measured by two probe method. The electrical conductivity of copolymers is found to be less than polyaniline but processability has been improved significantly in solvents like 1-methyl-2-pyrrolidone, dimethyl sulphoxide, and dimethyl formamide. The decrease in room temperature conductivity of the copolymers with increase of the *N*-phenylaniline is due to the introduction of phenyl groups in to the copolymer which reduces the conjugation of the polymer chain and reduces the mobility of charge carrier along the main chain. From temperature dependence of electrical conductivity, charge localization length and hopping distance are calculated and the effect of substituent and dopant on crystallinity is discussed. The temperature dependence of dielectric data and conductivity suggests that copolymers are quasi 1D-disordered state composed of 3D-metallic crystalline regions in 1D-localized amorphous regions.

2.1 INTRODUCTION

In recent years, much attention has been bestowed upon electrically conducting materials which may be useful for rechargeble batteries, electrochromic display devices, and modified electrodes.[1,2] However, most of the studies in the field are confined to conjugated organic polymers, prepared by chemical and electrochemical synthesis, including poly(pyrrole) and poly(aniline). These polymers can be doped with either electron acceptor or donors to yield conducting polymers.

Polyaniline (PANI) is made up of combination of fully reduced (B–NH–B–NH) and fully oxidized (B–N=Q=N–) repeating units, where B denotes a benzenoid and Q denotes a qsuinoid ring. Thus, different ratios of these fully reduced and fully oxidized units yield various forms of PANI, such as leucoemeraldine (100% reduced form), emeraldine base (50% oxidized form), and pernigraniline (fully oxidized form). However, all of these forms are electrically insulating in nature. Doping of emeraldine base with a protonic acid converts it into conducting form protonated emeraldine (emerdine salt). The main issue with PANI is processing difficulties due to its infusibility and relative insolubility in common organic solvents. It can be made processable/soluble either by polymerizing functionalized anilines[3] or by copolymerizing aniline with substituted monomers.[4] The copolymerization is a powerful method to improve processability of conducting polymers. In general, solubility of substituted PANIs in organic solvents is significantly higher than PANI. However, their thermal stability and electronic conductivities are substantially lower than doped PANI. In order to maintain balance between conductivity, stability, and processability copolymerization has been done.

In this chapter, poly(aniline-co-N-phenylaniline) copolymers have been synthesized by chemical peroxidation method. These copolymers are characterized by FTIR, UV–visible (UV–vis) spectroscopy. Their electronic conductivities have been measured by two-probe technique. Variable range hopping (VRH) model has been applied depending upon the nature of variation of conductivity with temperature.

2.2 EXPERIMENTAL SYNTHESIS OF HOMOPOLYMERS

a. PANI was chemically synthesized[5–7] using ammonium peroxodisulphate as an oxidant in an aqueous 1 M HCl in nitrogen atmosphere at 0–4°C.

b. Poly(N-phenylaniline) (PNPANI) was chemically synthesized[5–7] using ammonium peroxodisulphate as an oxidant in an aqueous 1 M HCl and CH_3CN (1:1 ratio) in nitrogen atmosphere at 0–4°C.

2.3 SYNTHESIS OF COPOLYMERS

Poly(aniline-co-N-phenylaniline) (PANI-co-PNPANI) copolymers were chemically copolymerized[8,9] from the monomers, aniline, and N-phenylaniline using ammonium peroxodisulphate as an oxidant in an aqueous 1 M HCl and CH_3CN (1:1 ratio) in nitrogen atmosphere at 0–4°C. The molar feed ratio of starting aniline monomer was varied to result in copolymers of different compositions. The homopolymers and copolymers obtained from reaction medium were filtered and washed with distilled water and methanol several times to remove unreacted monomers and then dried in an air oven at 70°C for 8 h.

2.4 CHARACTERIZATION

The solubility of the homopolymers and copolymers salt form was tested by dissolving each material in dimethyl formamide (DMF). The mixture was kept for 24 h at room temperature, after which the solution was filtered through sintered glass crucible G_4. The room temperature solubility was recorded. UV–vis spectra of homopolymers and copolymers were recorded at room temperature in 1-methyl-2-pyrrolidone (NMP), DMF, and dimethyl sulphoxide (DMSO) in 190–700 nm range using UV-240 Shimadzu Automatic Recording Double Beam Spectrophotometer. FTIR spectra of homopolymers and copolymers were recorded on 550 Series II, Nicolet, using KBr pellet technique in the range of 400–4000 cm^{-1}. DC electrical conductivity of polymer samples was measured by the two probe technique in the temperature range of 298–398 K. Dry powdered samples were made in to a pellet under hydraulic press IEBIG and placed between electrodes in a cell. Resistance was measured on a DC resistance bridge LCR Meter 926. The conductivity value was calculated from the measured resistance and sample dimensions.

2.5 RESULTS AND DISCUSSION

The solubility of homopolymers and copolymers in DMF are determined. PANI (0.0414 g/dl) and PNPANI (0.0732 g/dl) homopolymers show low and high solubility as compared to copolymers, which indicates that the incorporation of the substituted monomer units in the copolymer which gives a solubility intermediate between the corresponding homopolymers. The substituent introduces flexibility into the rigid PANI backbone structure; as a result, copolymers show higher solubility than PANI.

The empirical repeat unit for homopolymers and copolymers is shown in Figures 2.1 and 2.2.

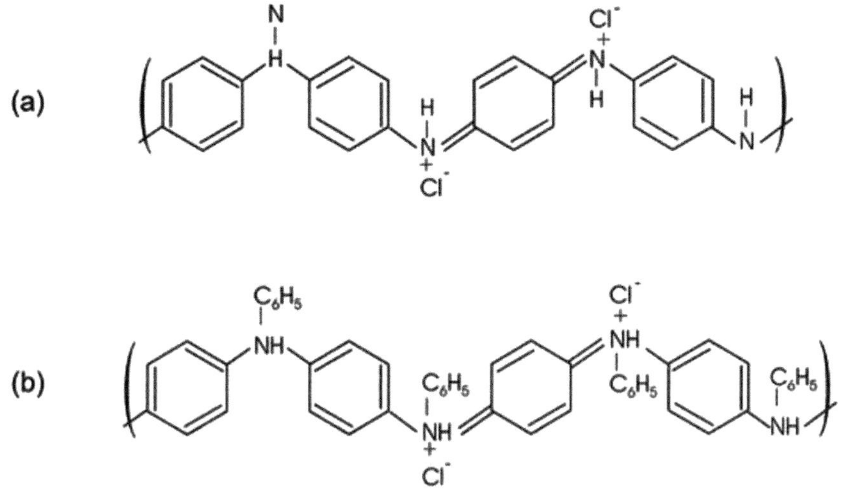

FIGURE 2.1 Repeat unit for homopolymers (a) PANI and (b) PNPANI.

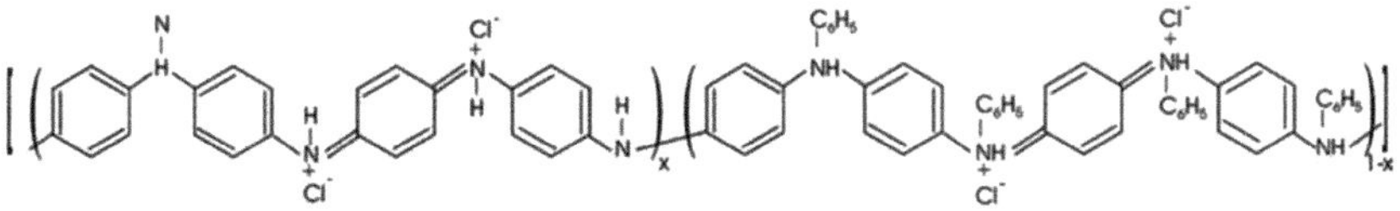

FIGURE 2.2 Repeat unit for copolymers.

The absorption bands of homopolymers and copolymers are recorded in NMP, DMF, and DMSO. The corresponding bands are given in Table 2.1.

TABLE 2.1 UV–Vis Absorption Bands of Homopolymer and Copolymers.

Polymer/copolymer	UV–vis absorption band (nm)					
	NMP		DMF		DMSO	
PANI	330	628	328	618	314	618
PNPANI	314	600	310	610	310	600
PANI-co-PNPANI (60:40)	292	560	298	560	294	580
PANI-co-PNPANI (30:70)	290	560	296	540	290	580

There are two absorption bands in the electronic spectra of homopolymers and copolymers. The bands around 290–330 nm (4.278–3.760 eV) are assigned to π–π* (bandgap) transition (which is related to the extent of conjugation between the adjacent rings in the polymer chain), and the band above 540 nm (exciton band) is due to inter band charge transfer associated with excitation of benzenoid to quinoid moieties (formation of exciton). These bands change with solvents. The π–π* band of copolymers shows hypsochromic shift with the increase in dielectric constant of the solvent.[10] The exciton band shows the bathochromic shift with an increase in dielectric constant of the solvent. The excitation leads to formation of molecular exciton (positive charge on the adjacent benzenoid units and negative charge centered on quinoid unit).[11] This interchain charge transfer from highest occupied molecular orbital (HOMO) to lowest unoccupied molecular orbital (LUMO) may lead to the formation of positive and negative polarons. A polymer in a solvent of high dielectric constant may exist in coil like conformation (decrease in conjugation) and a less polar solvent provides thermodynamically more stable chain conformation and restricts the polymer to lower energy high planarity state. Such shift may increase the conjugation of the system which yields a lower energy transition a red shift. The π–π* band in copolymers shifts to the lower wavelength as the percentage of N-phenylaniline in copolymer increase which may be due to addition of more phenyl groups which twist the torsion angle, which are expected to increase the average bandgap in the conjugated polymer chain. The continuous variation of the wavelength and intensity of the UV–visible bands may have resulted from the copolymerization effect of N-phenylaniline and aniline. In other words, the polymer formed by the oxidative polymerization of N-phenylaniline with aniline was the copolymer of two monomers rather than a mixture of two homopolymers.

The FTIR spectra of homopolymers and copolymers are shown in Figure 2.3 and spectral data are recorded in Table 2.2. The FTIR spectra of PANI show a broad band at 3431 cm^{-1} characteristics of N–H stretching. The band in the range 810–819 cm^{-1} assignable to 1,4 substituted aromatic ring indicating the bonding in polymers and copolymers was through 1,4 position. The bands of 1569 and 1492 cm^{-1} are assigned to quinoid nitrogen (N=Q=N) and benzoid (N–B–N) ring stretching. In the spectra of copolymer, there appears an absorption band at 1310 cm^{-1} (C–N stretching) and 1500 cm^{-1} (benzenoid stretching) indicating the existence of phenyl group on a benzene ring. The relative intensity of 1600–1500 cm^{-1} is between those of PNPANI and PANI, and so the coexistence of N-phenylaniline and aniline units in the copolymer was further confirmed.

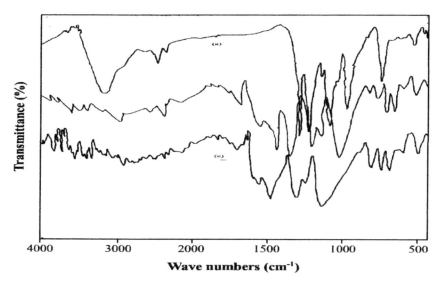

FIGURE 2.3 FTIR spectra of (a) PANI, (b) PNPANI, and (c) PANI-*co*-PNPANI (60:40).

TABLE 2.2 FTIR Spectral Data of Homopolymers and Copolymers.

Polymer/copolymer	Wave number (cm^{-1})						
PANI	819	1145	1246	1300	1492	1569	3431
PNPANI	810	1160	1250	1315	1510	1560	
PANI-*co*-PNPANI (60:40)	813	1141	1242	1310	1485	1556	

The temperature dependence of electrical conductivity data was fitted to an Arrhenius equation:

$$\sigma(T) = \sigma_0 \exp{-\frac{E_a}{2kT}} \quad (2.1)$$

and the measured values were plotted semi-logarithmically as a function of reciprocal of temperature (Fig. 2.4a). The conductivity is found to increase with temperature; however, there are deviations at lower temperature. In the present study, the temperature dependence of $\sigma(T)$ is fitted to the Zeller equation[12]:

$$\sigma(T) = \exp{-\left(\frac{T_0}{T}\right)^{1/2}} \quad (2.2)$$

where T_0 is the Mott characteristic temperature and is a measure of the hopping barrier.

Figure 2.4(b) describes the interchain conductivity where only the neighbor VRH of charge (which is quasi 1D) is considered.[13]

In Zeller equation, T_0 is related to delocalization length ($-^1$), most probable hopping distance (R) and hopping energy (w) by relations (2.3)–(2.5)[14]

$$T_0 = 8(/N(EF)ZK \qquad (2.3)$$

$$R = \left(\frac{T_0}{T}\right)^{1/2} (-^1/4) \qquad (2.4)$$

$$W = \frac{ZKT_0}{16} \qquad (2.5)$$

where Z is the number of nearest neighboring chains (≈ 4), K is the Boltzmann constant, and $N(E_f)$ the density of states at Fermi energy for sign of spin which is taken as 1.6 states per eV (2-ring unit) suggested for PANI.[15]

The value of T_0 is determined graphically from log $\sigma(T)$ versus $100/T^{1/2}$ (Fig. 2.4(b)) and other parameters are computed from the data (Table 2.3). It is observed that T_0 for PNPANI is more than PANI (maximum localization length). Therefore, correspondingly the localization length and average charge hopping distance decrease on increasing the amount of N-phenylaniline in copolymer chain. This increase in the electron localization is due to the presence of phenyl group in the copolymer chain. The temperature dependence of electrical conductivity fits (2.2) (Fig. 2.4(b)) for homopolymers and copolymers, which suggests that the charge conduction is quasi-1D VRH between nearest neighboring chains.[16]

In case of HCl-doped PANI Cl⁻ ions are small and the interchain separation is small, resulting in appreciable coupling interaction between the chains. Thus, charge carriers could easily hop from one chain to other to give of 3D-VRH conduction. However, in the case of copolymers, the charge carriers hop between 3D-VRH but interchain hopping has been strongly inhibited leading to 1D-VRH. The reduction in interchain hopping may be attributed to the presence of phenyl group in the copolymers that reduce the coupling interactions between the chains. Therefore, the net effect is the 1D-VRH conduction in case of copolymers.

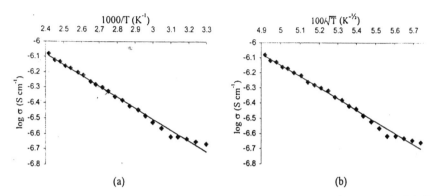

FIGURE 2.4 Temperature dependence of electrical conductivity of PANI-*co*-PNPANI (60:40): (a) Arrhenius equation and (b) Zeller equation.

TABLE 2.3 Transport Properties of Homopolymers and Copolymers.

Polymer/Copolymer	σ (S/cm)	T_0 (K)	ζ^{-1} (nm)	R (nm)	w (eV)	E_a (eV)
PANI	6.550×10^{-2}	3.315×10^3	6.995	5.736	0.071	0.047
PNPAN	1.026×10^{-9}	7.260×10^3	3.194	3.877	0.156	0.137
PANI-*co*-PNPANI (60:40)	2.191×10^{-7}	5.250×10^3	4.416	4.558	0.101	0.133
PANI-*co*-PNPANI (30:70)	1.517×10^{-8}	5.804×10^3	3.995	4.335	0.120	0.146

The electrical conductivity of copolymer salts are measured and compared to that of homopolymers the results are summarized in Table 2.3. The PANI shows conductivity of order of 6.55×10^{-2} S cm^{-1} and 1.026×10^{-7} S cm^{-1} for PNPANI while copolymer shows lower conductivity than PANI which indicates the ionization potential, bandgap, and bandwidth are affected by torsion angle between adjacent phenyl rings to relieve steric strain in polymer chain. The conductivity of copolymer decreases with increasing the content of *N*-phenylaniline in copolymers is due to the introduction of phenyl groups which reduce the conjugation of the polymer chain and reduces the mobility of charge carrier along the main chain.

2.6 CONCLUSIONS

The chemical copolymerization of aniline with *N*-phenylaniline has been carried out. The ratio of the two monomers has influence on the copolymerization process due to different reactivity of monomers. Elemental analysis data infer that one anion is substituted for every two phenyl rings for complete protonation. Spectroscopic information indicates the presence of both the comonomers in the polymer chain. The temperature dependence of conductivity suggests that copolymers are quasi 1D-disordered state composed of 3D-metallic crystalline regions in 1D-localized amorphous regions. The substituent group decreases the conductivity of copolymers and the coplanarity of the polymer chain gets disrupted. It also reduces the mobility of the charge carriers along the main chain. The solubility, electrical conductivity, and thermal stability can be modified by varying comonomer composition and it is depending on the substituent group. A soluble polymer is more easily processable than insoluble and is thus more attractive to industry.

ACKNOWLEDGMENTS

Thanks are due to the Director of RSIC Lucknow for recording the FTIR spectra.

KEYWORDS

- polyaniline
- copolymers
- poly(aniline-*co*-N-phenylaniline)
- transport properties
- conductivity

REFERENCES

1. Syed, A. A.; Dinesan, M. K.; Genies, E. M. *Electrochemistry* **1988**, *4*, 737–742.
2. MacDiarmide, A. G.; Mu, S. L.; Somasiri, N. L. D.; Wu, W. *Mol. Cryst. Liq. Cryst.* **1985**, *121*, 173–180.
3. Dao, L. D.; Leclerc, M.; Guay, J.; Chevalier, J. W. *Synth. Met.* **1989**, *29*, 377–385.
4. Savitha, P.; Rao, P. S.; Sathyanarayana, D. N. *Polym. Int.* **2005**, *54*, 1243–1250.
5. Gupta, M. C.; Umar, S. S. *Macromolecules* **1992**, *25*, 138–142.

6. Gupta, M. C.; Borkar, A. D. *Ind. J. Chem.* **1990**, *29A*, 613–634.
7. Umare, S. S.; Huque, M. M.; Gupta, M. C.; Viswanath, M. G. *Macromol. Rep.* **1996**, *33A*(Suppl. 7 & 8), 381–389.
8. Borkar, A. D.; Gupta, M. C.; Umare, S. S. *Polym. Plast. Technol. Eng.* **2001**, *40*(2), 225–234.
9. Borkar, A. D.; Umare, S. S.; Gupta, M. C. *Prog. Cryst. Grow. Charact. Mat.* **2002**, *44*, 201–209.
10. Ghosh, S. G.; Kalapagam, V. *Synth. Met.* **1989**, *1*, 33–40.
11. Ginder, J. M.; Epstein, A. J. *Phys. Rev.* **1990**, *41B*, 10674–10680.
12. Shante, V. K. S.; Verma, C. M.; Bloch, A. N. *Phys. Rev.* **1973**, *138*, 4885–4893.
13. Heeger, A. J.; Kivelson, S. A.; Schrieffer, J. R.; Su, W. P. *Rev. Mod. Phys. 60*, **1998**, 781–789.
14. Pouget, J. P.; Jozefowicz, M. E.; Epstein, A. J.; Tang, X.; MacDiarmid, A. G. *Macromolecules* **1991**, *24*, 779–789.
15. Jozefowicz, M. E.; Laversana, R.; Javadi, H. S.; Epstein, A. J.; Pouget, J. P.; Tang, X.; MacDiarmid, A. G. *Phys. Rev.* **1998**, *39B*, 12598–12605.
16. Wang, Z. H.; Li, C.; Scherr, E. M.; MacDiarmid, A. G.; A. J. Epstein, A. J. *Phys. Rev. Lett.* **1991**, *66*, 1745–1753.

CHAPTER 3

SURFACE-ALTERED POLY(METHYL METHACRYLATE) RESIN FOR ANTIFUNGAL DENTURES

ABHAY NARAYANE[1], AKSHAY MOHAN[2], and
SIDDHARTH MESHRAM[3*]

[1]*Department of Prosthodontics, SDKS Dental College and Hospital, Nagpur, India*

[2]*Department of Chemical Engineering, Laxminarayan Institute of Technology, RTM Nagpur University, Nagpur 440033, India*

[3]*Department of Applied Chemistry, Laxminarayan Institute of Technology, RTM Nagpur University, Nagpur 440033, India*

[*]*Corresponding author. E-mail: sidmesh2@gmail.com*

CONTENTS

Abstract	42
3.1 Introduction	42
3.2 Experimental Procedures	47
3.3 Results and Discussion	50
Keywords	52
References	52

ABSTRACT

Poly (methyl methacrylate) is the most popular acrylic resin used in denture fabrication, owing to its inert and colour retention properties. Denture stomatitis is a common oral disease, which is associated with the adherence of micro-organisms (*Candida albicans*) to denture surfaces. It was noticed that the localization of *C. albicans*, associated with about 90% of denture stomatitis cases mainly observed in patients affected by AIDS/HIV, results in deterioration of the denture. This chapter aims to induce antifungal properties through surface-altered acrylic denture resin. To achieve this, cationic surfactant such as such as hexadecyl trimethyl ammonium bromide was blended with PMMA resin to fabricate denture having antifungal properties. The microbial analysis clearly illustrates the adequate repellence of *C. albicans* due to the hydrophobic nature of QAS in hot cured denture with optimum dose. The novelty of the work lies in the use of surfactant to obtain antifungal denture following an economic route along with substituting the dependence on drugs used in customary treatment against candidiasis disease that may impart adverse effects on health. The application of the suggested technique avoids such adverse effects and provides as a viable solution.

3.1 INTRODUCTION

Denture-related stomatitis is probably the most common form of oral candidiasis and its reported prevalence varies widely ranging up to 65%.[1] The synthetic acrylic resins have a long, clinically proven history of use for dentures, since they exhibit adequate physical, mechanical, and esthetic properties;[2] however, they are susceptible to microbial adhesion, leading to denture stomatitis.[3] Every denture causes at least some harm to the underlying tissues either directly or indirectly. One important aspect of this harm is denture-related stomatitis. The presence of dentures may be the triggering local factor for diseased oral mucosa.[4] It was noticed that the localization of *Candida albicans*, associated with about 90% of denture stomatitis cases mainly observed in patients affected by AIDS/HIV, results in deterioration of the denture.

The use of polymers has revolutionized the biomedical industry ever since their discovery. Since time immemorial, the replacement of missing teeth has been a medical and cosmetic necessity for humans. Many prostheses and implants made from polymers have been in use for the last three decades, and there is a continuous search for more biocompatible and stronger polymer prosthetic materials.[5] Denture prosthetics has undergone many development stages since the first still preserved dentures were fabricated. The first sets of dentures were based on rubber and porcelain. With the emergence of polymer chemistry in the early 20th century, the foundation for the widespread use of removable dentures was laid. Quaternary Ammonium Compounds (QACs) are widely used as disinfectants in both medical and food environments.[6] Several authors have recently obtained encouraging results with antiseptic-coated devices using QACs which have proven efficient against microbial ecosystems, self-protected by a biofilm.[7–11] Several difficulties exist in producing a satisfactory denture material or designing a technique that is useful for its application. Conditions in oral cavity seem almost suited to annihilation. Biting stresses on dentures can be extremely high, temperatures may fluctuate between 25 and 45°C, and pH may change instantaneously from acidic to alkaline. The warm and moist oral environment, which is also enzyme and bacteria rich, is conducive to further decay. The soft tissues and structures in contact with the denture polymers may be injured from the toxic leaching or breakdown of the material. Synthetic plastic resins and lightweight metal alloys have made teeth more durable and natural looking. Most artificial dentures are made from high quality acrylic resins, which make them stronger and more attractive than was once possible.[12] Polymers such as polyamides, epoxy resin, polystyrene, vinyl acrylic, rubber graft copolymers, and polycarbonate have also been developed and tested as potential alternative denture-base materials. However, these have not generally proved successful.[13,14] Poly(methyl methacrylate) (PMMA) is the resin of choice for the fabrication of denture bases in clinical dentistry. It is inert, has color retention properties, and clearly defined polymerization process that is easy for modification. The surface properties of denture base materials are of paramount importance for maintaining the structural and functional integrity of the denture-covered oral mucosa. Control of denture plaque is facilitated if the denture surfaces, including the tissue surface, are well-polished as it will prevent the accumulation of debris.[15] The tissue surfaces are usually left as they were when removed from the cast; thus, it presents irregularities and microscopic pores that facilitate bacterial and fungal colonization.

 Many attempts have been made to modify PMMA taking advantage of the broad scope of modification available in polymer.[16] PMMA-based materials are available for the production of denture teeth on commercial scale too. Few of these materials contain organic and/or inorganic fillers, which improve the mechanical properties of the teeth. Alongside PMMA-based denture teeth, several composite teeth are also available among the marketable applications. The basic materials polymerize to a solid material during the production process of the denture teeth. The biocompatible properties of the solid material are different from those of the basic materials. The denture teeth consist of an insoluble polymer which is not accessible to the organism and can be regarded as inert. Only substances that may dissolve from the material could pose a risk of exposure to patients. Denture base acrylic resin is easily colonized by oral endogenous bacteria and *Candida* spp., and eventually by extraoral species such as *Staphylococcus* spp., Pseudomonadaceae, or members of Enterobacteriaceae with the duration of time over a few years. This microbial reservoir can be responsive for denture-related stomatitis and aspiration pneumonia, a life-threatening infection especially in geriatric patient.[17] In denture wearers, *Candida*, a yeast-like fungus, is a frequent cause of infections. One such infection is denture stomatitis.[18,19] *Candida*-associated denture stomatitis is a common recurring disease, observed in approximately 11–67% of otherwise healthy denture wearers.[20] Denture stomatitis is characterized by inflamed mucosa, particularly under the upper denture, and patients may complain of a burning sensation, discomfort, or bad taste, but in the majority of cases, they are unaware of the problem. Several excellent reviews discussing denture stomatitis in detail are available to the reader.[21–24] The major risk factor for the development of this condition is wearing an upper complete denture, particularly when it is not removed during sleep and cleaned regularly. Wearing dentures alters the oral microbiota. A microbial plaque composed of bacteria and/or yeast forms on the fitting surface of the denture (the surface which rests against the palate) and on the mucosa which is covered. Denture plaque is defined as a dense microbial layer comprising microorganisms and their metabolites, and it is known to contain more than 1011 organisms per gram in wet weight. Over time, this plaque may be colonized by *Candida* species. The local environment under a denture is more acidic and less exposed to the cleansing action of saliva, which favors high *Candida* enzymatic activity and may cause inflammation in the mucosa. *C. albicans* is the most commonly isolated organism, but occasionally bacteria are implicated. There are several factors, which affect *C. albicans* adhesion on acrylic resin surface;

quality and quantity of saliva, surface properties of acrylic resin[25] and oral bacteria.[26]

Mechanical irritation may possibly play a predisposing role by increasing the turnover of the epithelial cells, hence the barrier function of the epithelium is reduced and its penetration by microbial antigens is possibly enhanced. Due to the large size of *Candida*, the host phagocytes are unable to completely destroy it. Therefore, extracellular killing mechanisms are required to eliminate the *Candida*. The usual treatment of denture-related stomatitis involves elimination of the source of the infectious agent (denture cleansing and disinfection) and elimination of the oral tissues.[27] Moreover, the important aspect of treatment is improving denture hygiene, that is, removal of denture at night and disinfecting it by storing it in antiseptic solution. On the contrary, the medicinal treatment involves the resolution of the mucosal infection, for which topical antifungal medications are suggested, namely, nystatin, fluconazole/intraconazole, miconazole, amphotericin, and chlorhexidine as a mouth wash.[28] But, it is noteworthy that these medicines cause adverse effects on human health.[29–32] Some of the side effects are displayed in Table 3.1.

This is important as the denture is usually infected with *C. albicans* which will cause reinfection if it is not removed. The common antiseptic solution includes alkaline peroxides or hypochlorites which may overtime corrode the metal components of dental appliances. Quaternary ammonium salts (QASs) have been widely used in paint, water treatment, textiles, and food industry because of low toxicity and broad antifungal spectrum. They can also be chemically bound to the polymer carriers via active groups, thus integrating QAS monomers with composite matrix. As compared with the conventional antifungal agents of low molecular weight, the advantages of these polymerizable antibacterial agents include nonvolatility, chemical stability, and low permeation through skin. Hence, the application of QAS monomers to dentistry may provide more choices for development of dental antifungal materials.

Reports are also available on coating of cationic, anionic, and nonionic surfactants upon poly(ethylcyanoacrylate) nanoparticles to decrease the *C. albicans* both anti adherent and antifungal effects.[33] The requirement of dental resins is that they should be nontoxic, nonirritating, and otherwise nondetrimental to oral tissues. To fulfill these requirements, they should be preferably insoluble in saliva and all other body fluids. They should also not become insanitary or disagreeable in taste, odor or smell, and should be

TABLE 3.1 Side Effects of Drugs Commonly Prescribed for Denture Stomatitis.

Sr No.	Name of drug	Side effects
1	Nystatin	Breathing difficulty, closing of throat, swelling of lips and tongue, nausea, vomiting, diarrhea, skin irritation, rashes, itching, eczema, etc.
2	Amphotericin	Fever and chill, headache, irregular heartbeats and urination, nausea, vomiting, rigors, fever, hypertension or hypotension, and hypoxia (more common), nephrotoxicity, blurred or double vision, convulsions, weakness in hands or feet, shortness of breath, skin rash, itching, unusual bleeding or bruising (less common)
3	Miconazole	Local side effects, oral burning/discomfort, gingival pain and swelling, tongue unceleration, dry mouth, toothache, loss of taste in 12.1% HIV and 9.5% head and neck cancer patients. Gastrointestinal (9%), nausea (6.6%), nervous system (headache 7.6%), respiratory (cough 2.8%, respiratory infection 2.1%), hematologic (anemia 2.8%, lymphopenia 1.7%, neutropenia 0.7%), etc.
4	Fluconazole	Common (≥1% of patients): rash, headache, dizziness, nausea, vomiting, abdominal pain, diarrhea, and/or elevated liver enzymes Infrequent (0.1–1% of patients): anorexia, fatigue, constipation Rare (<0.1% of patients): oliguria, hypokalaemia, paraesthesia, seizures, alopecia, Stevens–Johnson syndrome, thrombocytopenia, other blood dyscrasias, serious hepatotoxicity including hepatic failure, anaphylactic/anaphylactoid reactions

highly stable. A quaternary ammonium silane-functionalized methacrylate (QAMS) represents an antifungal macro monomer synthesized by sol–gel chemistry and possesses flexible Si–O–Si bonds.[34] A partially hydrolyzed resin formulation containing QAMS copolymerized with bisphenol A glycidyl methacrylate retains kill-on-contact microbiocidal activities, which should contribute to prevent resin-based tooth filling failure caused by recurrence of caries. The drawback is that this method cannot be carried out on a laboratory scale and is quite expensive.[35] The results of the present study suggest that the modification of PMMA with methacrylic acid/methyl methacrylate and a surfactant HDTMABr having resistance to cationic microbial growth changes the physical properties of the resin, resilience and strength to biting, chewing, impact forces, and excessive wear under mastication. Hexadecyl trimethyl ammonium bromide being a QAC has been used in the denture making as an antifungal agent. It is stable under all conditions of service, including thermal and loading shocks although lighter in weight.[36,37]

3.2 EXPERIMENTAL PROCEDURES

3.2.1 MATERIALS AND METHODOLOGY

The materials generally used for heat-cured systems are as follows:

PMMA resin, benzoyl peroxide initiator, mercuric sulphide, cadmium sulphide—dyes; zinc oxide, titanium oxide—opacifiers; dibutyl phthalate—plasticizer; sodium alginate—anti-sticking agent; dyed particles—glasses; beads—for esthetics; dibutyl phthalate—plasticizer; glycol dimethacrylate (1–2%)—cross-linking agent; hydroquinone (0.006%)—inhibitor; acrylizing unit—for curing; and bench grinder—for finishing.

3.2.2 EXPERIMENTAL PROCESS

A mold was prepared with help of modeling wax sheet about 2 mm in thickness. The metal molds were initially thoroughly coated with a thin layer of sodium alginate, which was then allowed to dry prior to packing and processing for easy removal of the prepared denture. Polymer was mixed with QAS powder and monomer was added to it in the ratio of 3:1 by volume along with 0.2 g benzoyl peroxide as the initiator and hexadecyl trimethyl ammonium bromide salt as an antifungal agent. The mixture was then left to stand in a closed container until the dough stage had reached (approximately 10 min) and packed into the metallic mold half containing a deep coating of wax pattern void. A commercially available polythene sheet was placed over this and the opposing half of mold and investment then positioned and closed. A trial closure of the denture flask was performed by applying pressure. This pressure was increased slowly over a period of 1 min. Later, the mold was opened, the separating sheet was discarded and the excess flash was removed. The separating sheet was further reapplied and a check was made to ensure that the mold seal was intact. The acrylic material in the packed mold was then polymerized by heating for 90 min at 80–90°C followed by bench cooling overnight. The polymerized samples were deflasked and placed in the solution of sodium citrate to remove any residual stone adhered to it.

It was noticed that the denture thus formed was white in color as the hot-cured acrylic material had lost its pink color significantly and decolorized during curing due to the addition of excess amount (7.5%) of QAS as indicated in Figure 3.1. Such a white-colored denture may not be acceptable in the market as it resembles a different color than the natural pink denture and hence it indicates a drawback.

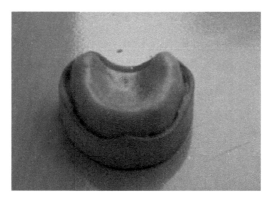

FIGURE 3.1 Decolorization of denture due to excess QAS.

To overcome this issue of decolorization of acrylic resin, studies have been conducted as per Table 3.2 toward the optimization of dose of QAS used while retaining satisfactory antifungal properties. The denture containing optimized QAS dose, after breaking open the mold, was found to satisfactorily retain its color as depicted in Figure 3.2.

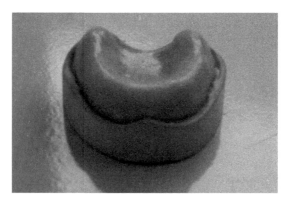

FIGURE 3.2 Color retention of denture with optimized quantity of QAS.

TABLE 3.2 QAS Dose Optimization Studies.

Sr. No.	Amount of polymer used (g)	% Conc. of QAMS	Amount of QAMS (g)
1 (Control)	6.40	Nil	Nil
2	6.41	1	0.064
3	6.40	2.5	0.16
4	6.40	5	0.32
5	6.39	7.5 (Excess)	0.48

3.2.3 MICROBIAL ANALYSIS

3.2.3.1 ANTIFUNGAL TEST

The denture was scrapped into fine powder with the help of metallic bur attached to lab micro motor as shown in Figure 3.3. And an amount of approximately 1.5 g was transferred to a test-tube containing 5 ml of xylene solvent and kept for 2–3 days. It was observed that the fine powder that was initially insoluble in the xylene solvent is completely dissolved in xylene after a lapse of few days.

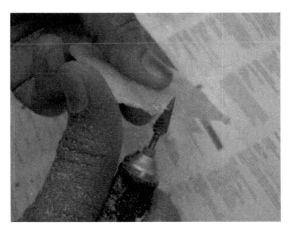

FIGURE 3.3 Sample grinding with the help of metallic bur attached to lab micro motor.

This solvent containing the denture base powder was then sent to the microbiology laboratory at Rajiv Gandhi Biotechnological Centre for testing under biological conditions to determine its antifungal properties. A volume of 25 ml potato dextrose agar was autoclaved at 121.5°C for 15 min, poured in sterile petri dish, and left for solidification. To this, 500 µl culture of *C. albicans* was taken and evenly distributed with the help of spreader. The testing procedure consisted of growth of the *C. albicans* microbes in broth and then the determination of antifungal activity by Kirby–Bauer method by maintaining a pH of 7 and temperature of 25–27°C was performed. The main steps involved are inoculum preparation either by growth method or direct colony suspension method. The samples were subjected for duration of 72 h to obtain the results. After completion of incubation time, zone of inhibition was evaluated on these plates.

3.3 RESULTS AND DISCUSSION

The incorporation of QAS in fabricating hot-cured denture for improved antifungal properties proved to be highly encouraging. However, the excess QAS results in decolorization of the traditional color using hot-cured method that may not be commercially acceptable. Also, it was noticed that if the QAS is added in an excess amount to the cold-cured acrylic material and heated then the outer appearance of the denture thus formed reflects white pores as shown in Figure 3.4.

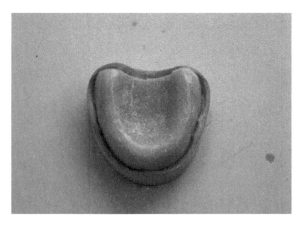

FIGURE 3.4 Cold-cured acrylic denture containing excess QAS.

To address this issue, experiments were conducted toward the incorporation of antifungal agent with hot curing in variable stoichiometry with resin material.

It was noticed from antifungal repellence studies that 5% QAS gave best results, but reappearance of white color was observed as shown in Figure 3.5. Hence, 2.5% was chosen as optimum dose as it maintained good color retention as well as antifungal properties.

QAS, the antifungal agent, can copolymerize with other monomers incorporated into the resinous material, the polymer attains antifungal property after curing. Therefore, the antimicrobial portion will not leach from the material and the composite resin, thus destroys any microbes that come in contact with the surface. Here, the presence of antifungal agent, that is, hexadecyl trimethyl ammonium bromide represents large number of carbon atoms in backbone chain, making it hydrophobic (water repelling) in nature.

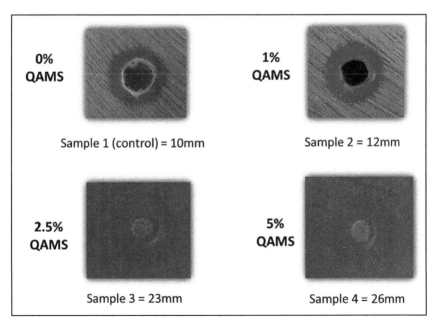

FIGURE 3.5 Repellence exhibited by hot-cured antifungal resin with varying dose of QAS.

The large surface provided by the polymeric denture results in the increased colonization of *C. albicans* microorganism. These organisms commute through the oral mucosa and reach the denture along with the saliva. Hence, the hydrophobic nature of the surface modified denture assists in the repellence of *C. albicans*, as the saliva is repelled due to the hydrophobicity. This nature of the proposed denture may cause a dryness feeling to the patient but may also prove effective against other diseases which similarly enter through the saliva.

The result achieved using the optimum dose of QAS as an antifungal agent is quite encouraging. This also helps in elimination of prescription dose of medicine to avoid the side effects on patients. The fabrication of the antifungal agent incorporated denture is nontedious and furthermore can be carried out on a commercial scale keeping it economically viable. The proposed future work includes, a detailed comparative patient study and analysis of the antifungal effect and shell life of material for practical use in patients suffering from diseases such as AIDS.

KEYWORDS

- **acrylic**
- **resin**
- **antifungal**
- **denture**
- *C. albicans*
- **quaternary ammonium salt**

REFERENCES

1. Jain, D.; Shakya, P. An *In Vitro* Study on Effect of Delmopinol Application on *Candida albicans* Adherence on Heat Cured Denture Base Acrylic Resin: A Thorough Study. *Indian J. Dent. Res.* **2013,** *24*(5), 645.
2. Peyton, F. A. History of Resins in Dentistry. *Dent. Clin. North. Am.* **1975,** *19*, 211–222.
3. Dhir, G.; Berzins, D. W.; Dhuru, V. B.; Periathamby, A. R.; Dentino, A. Physical Properties of Denture Base Resins Potentially Resistant to Candida Adhesion. *J. Prosthodont.* **2007,** *16*, 465–472.
4. Budtz-Jörgensen, E.; Lombardi, T. Antifungal Therapy in the Oral Cavity. *Periodontology* 2000 **1996,** *10*, 89–106.

5. Bhola, R. G.; Bhola, S. M.; Liang, H. J.; Mishra, B. Biocompatible Denture Polymers. *Trends Biomater. Artif. Organs* **2010**, *23*, 129–136.
6. Sundheim, G.; Langsrud, S.; Heir, E.; Holck, A. Bacterial Resistance to Disinfectants Containing Quaternary Ammonium Compounds. *Int. Biodeterior. Biodegrad.* **1998**, *41*, 235–239. DOI:10.1016/S0964-8305(98)00027-4.
7. Pesci-Bardon, C.; Fosse, T.; Madinier, I.; Serre, D. In Vitro New Dialysis Protocol to Assay the Antiseptic Properties of a Quaternary Ammonium Compound Polymerize with Denture Acrylic Resin. *Lett. Appl. Microbiol.* **2004**, *39*, 226–231. DOI:10.1111/j.1472-765X.2004.01569.x.
8. Chandra, J.; Mukherjee, P. K.; Leidich, S. D.; Faddoul, F. F.; Hoyer, L. L.; Douglas, L. J.; et al. Antifungal Resistance of Candidal Biofilms Formed on Denture Acrylic In Vitro. *J. Dent. Res.* **2001**, *80*, 903–908. DOI:10.1177/00220345010800031101.
9. Veenstra, D. L.; Saint, S.; Saha, S.; Lumley, T.; Sullivan, S. D. Efficacy of Antiseptic-impregnated Central Venous Catheters in Preventing Catheter-related Bloodstream Infection: A Meta-analysis. *J. Am. Med. Assoc.* **1999**, *281*, 261–267. DOI:10.1001/jama.281.3.261.
10. Lane, R. K.; Matthay, M. A. Central Line Infections. *Curr. Opin. Crit. Care* **2002**, *8*, 441–448. DOI:10.1097/00075198-200210000-00012.
11. Russell, A. D. Introduction of Biocides in Clinical Practice and the Impact on Antibiotic-resistant Bacteria. *J. Appl. Microbiol.* **2002**, *92*, 121–135. DOI:10.1046/j.1365-2672.92.5s1.12.x.
12. Park, S.; Chao, M.; Raj, P. Mechanical Properties by Surface Charged Poly(Methyl Methacrylate) as Denture Resins. *Int. J. Dentistry* **2009**, Article ID 841431, DOI:10.1155/2009/841431.
13. Monterio, D. Silver Distribution and Release from an Antimicrobial Denture Base Resin Containing Silver Colloidal Nanoparticles. *J. Prosthodont.* **2012**, *21*, 7–15. DOI:10.1111/j.1532-849X.2011.00772.x.
14. Silva, W. Poly(Methyl Methacrylate) Absorption and Releasing of Nystatin and Fluconazole. *Arch Oral Res.* **2011**, *7*(3), 259–266.
15. Sesma, N.; Laganá, D. C.; Morimoto, S.; Gil, C. Effect of Denture Surface Glazing on Denture Plaque Formation. *Braz. Dent. J.* **2005**, *16*, 129–134.
16. Beth, C. A Comparison of Polymeric Denture Base Materials, M.Sc. Thesis, University of Glasgow: Glasgow, 2010.
17. Holm, B. *Treatment of Denture Stomatitis: A Clinical, Microbiological, Histological Thesis*, Sweden, 1982.
18. Nikawa, H. A Review of In Vitro and In Vivo Methods to Evaluate the Efficacy of Denture Cleansers. *Int. J. Prosthodont.* **1999**, *12*, 153–159.
19. Paranhos, H. Effects of Mechanical and Chemical Methods on Denture Biofilm Accumulation. *J. Oral Rehabil.* **2007**, *34*, 606–612. DOI:10.1111/j.1365-2842.2007.01753.x.
20. Arendorf, T. M., Walker, D. M. Denture Stomatitis: A Review. *J. Oral Rehabil.* **1987**, *14*, 217–227. DOI:10.1111/j.1365-2842.1987.tb00713.x.
21. Iacopino, A. M.; Wathen, W. F. Oral Candidal Infection and Denture Stomatitis: A Comprehensive Review. *J. Am. Dent. Assoc.* **1992**, *123*, 46–51.
22. Budtz-Jorgensen, E. The Significance of *Candida albicans* in Denture Stomatitis. *Scand. J. Dent. Res.* **1974**, 82, 151–190. DOI:10.1111/j.1600-0722.1974.tb00378.x.

23. Radford, D. R.; Challacombe, S. J.; Walter, J. D. Denture Plaque and Adherence of *Candida albicans* to Denture-base Materials In Vivo and In Vitro. *Crit. Rev. Oral Biol. Med.* **1999**, *10*, 99–116. DOI:10.1177/10454411990100010501.
24. Webb, B. C.; Thomas, C. J.; Willcox, M. D.; Harty, D. W.; Knox, K. W. *Candida*-associated Denture Stomatitis. Aetiology and Management: A Review. Part 1. Factors Influencing Distribution of *Candida* Species in the Oral Cavity. *Aust. Dent. J.* **1998**, *43*, 45–50. DOI:10.1111/j.1834-7819.1998.tb00152.x.
25. Pereira-Cenci, T.; Del Bel Cury, A. A.; Crielaard, W.; Ten Cate, J. M. Development of *Candida*-associated Denture Stomatitis: New Insights. *J. Appl. Oral Sci.* **2008**, *16*, 86–94.
26. Webb, B. C.; Thomas, C. J.; Willcox, M. D.; Harty, D. W.; Knox, K. W. Candida-associated Denture Stomatitis. Aetiology and Management: A Review. Part 2. Oral Diseases Caused by *Candida* Species. *Aust. Dent. J.* **1998**, *43*, 160–166.
27. Yang, J.; Kim, H.; Chung, C. Photo-catalytic Antifungal Activity against *Candida albicans* by TiO_2 Coated Acrylic Resin Denture Base. *J. Korean Acad. Prosthodont.* **2006**, *44*, 284–294.
28. Crispian, S. *Oral and Maxillofacial Medicine: The Basis of Diagnosis and Treatment*, 2nd ed., Churchill Livingstone: Edinburgh, 2008; pp 201–203. ISBN 9780443068188.
29. John, P.; Facoep, D. Nystatin Oral Suspension Side Effects Center, 2014. Retreived from http://www.rxlist.com/nystatin-oral-suspension-side-effects-drug-center.htm (last accessed 20 April 2014).
30. Rossi, S. *Australian Medicines Handbook 2006*. Australian Medicines Handbook: Adelaide, 2006, ISBN 0-9757919-2-3.
31. Laniado, R.; Vargas, M. N. Amphotericin B: Side Effects and Toxicity. *Rev. Iberoam. Micol.* **2009**, *26*(4), 223–227. DOI:10.1016/j.riam.2009.06.003.
32. Oravig Side Effects. Retreived from http://www.drugs.com/sfx/oravig-side-effects.html (last accessed 22 April 2014).
33. McCarron, P.; Donnelly, R.; Marouf, W.; Calvert, D.; Anti-adherent and Antifungal Activities of Surfactant-coated Poly(ethylcyanoacrylate) Nanoparticles. *Int. J. Phar.* **2007**, *340*, 182–190. DOI:10.1016/j.ijpharm.2007.03.029.
34. Gottenbos, B. In Vitro and In Vivo Antimicrobial Activity of Covalently Coupled Quaternary Ammonium Silane Coatings on Silicone Rubber. *Biomaterials* **2002**, *23*, 1417–1423. DOI:10.1016/S0142-9612(01)00263-0.
35. Gong, S.; Niu, L.; Kemp, L.; Yiu, C.; Ryou, H.; Blizzard, J., et al. Quaternary Ammonium Silane-functionalized, Methacrylate Resin Composition with Antimicrobial Activities and Self-repair Potential. *Acta Biomater.* **2012**, *8*, 3270–3282. DOI:10.1016/j.actbio.2012.05.031.
36. Isquith, A.; Abbott, E.; Walters, P. Surface-bonded Antimicrobial Activity of an Organosilicon Quaternary Ammonium Chloride. *Appl. Microbiol.* **1972**, *24*, 859–863.
37. Tezel, U. Fate and Effect of Quaternary Ammonium Compounds in Biological Systems, Desertation Submitted to Department of Civil and Environmental Engineering, Georgia Institute of Technology, 2009.

CHAPTER 4

POLYBENZIMIDAZOLE AS MEMBRANES IN DIRECT METHANOL FUEL CELLS

P. P. KUNDU* and S. MONDAL

Advanced Polymer Laboratory, Department of Polymer Science and Technology, University of Calcutta, 92 APC Road, Kolkata 700009, West Bengal, India

*Corresponding author. E-mail: ppk923@yahoo.com.

CONTENTS

Abstract	56
4.1 Introduction	56
4.2 PBI Synthesis	58
4.3 Formation of PBI Membranes/Membrane Casting	60
4.4 Properties of PBI as PEM	61
4.5 DMFC Uses	67
Keywords	69
References	69

ABSTRACT

This is the era of advanced materials. Whether it is specialty polymers, composites, or nanomaterials, the interests are growing on demanding applications like space, power, and electronic devices. Not only the high temperature stability but also the material required in these applications needs a combination of physical, chemical, and mechanical properties. Aromatic polybenzimidazoles are one of the high-performance polymers which have drawn attention for research and development. The insolubility of wholly aromatic PBIs in common organic solvents is a big limitation for their processing and industrial uses. Despite the limitation, the outstanding properties of PBIs attract the researcher to investigate their use in gas separation, pervaporation, and polymer electrolyte membranes in fuel cell applications.

4.1 INTRODUCTION

This is the era of advanced materials. Whether it is specialty polymers, composites, or nanomaterials, the interests are growing on demanding applications like space, power, and electronic devices. Not only the high temperature stability, but also the material required in these applications needs a combination of physical, chemical, and mechanical properties.[1] Aromatic polybenzimidazoles (PBIs) are one of the high-performance polymers which have drawn attention for research and development. The insolubility of wholly aromatic PBIs in common organic solvents is a big limitation for their processing and industrial uses. Despite the limitation, the outstanding properties of PBIs attract the researcher to investigate their use in gas separation, pervaporation, and polymer electrolyte membranes (PEMs) in fuel cell applications.

Due to the increasing consumption of fossil fuels, increasing power demands and increasing pollution, alternative energy sources are very important nowadays. Among the clean and renewable energy, fuel cells are one of the most popular option. The first concept of fuel cell was built in 1829.[2] In 1960, NASA employed fuel cell in Gemini and Apollo Space Missions.[3] Till then, different types of fuel cells are reviewed in literature,[4–9] covering solid oxide fuel cells, molten carbonate fuel cell, phosphoric acid fuel cell, alkaline fuel cell, polymer electrolyte membrane fuel cells (PEMFCs), and direct methanol fuel cells (DMFCs). Among those, the most promising fuel cell system was PEMFC and DMFC. As shown in Figure 4.1, due to the inexpensive and easy storage of methanol fuel, simplified design, and high volumetric theoretical energy density of methanol, application of DMFC is ever-increasing.

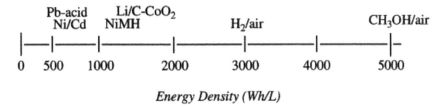

FIGURE 4.1 Theoretical energy density of batteries, H_2-PEMFCs and DMFCs.[10]

Protonic electrolyte-based DMFCs are directly fed with a mixture of methanol/water at the anode. Methanol is directly oxidized to carbon dioxide, although the possible formation of compounds such as formaldehyde, formic acid, or other organic molecules is not excluded. The formation of such organic molecules decreases the use of fuel for generation of energy. A scheme of the overall reaction process occurring in a DMFC equipped with a proton conducting electrolyte is outlined below:

$$CH_3OH + H_2O \rightarrow CO_2 + 6H^+ + 6e^- \quad \text{(anode)}$$

$$3/2\, O_2 + 6H^+ + 6e^- \rightarrow 3H_2O \quad \text{(cathode)}$$

$$CH_3OH + 3/2\, O_2 \rightarrow CO_2 + 2H_2O \quad \text{(overall)}$$

In the presence of an alkaline electrolyte, this process can be written as follows:

$$CH_3OH + 6OH^- \rightarrow CO_2 + 5H_2O + 6e^- \quad \text{(anode)}$$

$$3/2O_2 + 3H_2O + 6e^- \rightarrow 6OH^- \quad \text{(cathode)}$$

$$CH_3OH + 3/2O_2 \rightarrow CO_2 + 2H_2O \quad \text{(overall)}$$

In this chapter, we will discuss about synthesis, properties, and application of PBI membranes in DMFC.

4.2 PBI SYNTHESIS

In the early 1960s, a cooperative effort of the US Air Force Materials Laboratory with Dupont and the Celanese Research Company led to the synthesis of PBI fiber.[11] Poly(2,2'-m-phenylene-5,5'-bibenzimidazole) is the most widely investigated PBIs, commonly referred to as *m*-PBI as shown in Figure 4.2.

FIGURE 4.2 *m*-PBI.

Three types of procedures are available for synthesis of PBIs:

- thermally rearranged polymerization;
- solution polymerization; and
- catalytic polymerization.

4.2.1 THERMALLY REARRANGED POLYMERIZATION

In this method, 3,3'-diaminobenzidinetetrahydrochloride was dissolved in *N*-methyl-2-pyrrolidone (NMP), taken in a three-necked round-bottom flask and was stirred for 2 h under a nitrogen atmosphere at 60°C. The poly(amino amic acid) solution was obtained by a slow addition of equimolar amount of dianhydride, 4,4-(hexafluoroisopropylidene)diphthalic anhydride (dissolved in NMP), followed by heating to 80°C. After stirring for 12 h, the resulting solution was cast onto a glass plate, stored overnight at 80°C, slowly heated at less than 2°C min^{-1}, and thermally imidized at 250°C in a vacuum oven.[12] Resulting polyaminoimide membrane was thermally rearranged to polypyrrolone (PPL) at 300–450°C. The PPL membrane was then immersed in 1 M NaOH solution at 100°C for 3 h. The intermediate membrane was then treated with 0.1 M HCl in aqueous solution at room temperature, rinsed several times with water, and dried

12 h at 150°C under high vacuum. The final thermally rearranged PBI (TR-PBI) membrane was obtained through further heat treatment at 450°C for 1 h.[1,13] A schematic representations were given in Figure 4.3.

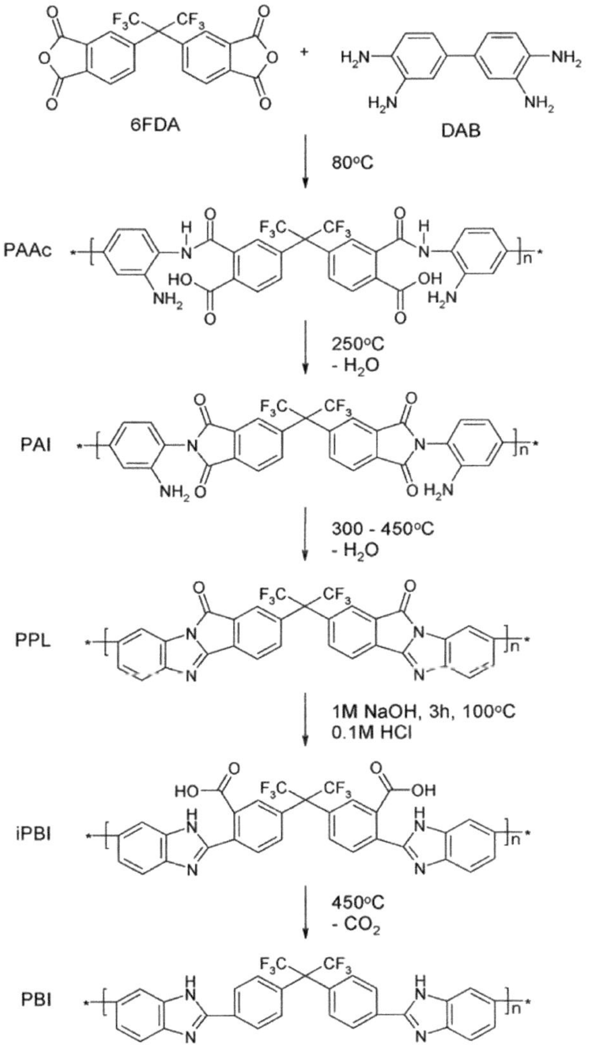

FIGURE 4.3 New synthesis route to microporous polybenzimidazole (TR-PBI) from polypyrrolone (PPL) using a thermal rearrangement concept.[13] (Reprinted from Han, S. H.; Lee, J. E.; Lee, K.-J.; Park, H. B.; Lee, Y. M. Highly Gas Permeable and Microporous Polybenzimidazole Membrane by Thermal Rearrangement. *J. Membr. Sci.* **2010**, *357*(1–2), 143–151. © 2010, with permission from Elsevier.)

4.2.2 CATALYTIC POLYMERIZATION

In a three-necked flask equipped with a mechanical stirrer, the tetra amine and solvent PPA were added. The solution was stirred at 110°C under nitrogen atmosphere for about 1.5 h to get a homogeneous solution. An equimolar amount of diacid was added to the solution, and the reaction was continued for 12 h at 140°C. Then, a catalytic amount of phosphorus pentoxide and triphenyl phosphite was added into the system. The solution became brownish and viscous. It was heated to about 230°C for another 24 h and then poured into deionized water, washed with deionized water several times, and neutralized by alkaline solution. Finally, the polymer was dried under vacuum for 24 h.

4.2.3 SOLUTION POLYMERIZATION

In this system, the polymerization is done in the arrangement state, in which polyphosphoric acid (PPA) or methane sulfonic acid/P_2O_5 (PPMA) are utilized as solvents. In a three-necked round bottomed flask furnished with a nitrogen channel and mechanical stirrer, the equimolar tetraamine, and dicarboxylic acids or their derivatives were initially included and afterward dissolved in PPA or PPMA at 140°C. The reaction mixture was then heated to 180–210°C for about 18–24 h with constant stirring and finally poured into the water and neutralized by alkaline solution. The fibrous polymers were then dried under vacuum. PPA acts as both the solvent and dehydrating agent at the same time during polyheterocyclization. PBI dissolves in strong acids, bases, and a few organic solvents and membranes can be cast from the solutions.

4.3 FORMATION OF PBI MEMBRANES/MEMBRANE CASTING

4.3.1 CASTING FROM DIMETHYL ACETAMIDE

In general, the most used solvent to dissolve PBI is *N,N*-dimethyl acetamide (DMAc). Dissolution takes place at a temperature above the boiling point (165°C) of the solvent and the presence of traces of oxygen should be excluded. Then, the membranes are casted on a clean glass plate. After that, the membranes are dried in an oven at 60–120°C temperature.

Finally, they are washed with hot water and again dried in a vacuum oven at 120°C.

4.3.2 CASTING FROM TRIFLUOROACETIC ACID AND PHOSPHORIC ACID

PBI is soluble in a mixture of trifluoroacetic acid (TFA) and phosphoric acid.[14] PBI is first refluxed with TFA for few hour. After that, required amount of H_3PO_4 is added for desired doping level. For complete dissolution of PBI, little amount of water is also added. Then, the membrane is casted under nitrogen atmosphere and dried at room temperature. TFA cast membranes are generally more rubbery than DMAc cast membranes.

4.3.3 CASTING FROM POLYPHOSPHORIC ACID

PPA is used as an efficient condensation reagent and solvent for PBI synthesis. Recently, Xiao et al.[15] have developed a sol–gel process to fabricate PBI–H_3PO_4 membranes directly from the PBI solution in PPA at around 200°C. This saved the time for isolation or redissolution of the polymer after synthesis. After casting, PPA hydrolyzed to phosphoric acid by moisture from the surrounding environment and induces a sol–gel transition, resulting in phosphoric acid-doped PBI membranes.

4.3.4 CASTING FROM ALKALINE SOLUTION

PBI membranes have also been prepared from a solution in a mixture of NaOH and ethanol.[16] The membrane is dried under nitrogen atmosphere. Thorough washing is needed until the pH reaches 7.

4.4 PROPERTIES OF PBI AS PEM

4.4.1 WATER UPTAKES AND CONDUCTIVITY

The water uptake values and proton conductivity of PEMs are very important properties for their application in DMFC. The affinity toward moisture is very well-known fact for PBIs. Due to its hydrophilic nature, PBI can absorb up to 15–19 wt% of water (3.2 water molecules per repeating units). This high water uptake values corresponds to the intermolecular hydrogen bonding between N and N–H groups of PBI and water molecules.[17,18] Earlier reports about the conductivity of pristine PBI were very contradictory. The values about 2×10^{-4}–8×10^{-4} S cm^{-1} was reported at relative humidity between 0% and 100%.[19] But other groups[20–22] published the proton conductivity value as 10^{-12} S cm^{-1} for pure PBI. These latter values are accepted for pristine PBI. But, these values are too low to be accepted in fuel cell applications. Thus, modification was necessary to improve the proton conduction properties of pure PBI. Two possible routes have been developed to improve the proton-conduction properties: (1) complexation with acids and (2) grafting of functional groups onto PBI. The reactions are possible because of the specific reactivity of PBI, arising from the –N= and –N–H groups of the imidazole ring.[23]

4.4.1.1 ACID DOPING

PBI readily reacts with inorganic and organic acids due to its basic character (pK_a value of ~5.5).[24,25] Different inorganic acids (H_2SO_4, H_3PO4, HCl, HNO_3, $HClO_4$)[26–30] have been used to provide good proton conductivity to the PBI membrane. After a high amount of doping by different concentrated acids, the order of conductivity of PBI membranes is observed as given in Table 4.1.

$$H_2SO_4 > H_3PO_4 > HNO_3 > HClO_4 > HCl$$

Although sulfuric acid shows the maximum conductivity, H_3PO_4 is the most popular acid among the researchers. Conductivity is directly proportional to the amount of acid present in the membrane. As the acid contents increases, the conductivity increases as shown in Figures 4.4 and 4.5. Some organic acids such as CH_3SO_3H, $C_2H_5SO_3H$, and aromatic phosphoric acids have also been investigated but the conductivity was less than inorganic acids.[32,33]

TABLE 4.1 Conductivity at Room Temperature after Immersion of PBI Film for 10 days in Acids of Indicated Concentration.[31]

	PBI/HCl (11.8 mol dm^{-3})	PBI/HClO$_4$ (11.6 mol dm^{-3})	PBI/HNO$_3$ (15.8 mol dm^{-3})	PBI/H$_3$PO$_4$ (14.4 mol dm^{-3})	PBI/H$_2$SO$_4$ (16 mol dm^{-3})
Conductivity (S cm^{-1})	1.4×10^{-3}	1.6×10^{-3}	1.8×10^{-3}	1.9×10^{-2}	6×10^{-2}

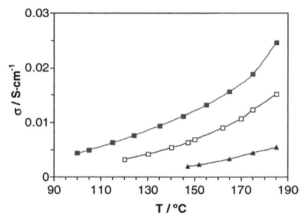

FIGURE 4.4 Conductivity of anhydrous ABPBI·2.7 H$_3$PO$_4$ (69%) (■), ABPBI·2.5 H$_3$PO$_4$ (68%) (□), and ABPBI·1.9 H$_3$PO$_4$ (62%) (▲) as a function of temperature. This plot shows the effect of the acid percentage present in the membrane.[34]

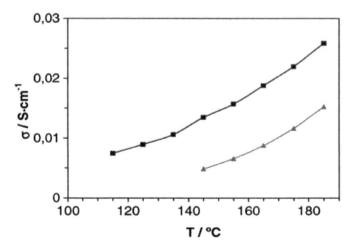

FIGURE 4.5 Conductivity of anhydrous PBI·6.7 H$_3$PO$_4$ (68.1% H$_3$PO$_4$) (■) and PBI·4.7 H$_3$PO$_4$ (59.7% H$_3$PO$_4$) (▲).[34]

Importantly, for the functioning of a fuel cell at temperatures above 100°C, the electroosmotic drag number of PBI/H_3PO_4 should be almost zero.[35] Slight gas hydration can therefore be used to prevent dehydration of the membrane, which may also assist in reducing reactant crossover. For this reason, PBI can be suitable as membrane for DMFC.[31]

4.4.2 MECHANICAL STRENGTH

The dry PBI membrane shows medium tensile strength about 60–70 MPa and a small elongation at brake about 2%. But, when saturated with water, both elongation at brake and tensile strength of PBI increase to 7–10% and 100–160 MPa, respectively.[36] Extensive hydrogen bonding between N and –NH– groups in PBI helps to form a close chain packing and therefore gives good mechanical strength of membranes. For acid doped membrane, mechanical strength reduced as the doping level increases. At low phosphoric acid-doping level (less than 2), intermolecular interaction between the chains of PBIs are decreased. On the other hand, the cohesion between N-atom and phosphoric acid are also observed. For these two opposite effects, no significant loss in strength is observed.[19,37] Free acids are present at the basic site of PBI, when the doping level is higher and these free acids increases the chain separation. As a result, a decrease in mechanical strength is observed. The strength is also strongly influenced by the average molecular weight. In a range from 20,000 to 55,000 g mol^{-1}, the tensile strength of acid-doped PBI membranes is found to increase from 4 to 12 MPa.[19,38]

Thus to consider both, conductivity and mechanical strength, a compromised doping level around 5–6 mol of H_3PO_4 per mole of PBI has been suggested for DMAc and TFA-cast membranes.[36] For high molecular weight PBIs, doping level extended to a higher range of 10–13 with sufficient mechanical strength given in Figure 4.6.

In case of PPA cast membranes, at an acid-doping level of 20–40 mol of phosphoric acid per mole of PB, a rather high tensile strength of 1–3.5 MPa is reported.[39,40] The possible reason may be the very high molecular weight obtained from PPA cast membranes.

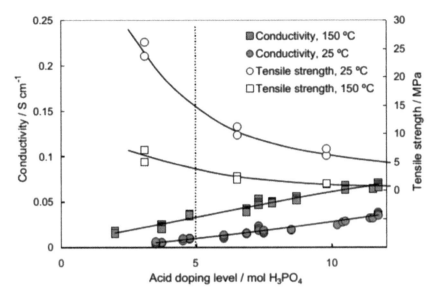

FIGURE 4.6 Conductivity and mechanical strength of acid doped PBI membranes as a function of the acid doping level.[36]

4.4.3 CHEMICAL AND THERMAL STABILITY

The thermal and chemical stability of a membrane is a major concern for its fuel cell application as long time durability is always desired. The instability of the PEM arises due to the harsh environment provided by the highly active catalyst. Chemical stability of membranes is of much concern to the lifetime of PEMFC. The principle degradation mechanism is assumed due to the in situ generation of H_2O_2 and ·OH and ·OOH radicals, which can attack the C– bonds of the membrane.[19] Usually, Fenton test is employed to determine the chemical stability of the membrane.[41,42] Most importantly, PBI membranes are less stable than Nafion 117. This is because the oxidative attacks occur preferentially to the weak linked nitrogen containing heterocyclic and adjacent aromatic rings.

But, it is interesting that ionical[43] and covalent[44] crosslinking of PBI can have comparable chemical stability to Nafion over the whole 120 h period of the test, shown in Figure 4.7.

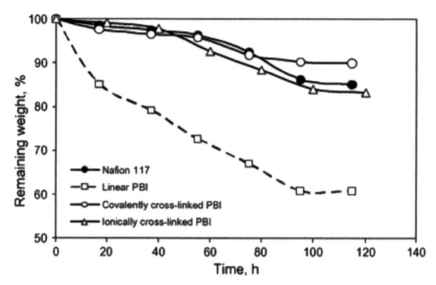

FIGURE 4.7 Membrane degradation in 3% H_2O_2 containing 4 ppm Fe^{2+} at 68°C. Solid lines indicate that the samples remained as a whole membrane, whereas dashed lines indicate that samples were broken into small pieces.[14] (Reprinted from Li, Q.; Jensen, J. O.; Savinell, R. F.; Bjerrum, N. J. High Temperature Proton Exchange Membranes Based on Polybenzimidazoles for Fuel Cells. *Progr. Polym. Sci.* **2009**, *34*(5), 449–477. © 2009, with permission from Elsevier.)

By thermogravimetric analysis, the thermal stability of PBI has been extensively studied.[45,46] At temperatures up to 150°C, a small weight loss (10–13%) is observed due to the absorbed water. But, no significant weight loss is found between 150 and 500°C. This shows the adequate thermal stability of this membrane for DMFC applications.

4.4.4 METHANOL CROSSOVER

A very simple two chambered diffusion cell separated by the membrane with liquid methanol solution in one side and pure water at the other side is used to evaluate the methanol crossover of the PEM.[47,48] Table 4.2 summarized the methanol crossover results as found in the literature. It seems that the undoped PBI membrane with the methanol crossover rate of 10^{-10} cm^{-2} s^{-1} shows less methanol crossover than that of the Nafion of about 10^{-8} cm^{-2} s^{-1}. These results indicate the usability of PBI as electrolyte for DMFCs.

TABLE 4.2 Methanol Crossover for Different PBIs and Nafion 117.[14]

Membranes	Conditions	Methanol crossover ($cm^2\ s^{-1}$)	Reference
Undoped PBI	1 M MeOH, 25°C	8.3×10^{-9}	[52]
Nafion 117		2.3×10^{-6}	
Undoped PBI	6 M MeOH, 25°C	1.5×10^{-9}	[50]
S-PBI		2.5×10^{-6}	
F6-PBI-5%MMT	1.8 M MeOH, 25°C	6×10^{-9}	[49]
SPOP-12%PBI	1 M MeOH, 60°C	1×10^{-7}	[51]
Nafion		3.5×10^{-6}	

(Reprinted from Li, Q.; Jensen, J. O.; Savinell, R. F.; Bjerrum, N. J. High Temperature Proton Exchange Membranes Based on Polybenzimidazoles for Fuel Cells. *Progr. Polym. Sci.* **2009**, *34*(5), 449–477. © 2009, with permission from Elsevier.)

4.5 DMFC USES

PBIs are generally used in mid-high temperature range in DMFC. Due to the poor conductivity of pure PBI, modified or doped PBIs are used for DMFC applications. The journey of the PBI membranes (doped with phosphoric acid) as a thermo-oxidative and stable proton conducting membranes for fuel cell applications began in 1994.[53] The membrane was tested by using various types of fuels including hydrogen,[54] methanol[24,25] and formic acid.[55] The results from Scopus database analysis based on the phrase "polybenzimidazole membranes for direct methanol fuel cells" are given in Figure 4.8.

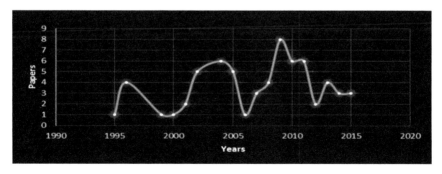

FIGURE 4.8 Number of paper published from 1995 to 2015.

Wainright et al. used an acid complexed PBI membrane in DMFC, prepared by doping of preformed membranes and by casting from a solution containing PBI and phosphoric acid.[24] With 4 mg cm^{-2} of Pt–Ru alloy electrode as anode and 4 mg cm^{-2} of Pt–black electrode as cathode, this DMFC produced power densities of 0.21 and 0.16 W cm^{-2}, respectively, at 500 mA cm^{-2} (feed of methanol/water mixture in 2/1 mole ratio and oxygen at atmospheric pressure).[24] At the temperature range 150–200°C, increased performance is attributed to the lower methanol crossover and higher electrolyte conductivity.

Due to the acid leaching problem at high temperature, researchers are trying to impregnate the sulfonic acid groups into PBI or blended PBI with some sulfonated/phosphonated polymers. BASF Fuel cells division has developed a membrane based on a blend of PBI and polyvinylphosphonic acid for its use in DMFCs called CeltecV.[56] Phosphonic acid-based electrolyte was immobilized in the PBI matrix and could not be leached out during operation. It was found that CeltecV MEAs showed ~50% lower methanol crossover than Nafion and at higher methanol concentrations (1.0 and 2.0 M), CeltecV showed superior performance over Nafion. Wycisk et al. prepared blend membranes of sulfonated poly[bis(phenoxy)phosphorene] (SPOP) and PBI and the DMFC performances were compared with Nafion 117. For an 82-µm thick membrane composed of 1.2 mmol g^{-1} IEC SPOP with 3 wt% PBI, the maximum power density was 89 mW cm^{-2} versus 96 mW cm^{-2} with Nafion 117, while the methanol crossover was 2.6 times lower than that with Nafion 117.[51]

The same group[57] prepared a membrane made of blends of PBI and Nafion for a DMFC operated at 60°C. Diaz et al.[58] mainly concentrated on the methanol permeability of acid-doped PBI and ABPBI membranes at the temperature range between 20 and 90°C. By comparison with the Nafion 117, the above works support the advantages of PBI-based membranes in increasing the cell performance, energy density, and also membrane selectivity.

The recent literatures are also focused on passive DMFC and alkaline DMFC by using PBI as an electrolyte membrane. Savadogo et al.[59] reported that the ionic conductivity of KOH-doped PBI membrane is in the range of 5×10^{-5}–10^{-1} S cm^{-1}; the highest ionic conductivity of 9×10^{-2} S cm^{-1} is achieved at 25°C.

Zhao et al. used H_3PO_4-doped PBI membrane as the electrolyte in passive DMFC and that can yield a peak power density of 37.2 and 22.1 mW cm^{-2} at 180°C when 16 M methanol solutions and neat methanol are used, respectively. In addition, the performance of this new DMFC was relatively stable during 132 h stability test.[60]

TABLE 4.2 Methanol Crossover for Different PBIs and Nafion 117.[14]

Membranes	Conditions	Methanol crossover ($cm^2\ s^{-1}$)	Reference
Undoped PBI	1 M MeOH, 25°C	8.3×10^{-9}	[52]
Nafion 117		2.3×10^{-6}	
Undoped PBI	6 M MeOH, 25°C	1.5×10^{-9}	[50]
S-PBI		2.5×10^{-6}	
F6-PBI-5%MMT	1.8 M MeOH, 25°C	6×10^{-9}	[49]
SPOP-12%PBI	1 M MeOH, 60°C	1×10^{-7}	[51]
Nafion		3.5×10^{-6}	

(Reprinted from Li, Q.; Jensen, J. O.; Savinell, R. F.; Bjerrum, N. J. High Temperature Proton Exchange Membranes Based on Polybenzimidazoles for Fuel Cells. *Progr. Polym. Sci.* **2009**, *34*(5), 449–477. © 2009, with permission from Elsevier.)

4.5 DMFC USES

PBIs are generally used in mid-high temperature range in DMFC. Due to the poor conductivity of pure PBI, modified or doped PBIs are used for DMFC applications. The journey of the PBI membranes (doped with phosphoric acid) as a thermo-oxidative and stable proton conducting membranes for fuel cell applications began in 1994.[53] The membrane was tested by using various types of fuels including hydrogen,[54] methanol[24,25] and formic acid.[55] The results from Scopus database analysis based on the phrase "polybenzimidazole membranes for direct methanol fuel cells" are given in Figure 4.8.

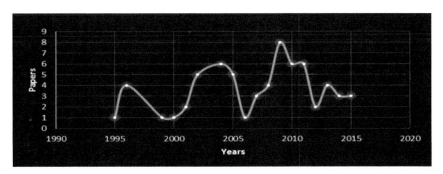

FIGURE 4.8 Number of paper published from 1995 to 2015.

Wainright et al. used an acid complexed PBI membrane in DMFC, prepared by doping of preformed membranes and by casting from a solution containing PBI and phosphoric acid.[24] With 4 mg cm^{-2} of Pt–Ru alloy electrode as anode and 4 mg cm^{-2} of Pt–black electrode as cathode, this DMFC produced power densities of 0.21 and 0.16 W cm^{-2}, respectively, at 500 mA cm^{-2} (feed of methanol/water mixture in 2/1 mole ratio and oxygen at atmospheric pressure).[24] At the temperature range 150–200°C, increased performance is attributed to the lower methanol crossover and higher electrolyte conductivity.

Due to the acid leaching problem at high temperature, researchers are trying to impregnate the sulfonic acid groups into PBI or blended PBI with some sulfonated/phosphonated polymers. BASF Fuel cells division has developed a membrane based on a blend of PBI and polyvinylphosphonic acid for its use in DMFCs called CeltecV.[56] Phosphonic acid-based electrolyte was immobilized in the PBI matrix and could not be leached out during operation. It was found that CeltecV MEAs showed ~50% lower methanol crossover than Nafion and at higher methanol concentrations (1.0 and 2.0 M), CeltecV showed superior performance over Nafion. Wycisk et al. prepared blend membranes of sulfonated poly[bis(phenoxy)phosphorene] (SPOP) and PBI and the DMFC performances were compared with Nafion 117. For an 82-μm thick membrane composed of 1.2 mmol g^{-1} IEC SPOP with 3 wt% PBI, the maximum power density was 89 mW cm^{-2} versus 96 mW cm^{-2} with Nafion 117, while the methanol crossover was 2.6 times lower than that with Nafion 117.[51]

The same group[57] prepared a membrane made of blends of PBI and Nafion for a DMFC operated at 60°C. Diaz et al.[58] mainly concentrated on the methanol permeability of acid-doped PBI and ABPBI membranes at the temperature range between 20 and 90°C. By comparison with the Nafion 117, the above works support the advantages of PBI-based membranes in increasing the cell performance, energy density, and also membrane selectivity.

The recent literatures are also focused on passive DMFC and alkaline DMFC by using PBI as an electrolyte membrane. Savadogo et al.[59] reported that the ionic conductivity of KOH-doped PBI membrane is in the range of 5×10^{-5}–10^{-1} S cm^{-1}; the highest ionic conductivity of 9×10^{-2} S cm^{-1} is achieved at 25°C.

Zhao et al. used H_3PO_4-doped PBI membrane as the electrolyte in passive DMFC and that can yield a peak power density of 37.2 and 22.1 mW cm^{-2} at 180°C when 16 M methanol solutions and neat methanol are used, respectively. In addition, the performance of this new DMFC was relatively stable during 132 h stability test.[60]

44. Li, Q.; Pan, C.; Jensen, J. O.; Noyé, P.; Bjerrum, N. J.; Cross-linked Polybenzimidazole Membranes for Fuel Cells. *Mater. Chem.* **2007**, *19*, 350–352.
45. Jaffe, M.; Haider, M. I.; Menczel, J.; Rafalko, J. Thermal Characterization of high-performance PBI and 6F Polymers and their Alloys. *Polym. Eng. Sci.* **1992**, *32*, 1236–1241.
46. Samms, S. R.; Wasmus, S.; Savinell, R. F. Thermal Stability of Proton Conducting Acid Doped Polybenzimidazole in Simulated Fuel Cell Environments. *J. Electrochem. Soc.* **1996**, *143*, 1225–1232.
47. Asensio, J. A.; Borro, S.; Romero, G. P. Polymer Electrolyte Fuel Cells Based on Phosphoric Acid-impregnated Poly(2,5-benzimidazole) Membranes. *J. Electrochem. Soc.* **2003**, *151*, A304–A310.
48. Mondal, S.; Soam, S.; Kundu, P. P. Reduction of Methanol Crossover and Improved Electrical Efficiency in Direct Methanol Fuel Cell by the Formation of a Thin Layer on Nafion 117 Membrane: Effect of Dip-coating of a Blend of Sulphonated PVdF-*co*-HFP and PBI. *J. Membr. Sci.* **2015**, *474*, 140–147.
49. Chuang, S. W.; Hsu, S. L. C.; Hsu, C. L.; Synthesis and Properties of Fluorinecontaining Polybenzimidazole/Montmorillonite Nanocomposite Membranes for Direct Methanol Fuel Cell Applications. *J. Power Sources* **2007**, *168*, 172–177.
50. Pu, H. T.; Liu, Q. Z.; Liu, G. H. Methanol Permeation and Proton Conductivity of Acid-doped Poly(*N*-Ethylbenzimidazole) and Poly(N-Methylbenzimidazole). *J. Membr. Sci.* **2004**, *241*, 169–175.
51. Wycisk, R.; Lee, J. K.; Pintauro, P. N. Sulfonated Polyphosphazene–Polybenzimidazole Membranes for DMFCs. *J. Electrochem. Soc.* **2005**, *152*(5), A892–A898.
52. Pivovar, B. S.; Wang, Y. X.; Cussler, E. L. Pervaporation Membranes in Direct Methanol Fuel Cells. *J. Membr. Sci.* **1999**, *154*, 155–162.
53. Wainright, J. S.; Wang, J. T.; Savinell, R. F.; Litt, M.; Moaddel, H.; Rogers, C. Acid Doped Polybenzimidazoles, a New Polymer Electrolyte. *Proc. Electrochem. Soc.* **1994**, *94*, 255.
54. Wang, J. T.; Savinell, R. F.; Wainright, J. S.; Litt, M.; Yu, H.; A H_2/O_2 Fuel Cell Using Acid Doped Polybenzimidazole as Polymer Electrolyte. *Electrochim. Acta* **1996**, *41*, 193–197.
55. Weber, M.; Wang, J. T.; Wasmus, S.; Savinell, R. F. Formic Acid Oxidation in a Polymer Electrolyte Fuel Cell: A Real-time Mass-spectrometry Study. *J. Electrochem. Soc.* **1996**, *143*, L158–L160.
56. Gubler, L.; Kramer, D.; Belack, J.; Unsal, O.; Schmidt, T. J; Scherer, G. G.; Celtec-V a Polybenzimidazole-based Membrane for the Direct Methanol Fuel Cell. *J. Electrochem. Soc.* **2007**, *154*, B981–B987.
57. Wycisk, R.; Chisholm, J.; Lee. J.; Lin, J.; Pintauro, P. N. Direct Methanol Fuel Cell Membranes from Nafione Polybenzimidazole Blends. *J. Power Sources* **2006**, *163*, 9–17.
58. Diaz, A. Z.; Abuin, C.; Corti, H. R. Water and Phosphoric Acid Uptake of Poly[2,5-benzimidazole] (ABPBI) Membranes Prepared by Low and High Temperature Casting. *J. Power Sources* **2009**, *188*, 45–50.
59. Xing, B.; Savadogo, O. Hydrogen/Oxygen Polymer Electrolyte Membrane Fuel Cells (PEMFCs) Based on Alkaline-doped Polybenzimidazole (PBI). *Electrochem. Commun.* **2000**, *2*(10), 697–702.

27. Kawahara, M.; Morita, J.; Rikukawa, M.; Sanui, K.; Ogata, N. *Electrochim. Acta* **2000**, *45*, 1395–1398.
28. Bouchet, R.; Siebert, E. Proton Conduction in Acid Doped Polybenzimidazole. *Solid State Ionics* **1999**, *118*, 287.
29. Bouchet, R.; Siebert, E.; Vitter. G. Acid-doped Polybenzimidazole as the Membrane of Electrochemical Hydrogen Sensors. *J. Electrochem. Soc.* **1997**, *144*, L95.
30. Glipa, X.; Bonnet, B.; Mula, B.; Jones, D. J.; Rozière, J. Investigation of the Conduction Properties of Phosphoric and Sulfuric Acid Doped Polybenzimidazole. *J. Mater. Chem.* **1999**, *9*, 3045.
31. Jones, D. J.; Roziere, J. Recent Advances in the Functionalisation of Polybenzimidazole and Polyetherketone for Fuel Cell Applications. *J. Membr. Sci.* **2001**, *185*(1), 41–58.
32. Kawahara, M.; Morita, J.; Rikukawa, M.; Sanui, K.; Ogata, N. Synthesis and Proton Conductivity of Thermally Stable Polymer Electrolyte: Poly(benzimidazole) Complexes with Strong Acid Molecules. *Electrochem. Acta* **2000**, *45*, 1395–1398.
33. Akita, H.; Ichikawa, M.; Nosaki, K.; Oyanagi, H.; Iguchi, M. US Patent 2000, *6*, 124.
34. Asensio, J. A.; Sánchez, E. M.; Romero, P. G.; Proton-conducting Membranes Based on Benzimidazole Polymers for High-temperature PEM Fuel Cells. A Chemical Quest. *Chem. Soc. Rev.* **2010**, *39*, 3210–3239.
35. Weng, D.; Wainright, J. S.; Landau, U.; Savinell, R. F. Electroosmotic Drag Coefficient of Water and Methanol in Polymer Electrolytes at Elevated Temperatures. *J. Electrochem. Soc.* **1996**, *143*, 1260–1263.
36. Li, Q. F.; He, R. H.; Jensen, J. O.; Bjerrum, N. J. PBI-based Polymer Membranes for High Temperature Fuel Cells—Preparation, Characterization and Fuel Cell Demonstration. *Fuel Cells* **2004**, *4*, 147–159.
37. Litt, M.; Ameri, R.; Wang, Y.; Savinell, R.; Wainwright, J.; Polybenzimidazoles/Phosphoric Acid Solid Polymer Electrolytes: Mechanical and Electrical Properties. *Mater. Res. Soc. Symp. Proc.* **1999**, *548*, 313–323.
38. He, R. H.; Li, Q. F.; Bach, A.; Jensen, J. O.; Bjerrum, N. J. Physicochemical Properties of Phosphoric Acid Doped Polybenzimidazole Membranes for Fuel Cells. *J. Membr. Sci.* **2006**, *277*, 38–45.
39. Yu, S.; Xiao, L.; Benicewicz, B. C. Durability Studies of PBI-based High Temperature PEMFCs. *Fuel Cells* **2008**, *8*, 156–174.
40. Chuang, S. W.; Hsu, S. L. C. Synthesis and Properties of a New Fluorine-containing Polybenzimidazole for High-temperature Fuel Cell Applications. *J. Polym. Sci. A* **2006**, *44*, 4508–4513.
41. LaConti, A. B.; Hamdan, M.; McDonald, R. C. Mechanisms of Membrane Degradation. In: *Handbook of Fuel Cells*, vol. 3; Vielstichm, W., Lamm, A., Gasteiger, H. A.; John Wiley & Sons Ltd., 2003.
42. Noyé, P.; Li, Q. F.; Pan, C.; Bjerrum, N. J. Cross-linked Polybenzimidazole Membranes for High Temperature Proton Exchange Membrane Fuel Cells with Dichloromethyl Phosphoric Acid as a Cross-linker. *Polym. Adv. Technol.* **2008**, *19*, 1270–1275.
43. Li, Q.; Jensen, J. O.; Pan, C.; Bandur, V.; Nilsson, M.; Schönberger, F. Partially Fluorinated Arylene Polyethers and their Ternary Blends with PBI and H_3PO_4, Part II. Characterizations and Fuel Cell Tests of the Ternary Membrane. *Fuel Cells* **2008**, *8*, 188–199.

11. Kreuer, K. D. Fuel Cells. In: *Selected Entries from the Encyclopedia of Sustainability Science and Technology*. Springer: New York, 2013.
12. Marvel, C. S. Polymers from *Ortho*-Aromatic Tetraamine and Aromatic Dianhydride. *J. Polym. Sci. A* **1965**, *3*, 3549–3571.
13. Han, S. H.; Lee, J. E.; Lee, K.-J.; Park, H. B.; Lee, Y. M. Highly Gas Permeable and Microporous Polybenzimidazole Membrane by Thermal Rearrangement. *J. Membr. Sci.* **2010**, *357*(1–2), 143–151.
14. Li, Q.; Jensen, J. O.; Savinell, R. F.; Bjerrum, N. J. High Temperature Proton Exchange Membranes Based on Polybenzimidazoles for Fuel Cells. *Progr. Polym. Sci.* **2009**, *34*(5), 449–477.
15. Xiao, L.; Zhang, H.; Scanlon, E.; Ramanathan, L. S.; Choe, E.-W.; Rogers, D.; Benicewicz, B. C. High-temperature Polybenzimidazole Fuel Cell Membranes via a Sol–Gel Process. *Chem. Mater.* **2005**, *17*(21), 5328–5333.
16. Litt, M.; Ameri, R.; Wang, Y.; Savinell, R.; Wainwright, J. Polybenzimidazoles/Phosphoric Acid Solid Polymer Electrolytes: Mechanical and Electrical Properties. *MRS Proc.* **2011**, *548*, 313.
17. Li, Q. F; He, R. H.; Berg, R. W.; Hjuler, H. A.; Bjerrum, N. J. Water Uptake and Acid Doping of Polybenzimidazoles as Electrolyte Membranes for Fuel Cells. *Solid State Ionics* **2004**, *168*, 177–185.
18. Brooks, N. W.; Duckett, R. A.; Rose, J.; Ward, I. M.; Clements, J. An NMR-study of Absorbed Water in Polybenzimidazole. *Polymer* **1993**, *34*, 4039–4042.
19. Li, Q.; Jensen, J. O.; Savinell, R. F.; Bjerrum, N. J. High Temperature Proton Exchange Membranes Based on Polybenzimidazoles for Fuel Cells. *Progr. Polym. Sci.* **2009**, *34*(5), 449–477.
20. Glipa, X.; Haddad, M. E.; Jones, D. J.; Rozière, J. Synthesis and Characterisation of Sulfonated Polybenzimidazole: A Highly Conducting Proton Exchange Polymer. *Solid State Ionics* **1997**, *97*, 323.
21. Powers, E. J.; Serad, G. A. History and Development of Polybenzimidazoles. In: *High Performance Polymers: Their Origin and Development*; Seymour, R. B., Kirschenbaum, G. S., Eds.; Elsevier: Amsterdam, 1986.
22. Aharoni, S. M.; Litt, M. H. Synthesis and Some Properties of Poly(2,5-Trimethylenebenzimidazole) and Poly(2,5-Trimethylenebenzimidazole Hydrochloride). *J. Polym. Sci., Polym. Chem.* **1974**, *12*, 639.
23. Jones, D. J.; Jacques Rozière, J. Recent Advances in the Functionalisation of Polybenzimidazole and Polyetherketone for Fuel Cell Applications. *J. Membr. Sci.* **2001**, *185*, 41–58.
24. Wainright, J. S.; Savinell, R. F.; Litt, M. H. Acid Doped Polybenzimidazole as a Polymer Electrolyte for Methanol Fuel Cells. In: *Proceedings of the Second International Symposium on New Materials for Fuel Cells and Modern Battery Systems*; Savadogo, O., Roberge, P. R., Eds.; Montreal, Canada, 6–10 July 1997.
25. Wainright, J. S.; Wang. J.-T.; Weng, D.; Savinell, R. F.; Litt, M. H. Acid-doped Polybenzimidazoles: A New Polymer Electrolyte. *J. Electrochem. Soc.* **1995**, *142*(7), L121–L123.
26. Xing, B.; Savadogo, O. The Effect of Acid Doping on the Conductivity of Polybenzimidazole (PBI). *J. New. Mater. Electrochem. Syst.* **1999**, *2*, 95.

Hasani-Sadrabadi et al. found that nanocomposite of organically modified montmorillonite (OMMT)/PBI polyelectrolyte membranes doped with phosphoric acid performed better than Nafion 117. The membrane composed of 500 mol% doped acid and 3.0 wt% OMMT showed a membrane selectivity of approximately 109,761 in comparison with 40,500 for Nafion 117 and also a higher power density (186 mW cm^{-2}) than Nafion 117 (108 mW cm^{-2}) for a single-cell DMFC at a 5 M methanol feed.[61]

KEYWORDS

- polybenzimidazole
- DMFC
- polymer electrolyte membrane
- fuel cell

REFERENCES

1. Banerjee, S. *Handbook of Specialty Fluorinated Polymers Preparation, Properties, and Applications*. Elsevier: Oxford, 2015.
2. Bossel, U. *The Birth of the Fuel Cell*. European Fuel Cell Forum: Oberrohrdorf, 2000.
3. Mittal, V. *High Performance Polymers and Engineering Plastics*. Wiley and Scrivener Publishing: Lowell, MA, 2011.
4. Perry, M. L.; Fuller, T. F. A Historical Perspective of Fuel Cell Technology in the 20th Century. *J. Electrochem. Soc.* **2002**, *149*, S59–S67.
5. Baladauf, M.; Preidel, W.; Status of the Development of a Direct Methanol Fuel Cell. *J. Power Sources* **1999**, *84*, 161–166.
6. Appleby, A. J. Issues in Fuel Cell Commercialization. *J. Power Sources* **1996**, *59*, 153–176.
7. Carrette, L.; Friedrich, K. A.; Stimming, U. Fuel Cells—Fundamentals and Applications. *Fuel Cell* **2001**, *1*, 5–39.
8. Cameron, D. S.; Hards, G. A.; Harrison, B.; Potter, R. J. Direct Methanol Fuel Cells. *Platinum Metals Rev.* **1987**, *31*, 173–181.
9. Waidhas, M.; Drenckhahn, W.; Preidel, W.; Landes, H.; Direct-fuelled Fuel Cells. *J. Power Sources* **1996**, *61*, 91–97.
10. McGrath, K. M.; Surya Prakash, G. K.; Olah, G. A.; Direct Methanol Fuel Cells. *J. Ind. Eng. Chem.* **2004**, *10*(7), 1063–1080.

60. Zhao, X.; Yuan, W.; Wu, Q.; Sun, H.; Luo, Z.; Fu, H. High-temperature Passive Direct Methanol Fuel Cells Operating with Concentrated Fuels. *J. Power Sources* **2015,** *273*, 517–521.
61. Hasani-Sadrabadi, M. M.; Dorri, N. M.; Ghaffarian, S. R.; Dashtimoghadam, E.; Sarikhani, K.; Majedi, F. S. Effects of Organically Modified Nanoclay on the Transport Properties and Electrochemical Performance of Acid-doped Polybenzimidazole Membranes. *J. Appl. Polym. Sci.* **2010,** *117*, 1227–1233.

CHAPTER 5

APPLICATIONS OF SMART POLYMERS IN EMERGING AREAS

U. V. GAIKWAD[1*], A. R. CHAUDHARI[2], and S. V. GAIKWAD[3]

[1]Department of Physics, Priyadarshini Bhagwati College of Engineering, Nagpur, India

[2]Department of Chemistry, Priyadarshini Bhagwati College of Engineering, Nagpur, India

[3]Department of Chemistry, Dr. Babasaheb Ambedkar College of Engineering & Research, Nagpur, India

*Corresponding author. E-mail: umagaikwad353@gmail.com

CONTENTS

Abstract 76
5.1 Introduction 76
5.2 Applications in Various Fields 77
5.3 Conclusion 88
Keywords 88
References 89

ABSTRACT

Smart polymers (SPs) are one of the important classes of polymers and its applications have been increasing significantly. SPs are materials that respond to small external stimuli. These are also referred as "stimuli responsive" materials or "intelligent" materials. Last two to three decades shows explosive growth in this subject.

Smart materials field of research is wide and complex not only relative to the terms but also regarding to its technical aspects and applications. These polymers show important changes in their properties with environmental stimulation. The stimuli include salt, UV irradiation, temperature, pH, magnetic or electric field, ionic factors, etc. SPs have very promising applications in drug delivery, tissue engineering, cell culture, nanocarriers, textile engineering, bioseperation, optical storage device, and so on.

SPs including thermal, moisture, light-responsive polymers, and pH-responsive hydrogels have been applied to improve textile functionalities. SPs based on photoresponsive azobenzene moieties have been extensively explored as potential materials for high-capacity optical storage.

Smart polymeric nanocarriers are an important emerging area in drug delivery research. Conductive polymers have been recently used in fuel cells, computer displays and microsurgical tools, and also in biomaterials. This chapter is focused on the entire features of SPs and their most recent and relevant applications.

5.1 INTRODUCTION

The synthetic polymers can be classified into different categories based on their chemical properties. Out of these, some special types of polymers have emerged as a very useful class of polymers and have their own special chemical properties and applications in various areas. These polymers are coined with different names, based on their physical or chemical properties like, "stimuli-responsive polymers"[1] or "smart polymers (SPs)" or 'intelligent polymers" or "environmental-sensitive" polymers. The characteristic feature that actually makes them "smart" is their ability to respond to very slight changes in the surrounding environment.[1,15]

SPs classify on the basis of following three perspectives:

- the type of polymeric material,
- external stimuli, and
- their given response.

The uniqueness of these materials lies not only in the fast macroscopic changes occurring in their structure but also these transitions being reversible. SPs may also be classified depending on their response to external stimuli. The responses are manifested as changes in one or more of the following—shape, surface characteristics, solubility, formation of a complex molecular assembly, a sol-to-gel transition, and others. The environmental trigger behind these transitions can be either change in temperature or pH shift,[1] increase in ionic strength, presence of certain metabolic chemicals, addition of an oppositely charged polymer and polycation–polyanion complex formation.[1] More recently, changes in electric and magnetic field[1,16] light or radiation forces have also been reported as stimuli for these polymers. The physical stimuli, such as temperature, electric or magnetic fields, and mechanical stress, will affect the level of various energy sources and alter molecular interactions at critical onset points. They undergo fast, reversible changes in microstructure from a hydrophilic to a hydrophobic state.[1] These changes are apparent at the macroscopic level as precipitate formation from a solution or order-of-magnitude changes in the size and water content of stimuli-responsive hydrogels.[1] An appropriate proportion of hydrophobicity and hydrophilicity in the molecular structure of the polymer is believed to be required for the phase transition to occur.[2,17]

Most frequent and promising applications of SPs are focused on biotechnology, medicine, and electronic technologies.[1] There is not a single area left where SPs are unreachable. In this review, we are focusing various applications of biopolymers.

5.2 APPLICATIONS IN VARIOUS FIELDS

5.2.1 BIOMATERIAL

Living systems respond to external stimuli adapting themselves to changing conditions. Polymer scientists have been trying to mimic this behavior for the last 20 years creating the so-called SPs. SPs have very promising applications in the biomedical field as delivery systems of therapeutic agents, tissue engineering scaffolds, cell culture supports, bioseparation devices, sensors, or actuators systems.[3]

Now, we will focus on pH and temperature sensitive polymers and their most recent and relevant applications as biomaterials in drug delivery and tissue engineering. Dual-stimuli-responsive materials will also be presented because of their high potential in the biomedical field.[3]

5.2.1.1   pH-SENSITIVE POLYMERS

pH-sensitive polymers are polyelectrolytes that bear in their structure weak acidic or basic groups that either accept or release protons in response to changes in environmental pH. The pendant acidic or basic groups on polyelectrolytes undergo ionization just like acidic or basic groups of monoacids or monobases.[3]

pH-sensitive polymers have been used in several biomedical applications, the most important being their use as drug and gene delivery systems, and glucose sensors. Between all the systems described in the literature, we report in this section the most attractive examples reported in the last years.[3]

5.2.1.2   DRUG DELIVERY SYSTEMS

pH varies along the gastrointestinal tract between 2 (stomach) and 10 (colon).[3] This condition makes pH-sensitive polymers ideal for colon-specific drug delivery. The most common approach utilizes enteric polymers that resist degradation in acidic environment and release drug in alkaline media due to the formation of salt.[3] There are a large number of polysaccharides, such as amylose, guar gum, pectin, chitosan, inulin, cyclodextrin, chondroitin sulfate, dextran, and locust beam gum, have been also investigated for colon specific drug release.[3]

5.2.1.3   GLUCOSE SENSORS

One of the most popular applications of pH-sensitive polymers is the fabrication of insulin-delivery systems for the treatment of diabetic patients. Delivering insulin is different from delivering other drugs, since insulin has to be delivered in an exact amount at the exact time of need. Many devices have been developed for this purpose and all of them have a glucose sensor built into the system.[3] In a glucose-rich environment, such as the bloodstream after a meal, the oxidation of glucose to gluconic acid catalyzed by glucose oxidase can lower the pH to approximately 5.8. This enzyme is probably the most widely used in glucose sensing and makes possible to apply different types of pH-sensitive hydrogels for modulated insulin delivery.[3,20,21]

5.2.1.4 TEMPERATURE-RESPONSIVE POLYMERS

Polymers sensitive to temperature changes are the most studied class of environmentally sensitive polymers as they have potential applications in the biomedical field.[3] This type of systems exhibit a critical solution temperature (typically in water) at which the phase of polymer and solution is changed in accordance with their composition. Those systems exhibiting one phase above certain temperature and phase separation below it possess an upper critical solution temperature. On the other hand, polymer solutions that appear as monophasic below a specific temperature and biphasic above it generally exhibit the so-called lower critical solution temperature (LCST). These represent the type of polymers with most number of applications.[3,18] The typical example is poly(N-isopropylacrylamide) (PNIPAAm) that presents a LCST at 32°C in water solution. Below that temperature, the polymer is soluble as the hydrophilic interactions, due to hydrogen bonding, are predominant, whereas a phase separation occurs above the LCST (cloud point) due to predomination of hydrophobic interactions. Other type of temperature sensitivity is based on the intermolecular association as in the case of pluronics or poloxamers (PEO–PPO–PEO)[3] where hydrophobic associations of PPO blocks lead to the formation of micelle structures above critical micelle temperature.[19]

Thus pH-sensitive polymer systems characteristics and applications as drug delivery systems, in particular as nanocarriers, have been exposed as one of the most promising applications. Temperature-responsive polymers from natural and synthetic origin as well as polymer–protein bioconjugates, and their temperature behavior, LCST, and gelation mechanisms, have been described paying special attention to their applications in tissue engineering processes.[3] Finally, the dual stimuli responsiveness of new synthetic polymers such as elastin-like systems and acrylic homopolymers and copolymers are discussed in terms of characteristics and possible applications in the biomedical field.[3]

There is one example when hydrogels are prepared by crosslinking T-sensitive polymers the temperature sensitivity in water results in changes in the polymer hydration degree. Below the transition temperature, the polymer swells up to equilibrium hydration degree being in an expanded state. By increasing the temperature above the transition, hydrogel deswells to a collapsed state. This process is generally reversible and can be applied in a pulsatile manner making the polymer to behave as an on-off system when the stimulus is applied or removed.[3]

5.2.2 TISSUE ENGINEERING

Conductive polymers are already used in fuel cells, computer displays, and microsurgical tools. These versatile polymers can be synthesized alone, as hydrogels, combined into composites or electrospun into microfibers. They can be created to be biocompatible and biodegradable. Their physical properties can easily be optimized for a specific application through binding biologically important molecules into the polymer using one of the many available methods for their fictionalizations. Their conductive nature allows cells or tissue cultured upon them to be stimulated, the polymers' own physical properties to be influenced post-synthesis and the drugs bound in them released, through the application of an electrical signal. It is thus little wonder that these polymers are becoming very important materials for biosensors, neural implants, drug delivery devices, and tissue-engineering scaffolds.[4]

Electroactive biomaterials are a part of a new generation of "smart" biomaterials that allow the direct delivery of electrical, electrochemical, and electromechanical stimulation to cells.[4] The family of electroactive biomaterials includes conductive polymers, electrets, piezoelectric, and photovoltaic materials.[4]

Electrets and piezoelectric materials allow the delivery of an electrical stimulus without the need for an external power source, but the control over the stimulus is limited.[4] Conductive polymers, on the other hand, allow excellent control of the electrical stimulus, possess very good electrical and optical properties, have a high conductivity/weight ratio and can be made biocompatible, biodegradable, and porous.[4] Furthermore, a great advantage of conductive polymers is that their chemical, electrical, and physical properties can be tailored to the specific needs of their application by incorporating antibodies, enzymes, and other biological moieties.[4] Additionally, these properties can be altered and controlled through stimulation (e.g., electricity, light, pH) even after synthesis.[4]

Considering the vast amount of new possibilities conductive polymers offer, we believe they will revolutionize the world of tissue engineering. Thus, we chose to gather together in this review all of the available information on the most commonly used conductive polymers, their biocompatibility, conductivity, synthesis, biomolecule doping, and drug-release applications.

Conductive polymers merge the positive properties of metals and conventional polymers the ability to conduct charge, great electrical and optical properties with flexibility in processing and ease of synthesis.[4] The early work on conductive polymers was triggered by the observation that the conductivity of polyacetylene, a polymer that is normally only semi-conducting at best, increases by 10 million-fold when polyacetylene is oxidized using iodine vapor.[4] The underlying phenomenon "doping" and is essential for the conductivity of polymers, as only through this process do they gain their high conductivity. Doping process introduces the charge carriers (polarons and bipolarons) into the polymer and renders it conductive.[4] Similarly to what happens in semiconductor technology, this can happen in two ways: p-doping, where the polymer is oxidized and will have a positive charge, and n-doping, where the polymer is reduced and will possess a negative charge.[4] The doping process occurs during synthesis and can be carried out chemically, electrochemically or via photodoping.[4]

In most cases, when the dopant is a biological molecule, as many of them are not capable of the redox chemistry that is necessary for chemical synthesis, the electrochemical method has to be used. In this case, the biological molecule has to be charged and placed with the monomer when the electrochemical synthesis occurs.[4] There is a proportional relationship between the amount of dopant used and the conductivity of the doped polymer.[4] The conductivity can be further increased by choosing a different dopant, but this will affect the surface and bulk structural properties (e.g., color, porosity, volume) of the polymer.[4] The doping is reversible, as an electrical potential applied through the polymer will cause the dopant to leave or reenter the polymer, switching it between its conductive and insulating redox states.

Dopants can be separated into two categories based on their molecular size: small dopants (e.g., Cl⁻) and large dopants (e.g., sodium polystyrenesulfonate) will behave and affect the polymer itself differently.[4] Both will affect the conductivity and structural properties of the polymer.[4] As polyacetylene was difficult to synthesize and is unstable in air, the search for a better conductive polymer began.[4] Polyheterocycles since then have emerged as a family of conductive polymers with both good stability and high conductance.[4]

There are currently two main methods for synthesizing conductive polymers: chemical and electrochemical.[4] During chemical synthesis, the monomer solution is mixed with an oxidizing agent (e.g., ferric chloride, ammonium persulfate).[4,11] This process creates a powder or a thick film of the polymer and allows its bulk production, which makes it the method of choice for commercial applications.[4,13,14] An additional advantage of chemical polymerization is that it can be used to create all types of conductive polymers, including some novel conducting polymers that cannot be synthesized with the electrochemical method.[4,10,12]

The choice of dopant defines the properties of the polymer and allows its functionalization for a specific application.[4]

Applying an electrical signal through a conductive polymer substrate allows the behavior of the cells or tissue cultured upon it to be influenced.[4] These multiple levels, where conductive polymers can be easily modified, grant a degree of control over the properties of the material, before and after synthesis, that no other material can provide. Thus, conductive polymers have lot of applications in tissue engineering.

5.2.3 TEXTILE APPLICATIONS

SPs are materials which can show noticeable changes in their properties with environmental stimulation. Novel functionalities can be delivered to textiles by integrating SPs into them. SPs including thermal-, moisture-, and light-responsive polymers, and thermal- and pH-responsive hydrogels have been applied to textiles to improve or achieve textile smart functionalities. The functionalities include esthetic appeal, comfort, drug release, wound monitoring, fantasy design (color changing), smart wetting properties, and protection against extreme environmental variations.[5]

SPs, such as shape memory polymers (SMPs), phase change materials (PCMs), color change polymers, and intelligent polymer hydrogels, have developed rapidly in the past few decades. Because they are easy to process, and are light and flexible, SPs have been an important material in textile processing. SMPs have the capability to memorize a permanent shape and to be programed to assume one or many temporary shapes; upon exposure to an external stimulus, they spontaneously recover their original permanent shape. For example, a closed flower (temporary shape) made of a SMP is fixed at a lower temperature and recovers from the closed state to an open flower (original shape) when the temperature is increased above its switch temperature.[5]

PCMs have the ability to absorb and emit heat energy without changing temperature themselves. These waxes include eicosane, octadecane, nonadecane, heptadecane, and hexadecane, which all have different freezing and melting points, and when combined in a microcapsule will store and emit heat energy and maintain their temperature range of 30–34°C, which is comfortable for the body.[5]

SP hydrogels undergo reversible volume change responding to a small variation in solution conditions (external stimuli), such as temperature, 2–7 pH[2,8,9] and solvent compositions. Some hydrogels such as poly(*N*-substituted acrylamide), poly(*N*-vinyl alkyl amide), poly(vinyl methyl ether), and poly(ethylene glycol-*co*-propylene glycol) have been studied and utilized for diverse textile applications. Poly(*N*-isopropyl acrylamide)[5] (PNIPAAm) is an intensively concentrated temperature-sensitive polymer which has a simultaneously hydrophilic and hydrophobic structure. Because of its sharp temperature-induced transition, PNIPAAm (and in particular the PNIPAAm hydrogel) has been developed into stimuli-sensitive textiles.[5]

Polymers which change their visible optical properties in response to external stimuli have aroused the growing interest of researchers. According to the external stimulus, these polymers are classified as

- thermochromic (stimulus: temperature);
- photochromic (stimulus: light);
- electrochromic (stimulus: electric field);
- piezochromic (stimulus: pressure);
- ionochromic (stimulus: ion concentration); and
- biochromic (stimulus: biochemical reaction).

Most SPs can be triggered in a variety of ways such as the response of SMPs to thermal, chemical, magnetic, and water stimuli. Another example is the way a hydrogel can respond to pH, heat, light, magnetic fields, etc.

These diverse stimuli make it possible to use SPs in different applications. SPs used in textiles usually appear in various forms such as film, fiber, solution, or gel to meet different processing requirements in textiles.[5]

Different types of SPs have significant effects on the applications used in textile technology. There are a variety of SPs that can be used with specific processing techniques, such as finishing, spinning, weaving, or laminating. It is important to use the correct type of SP for SPs for textile applications 439 specific textile applications. There are five common forms or shapes used to describe the SPs mentioned in the introduction. The five forms are

- solution;
- microcapsules;
- gel;
- film/foam;
- fiber; and
- nonwoven.

These forms are either in direct or intermediate states when incorporated into textiles. I would like to focus some amazing facts of these polymers in textile applications.[5]

5.2.3.1 COLOR CHANGE FIBER

The fiber method was developed later than the printing and dyeing method for color change materials. Color-change fibers have better washing durability than solution finishing and film lamination on textiles. The main manufacturing methods of color-change fibers include solution spinning, melt spinning, treatment, and graft copolymerization. The polymer and antitransfer agent are added in the solution and spun directly.[5]

The melt spinning method includes the polymerization, polyblending, and sheath–core compound spinning methods. Posttreatment also gives the fiber a chromic property. Common fibers or fabrics are dipped into solution which includes styrene monomer with pyrone, and the fiber and fabrics gain photochromic properties due to the monomer polymerizing. Chromic groups can also be introduced into polymers during polymerization. These polymers can be spun to produce chromic fibers. This method is a polymerization method. The polyblending method uses chromic polymers and polyester, polyamide, and other polymers blended and melt spun. It is simple but has a high demand for the chromic polymer. Sheath–core compound spinning is the main method for chromic fiber manufacturing. The core component is a chromic agent and the sheath is a common fiber. The core contains 1–40% chromic agent and the melting point is below 230°C. This chromic fiber has good hand-feeling, washability, and color reaction. Posttreatment and graft copolymerization methods are also used in chromic fiber production.

This method is simple and easy to control and extend. The simplest methods of combining chromic material and fabrics are printing and dyeing.

The chromic materials should be encapsulated first and mixed into resin solution. The solution is printed onto the fabric which gains the chromic qualities.

5.2.3.2 MEDICAL TEXTILE

The expanding field of medical textiles comprise all textile products that contribute to improving human health and well-being, protecting us against bacteria and infection, providing external support for injured skin, promoting the healing of wounds, and replacing injured and diseased tissues and organs. SPs can provide diversified and multifunctional options for design and fabrication. Some important advantages of these medical textiles are discussed in the following sections.

5.2.3.3 SKINCARE TEXTILE

The controlled release behavior and thermally regulated pore size control characteristics of hydrogel modified fabrics may be used for wound dressing, dialysis membranes, drug delivery carriers, separation membranes, skincare, and cosmetic materials.[5] With suitable design, the skincare products prepared from stimuli-responsive hydrogel-treated textiles can bring a moisturizing, whitening, brightening, or even antiaging effect to human skin. A facial mask made of the TRPG-treated nonwoven fabric, which is sensitive to the temperature of human skin, can act as a carrier media with controlled release of nutritious ingredients, perfumes, or other drugs to human skin in response to changes in human skin temperature.[5] The study has shown that the release of vitamin C could be controlled by varying the surrounding temperature. At present, there are still some challenges that need to be solved, such as the high stiffness and brittleness especially when the material is in a dry state.[5]

5.2.3.4 WOUND DRESSING TEXTILE

Chitin/chitosan and chitosan derivatives have excellent antibacterial properties and a good wound-healing effect. Chitosan hydrogel as a wound dressing can aid in the reestablishment of skin architecture. Alginate filaments and cotton fabric coated with chitosan have been developed for advanced wound dressings. In addition to chitin/chitosan, many hydrogel products for wound dressings have been developed from biopolymers.[5,8]

Wound dressings with stimuli-responsive hydrogels can provide a novel drug release system in response to variations in pH/temperature causing the wounds to heal at a faster rate. SPs have been providing enormous opportunities and potential in the textiles industry. By integrating SMPs into textile structures, we can obtain novel functional textiles with profound properties in terms of esthetic appeal, moisture/temperature management, protection against extreme climatic conditions, textile soft display, wound monitoring, skincare, fantasy design with color change, controlled release, and smart-wetting properties on textile surfaces.[5,9]

5.2.4 OPTICAL DATA STORAGE

SPs based on photoresponsive azobenzene moieties have been extensively explored as potential materials for high capacity optical storage.[6] The expansion of Information and Communication Technology has had an enormous impact on the socioeconomic progress of humankind.[6]

Modern digital society requires continuous development of the means to generate, distribute, and store increasing amounts of information in a fast and safe fashion.[6] This progress has been possible in part due to advances in different areas of material science and technology. For example, the continuous development of photolithographic techniques and polymeric photoresists has kept pace with Moore's law, doubling the number of transistors per unit area on an integrated circuit every two years. Another key element in this information revolution has been the development of information storage technologies.[6] Nowadays, there are several types of information storage systems, with the most common ones being magnetic hard drives, magnetic tapes, optical discs and solid-state drives (flash memory). In all cases, information is stored as a change in a physical property of a material or device.[6]

Flash memory stores information as electronic charge in a solid-state memory cell, while changes in magnetic or optical properties are used to store data in magnetic and optical discs. All of these technologies coexist, each having pros and cons with respect to access speed, durability, capacity, and price, and so the needs of each application holographic storage is an alternative optical technology with significantly improved storage capacity and information transfer rates. Using a completely different approach, holography makes use of the whole volume dictate the choice of equipment.

Holographic storage is an alternative optical technology with significantly improved storage capacity and information transfer rates. Using a completely different approach, holography makes use of the whole volume of the recording material, boosting the storage capacity of a single disc. In addition, it is possible to record or read a complete page of information (containing thousands of bits) in one single event, enormously increasing the transfer rate compared to conventional, bit by bit reading, optical discs.

The idea for holographic data storage arose in the early 1960s, but its commercial introduction has been a slow process. Despite the advances in systems design and materials, mainstream storage technologies have evolved even faster, and they have narrowed the market for holographic storage.

Materials development is a major issue in developing holographic optical storage technology due to the demanding storage media requirements, the most important being:

good optical quality with low scattering losses;

- good sensitivity, which allows rapid recording of holograms and is energy-efficient;
- a large dynamic range, with the medium able to multiplex several holograms in the same volume;
- low optical absorption, so that the whole volume of the recording media can be sensitized; and
- stability of the recorded holograms.[6]

If the recording medium needs to be rewritable, it should be possible to erase recorded information, preferentially in a local fashion, for example, using light.[6] The search for volume holographic materials that fulfill all the previously described properties has been an active topic of research for several decades. However, despite these intensive efforts, further research is still needed. Inorganic materials showing photo refractivity (e.g., $LiNiO_3$, $KNiO_3$) have been explored as suitable reversible volume holographic materials. Polymeric systems have also been investigated, especially those based on photopolymers in which light triggers a polymerization reaction leading to a change in the refractive index of the material.[6] SPs containing photochromic units, which experience reversible photo-induced transformations between two states with different absorption spectra, have also been studied as rewritable holographic materials.[6]

For example, light-induced pericyclic reactions have proven suitable for optical data storage Thus, smart photo addressable polymers containing azobenzene units have been actively investigated over more than two decades as media for optical data storage. Information can be recorded in these materials, through the effect of polarized light, as local changes in birefringence. This information can later be erased by optical means, making these materials potential candidates for rewritable optical storage.[6]

5.3 CONCLUSION

Thus from the above discussions, we can say that there is not a single area in which SPs cannot be utilized. From drug delivery to textile engineering, tissue engineering to optical storage device, everywhere these polymers are not only useful but also advantageous than the earlier product which were used. They really work as intelligent polymers.

KEYWORDS

- **smart polymers**
- **stimuli responsive**
- **hydrogels**
- **temperature responsive**
- **moisture-responsive polymer**

- **tissue engineering**
- **optical storage**
- **glucose sensor**
- **chitosan**

REFERENCES

1. Encarnacion, C. S. Smart Polymers and Applications. *Technology Watch Report*, 2009; pp 1–3.
2. Kumara, A.; Srivastava, A.; Galaevb, I. Y.; Mattiassonb, B. Smart Polymers: Physical Forms and Bioengineering Applications. *Prog. Polym. Sci.* **2007**, *32*, 1205–1237.
3. Aguilar, R. M.; Elvira, S. M.; Gallardo, A.; Vázquez, B.; Román, J. S. In: *Smart Polymers and their Applications in Biomaterials, Topics in Tissue Engineering*, Vol. 3; Ashammakhi, N., Eds.; R Reis & E Chiellini, 2007; pp 1–27.
4. Balint, R.; Cassidy, N. J.; Cartmell, H. Conductive Polymers: Towards a Smart Biomaterial for Tissue Engineering. *Acta Bio* **2014**, *10*, 2341–2353.
5. Hu, J.; Lu, J. *Smart Polymers for Textile Applications*; Woodhead Publishing Limited, 2014; pp 437–475.
6. Blasco, E.; Pinol, M.; Berges, C.; Sánchez-Somolinos, L.; Oriol, L. *Smart Polymers for Optical Data Storage*, Woodhead Publishing Limited, 2014; pp 510–548.
7. George, P. M.; et al. Fabrication and Biocompatibility of Polypyrrole Implants Suitable for Neural Prosthetics. *Biomaterials* **2005**, *26*, 3511–3519.
8. Gizdavic-Nikolaidis, M.; Ray, S.; Bennett, J.; Swift, S.; Bowmaker, G.; Easteal, A. Electrospun Poly(aniline-*co*-ethyl 3-aminobenzoate/Poly(Lactic Acid) Nanofibers and their Potential in Biomedical Applications. *J. Polym. Sci. A: Polym. Chem.* **2011**, *49*, 4902–4910.
9. Gizdavic-Nikolaidis, M.; Ray, S.; Bennett, J. R.; Easteal, A. J.; Cooney, R. P. Electrospun Functionalized Polyaniline Copolymer-based Nanofibers with Potential Application in Tissue Engineering. *Macromol. Biosci.* **2010**, *10*, 1424–1431.
10. Huang, L.; et al. Synthesis of Biodegradable and Electroactive Multiblock Polylactide and Aniline Pentamer Copolymer for Tissue Engineering Applications. *Biomacromolecules* **2008**, *9*, 850–858.
11. Bettinger, C. J.; Bruggeman, J. P.; Misra, A.; Borenstein, J. T.; Langer, R. Biocompatibility of Biodegradable Semiconducting Melanin Films for Nerve Tissue Engineering. *Biomaterials* **2009**, *30*, 3050–3057.
12. Blinova, N. V.; Stejskal, J.; Trchova, M.; Prokes, J. Control of Polyaniline Conductivity and Contact Angles by Partial Protonation. *Polym. Int.* **2008**, *57*, 66–69.
13. Cullen, D. K.; Patel, A. R.; Doorish, J. F.; Smith, D. H.; Pfister, B. J. Developing a Tissue-engineered Neural-electrical Relay Using Encapsulated Neuronal Constructs on Conducting Polymer Fibers. *J. Neural Eng.* **2008**, *5*, 374–384.

14. Dai, L. Conducting Polymers. In: *Intelligent Macromolecules for Smart Devices: From Materials Synthesis to Device Applications*; Dai, L., Ed.; Springer: London, 2004; pp 41–80.
15. Jeong, B.; Gutowska, A. Lessons from Nature: Stimuli Responsive Polymers and their Biomedical Applications. *Trends Biotechnol.* **2002**, *20*, 305–311.
16. Hoffman, A. S.; Stayton, P. S.; Bulmus, V.; Chen, G.; Jinping, C.; Chueng, C.; et al. Really Smart Bioconjugates of Smart Polymers and Receptor Proteins. *J. Biomed. Mater. Res.* **2000**, *52*, 577–586.
17. Galaev, I. Yu.; Mattiasson, B. 'Smart' Polymers and What They Could Do in Biotechnology and Medicine. *Trends Biotechnol.* **2000**, *17*, 335–340.
18. Singh, S.; Webster, D. C.; Singh, J. Thermosensitive Polymers: Synthesis, Characterization, and Delivery of Proteins. *Int. J. Pharm.* **2007**, *341*, 68–77.
19. Aguilar, M. R.; Elvira, C.; Gallardo, A.; Vázquez, B.; Román, J. S. Smart Polymers and their Applications as Biomaterials. In: *Topics in Tissue Engineering*. Vol. 6; Ashammakhi, N., Reis, R., Chiellini, E., Eds.; CiteSeer: Pennsylvania, 2007; pp 1–27.
20. Ravaine, V.; Ancla, C.; Catargi, B. Chemically Controlled Closed-loop Insulin Delivery. *J. Control. Release* **2008**, *132*, 2–11.
21. Roy, V. D.; Cambre, J. N.; Sumerlin, B. S. Future Perspectives and Recent Advances in Stimuli-responsive Materials. *Prog. Polym. Sci.* **2010**, *35*, 278–301.

PART II
Polymer Nanocomposites and Their Applications

CHAPTER 6

ADVANCEMENTS IN POLYMER NANOTECHNOLOGY

SATYENDRA MISHRA[*] and DHARMESH HANSORA

University Institute of Chemical Technology, North Maharashtra University, Jalgaon 425001, Maharashtra, India

[*]*Corresponding author. E-mail: profsm@rediffmail.com*

CONTENTS

Abstract 94
6.1 Introduction 94
6.2 Types of PNCs 95
6.3 Preparation Methods for PNCs 97
6.4 Properties of PNCs 103
6.5 Applications of PNCs 112
6.6 Summary 115
Keywords 116
References 116

ABSTRACT

Nanotechnology seems to be gaining importance rapidly as a powerful technology. Its immense potential promises the possibility of significant changes in near future. Today, the products made from polymer nanocomposites (PNCs) have shown their potential applications in gas sensing, drug delivery, nanoelectronics, energy harvesting for self-powered nanosystems, solar cells and batteries, aerospace materials, etc. In the last two decades, there has been proliferation in the research field of polymer nanotechnology. This research area includes nanopolymers, nanomaterials, nanocomposites, nanohybrids, nanodevices, nanocrystals, nanofibers, nanoclays, nanotubes, nanofilters, nanohorn, nanowires, nanosprings, nanorods, and nanomembranes, which have been explored. The essence of nanoscience is the ability to work at molecular level, atom by atom to create macrostructures with fundamentally new molecular organization. Nanotechnology is the application of these structured materials at nanoscale in solving problems of real life. As a result of molecular manufacturing and advanced form of nanotechnology, various types, preparation methods, properties, and applications of PNC are discussed in this chapter, which aims to review the previous work done and recent advancements in the field of polymer and plastics nanotechnology.

6.1 INTRODUCTION

Polymer nanocomposites (PNCs) are extensively utilized as numerous products across a wide variety of industries. In essence, the difference, between a pure polymer and a PNC as the choice of use in applications, is the addition of a suitable nanofiller (inorganic, conducting, stacked layers, clays, and biomaterials) at the nanoscale. The purpose behind addition of these nanofillers is to enhance the various properties of "pure and neat polymer" which make them "PNC matrix" embedded by nanofillers with improved performance. This is achieved by creating an interfacial phase due to the nanofiller dispersed within the "bulk" phase of the polymer. For optimal results, the nanomaterials need to be completely separated and dispersed uniformly throughout. In practice, this is challenging because of the inorganic/organic incompatibility between the phases that may result in aggregation. PNCs possess superior characteristic properties over conventional micro-composites; therefore, PNCs can offer new potential for various applications.[1–71]

6.2 TYPES OF PNCs

6.2.1 NATURAL FILLER-BASED POLYMER COMPOSITES

In early 1990s, wood polymer composites have been researched in plastics and wood technology because of their superior mechanical properties that are better than the pure wood. Wheat straw, cane bagasse, and teak sawdust (agro waste) can be reinforced with and without maleic anhydride (MA) treatment using Novolac resin to prepare composite sheet.[1] Natural filler like potato starch and urea have shown good degradable properties in linear low density polyethylene (LLDPE) composite.[2] Cane bagasse pith is also useful natural filler which can be esterified with MA, succinic anhydride, and phthalic anhydrides. These esterified natural fillers strengthen the mechanical properties of composites.[3,4] Natural and MA esterified fibers of banana, hemp, and sisal are the reinforcing agents in Novolac resin which can be useful in study of water and steam absorption.[5]

6.2.2 INORGANIC NANOMATERIAL-FILLED PNCs

Inorganic hydrates are commonly used in formulation of flame-retardant composite due to their ability to dehydrate under endothermic fire conditions. Inorganic nanofillers, like $Mg(OH)_2$,[6–8] $CaCO_3$,[9–28,49] $CaSO_4$,[29–36] $Ca_3(PO_4)_2$,[29,32,37–39] $BaSO_4$,[40,41] $Al(OH)_3$,[42] and $BaCO_3$[43] have been used for preparation of PNCs.[36,44] Inorganic filled PNCs have been prepared by various polymers like PBR,[9,16,20] epoxy resin (ER),[10,21,24,41] silicon rubber (SR),[22] polymethyl methacrylate (PMMA),[26,27] polyurathene rubber (PU),[8,38,42] polypropylene (PP),[13,27,43] ethylene propylene diene monomer (EPDM) rubber,[17,43] polyamide,[7,19,33,39,45,46] poly vinyl chloride (PVC),[18,34,47,48] styrene–butadiene rubber (SBR),[11,14,31,49] and polystyrene (PS).[23,35,50]

6.2.3 ELASTOMERIC NANOCOMPOSITES

Rubbers like PBR,[9,16,20] SBR,[11,14,49] PU,[8,38,42] EPDM,[17,43] SR,[22,50,51] viton rubber (VR)[52] are widely used for preparation of elastomeric PNCs. Various inorganic nanofillers, nanoclays,[50–52] and polymeric, conducting and bio-based materials have been used to prepare elastomeric PNCs.

6.2.4 CLAY AND NANOLAYERS-BASED PNCs

Various nanoclays like, organically modified montmorillonite (OMMT) nanoclays and MMT,[45,47,48,50–56] natural bentonite,[46] aluminosilicate clay (halloysite),[57] and surfactants (modified by bio and organic) have been used to prepare PNCs. Polymers like polyamide,[45] PVC,[47,48] SR,[50,51] VR,[52] polylactic acid (PLA),[53,55] polyvinyl alcohol,[56] and polyacrylic acid (PAA)[46,57] are widely used to prepare clay-based PNCs.

6.2.5 POLYMER NANOBLENDS

Core–shell nanoparticles (CSNPs) of nano-$CaCO_3$/PMMA were successfully blended with commodity polymer, that is, PP.[27] An isotactic PP-EPDM filled with $nBaCO_3$ is also a useful blend.[43] PS can be used to prepare the blend of PS:PLA and PS:PLA:OMMT nanocomposites,[53] which possess biodegradability. Polystyrene nanoparticles (nPS) can be blended with copolymers nanoparticles of polyacrylonitrile and poly(styrene/acrylonitrile))[58–61] and PP,[59–61] which were synthesized by microemulsion process using different monomer ratios.[58–61] Similarly, blend which includes polymer nanoparticles nPMMA/LLDPE is also useful.[62]

6.2.6 CONDUCTING MATERIAL OR POLYMER-BASED PNCs

Nanofillers like carbon nanotbes (CNTs),[63] silver nanoparticles,[64] γ-ferric oxide (γ-Fe_2O_3) nanoaparticles are most widely useful conducting nanomaterials to prepare their PNCs. Various conducting polymers like polyethylene oxide (PEO)[63] and polyaniline (PANI)[64,65] are generally used to prepare conducting PNCs.[66,67]

6.2.7 BIO-MATERIAL OR BIOPOLYMER-BASED PNCs

Most widely used biopolymer, that is, PLA has been used to prepare PS:PLA and PS:PLA:OMMT[53,55] PNCs for their biodegradable study. Biopolymer like gum ghatti can be also used to prepare biosurfactant-assisted $CaSO_4$/PS nanocomposites. Biosurfactants like rhamnolipid and surfactin have been used to modify inorganic various nanofillers.[35] Various biomaterials like chitosan and CNT-grafted chitosan are generally used. PNC of polycaprolactone grafted with chitosan can be used for vapor-sensing purpose.[68] Biosurfactant, like trehalose lipid and rhamnolipid, modified nanoparticles have been used to prepare PMMA nanocomposites for biodegradable study.[69]

6.2.8 CORE–SHELL MATERIAL-BASED PNCs

PS/$CaCO_3$ nanoparticles (having size <100 nm) with core–shell structure were synthesized by atomized microemulsion technique. The polymer chains were anchored onto the surface of nano-$CaCO_3$ using silane as a coupling agent.[23,25] The nanocontainers aluminosilicate clay (halloysite) can be used in nanocylindrical shape for loading of fragrance of rosewater. The fragrance-loaded nanocontainers were coated with a thin layer of polyelectrolyte, that is, PAA.[57]

6.3 PREPARATION METHODS FOR PNCs

6.3.1 MELT-EXTRUSION PROCESSING

Variable quantity of inorganic nanofiller was premixed into a polymer matrix, which was prepared by melt-extrusion processing or rheomix mixing (Fig. 6.1).

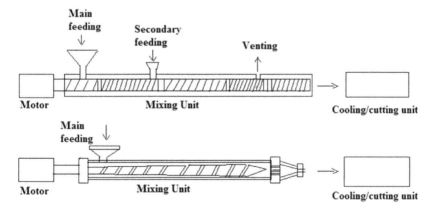

FIGURE 6.1 Preparation of PNC by melt-extrusion process.

PNCs of PP nanocomposites were injection-molded by reinforcing variable amount of inorganic nanofillers like $Mg(OH)_2$,[6] $CaCO_3$,[13,15] $CaSO_4$, and $Ca_3(PO_4)_2$[29,30,32,37] nanoparticles in an injection-molding machine by keeping temperature at 210°C in the feed zone, 220°C in the compression zone, 225°C in the metering zone, and 230°C in the nozzle. An isotactic nanoblend of PP/EPDM filled with $nBaCO_3$ was prepared in twin-screw extruder, which was then subjected to injection molding to get dumbbell-shaped specimens.[43]

6.3.2 MELT MIXING AND COMPOUNDING

Elastomeric nanocomposites can be compounded in two roll mill (Fig. 6.2) and molded in compression molding machine. For the preparation of elastomeric PNC, ingredients like rubber, stearic acid, zinc diethyl dithiocarbamate 2,2-dibenzothiazyl disulfide (MBTS), zinc oxide (ZnO) as accelerators, vulconex as antioxidant, sulfur, and the variable amount of nanofillers (nano-$CaCO_3$, commercial $CaCO_3$)[9,11,12,14,16,17,49] are generally added. Linseed oil can be used as extenders in rubber composites for homogeneous mixing of nanoparticles.[12]

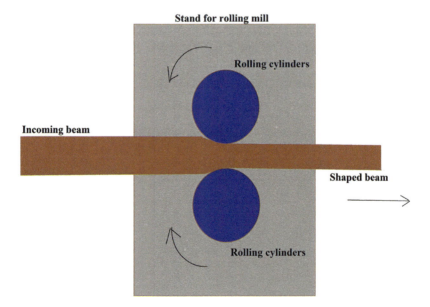

FIGURE 6.2 Preparation of PNCs by two roll mill.

Various rubber-based nanocomposites like millable PU/nano-Mg(OH)$_2$,[8] PBR/nano-CaCO$_3$,[9,16,20] ER/nano-CaCO$_3$,[10,21] SBR/nano-CaCO$_3$,[11,14,49] SR/nano-CaSO$_4$,[22] SBR/nano-CaSO$_4$,[31] PU/nAl(OH)$_3$,[42] SR/OMMT,[50,51] and VR/OMMT[52] have been masticated by compounding in two roll mill. After complete mixing of filler, curing agent was generally added and further mastication can be performed for 5 min to get a uniform dispersed soft sheet. For curing, the compounded rubber was compressed in a compression molding machine for optimum cure time at cure temperature under 100-kg/cm^2 pressure to get a cure sheet. The isotactic PP/EPDM/nBaCO$_3$ rubber nanoblend[43] was prepared by melt mixing using Brabender counter-rotating twin-screw plasticorder unit equipped with helical die head.[43]

Polyamide nanocomposites filled with nano-Mg(OH)$_2$,[7] nano-CaSO$_4$,[33] nano-Ca$_3$(PO$_4$)$_2$,[39] and OMMT[45] were compounded via melt intercalation on twin-screw extruder. PVC nanocomposites filled with nano-CaCO$_3$,[18] OMMT,[47,48] and nano-CaSO$_4$[51] were prepared using ingredients by dry blending with other additives in a conical twin-screw high intensive extruder/plasticorder.[51] The dry mix was fed into the conical twin-screw extruder through the vibrating pad hopper. The dry blending was carried out in a high intensive mixer with rotating speed of 50 rpm for 15 min at the temperatures of the feed zone, compression zone, and metering zone of 150, 160, and 170°C, respectively, while the temperature of die was kept at 175°C. The compounded materials were extruded through the slit die and the PVC nanocomposites sheets were prepared.[18,46,48]

6.3.3 SOLUTION BLENDING

PU/Ca$_3$(PO$_4$)$_2$ nanocomposites can be prepared by solution blending by addition of caster oil at ambient temperature for various periods of mixing. On complete mixing of nanoparticles in caster oil, methylene di-isocyante and polyol were mixed in the complex with various stirring periods. Foam was formed after proper mixing, which was poured in a well-defined mold to obtain sheet.[38] SR/OMMT nanocomposites were prepared through solution blending method by dissolving SR in chloroform followed by dispersion of OMMT in chloroform under controlled stirring. After complete mixing, solvent was evaporated at room temperature.[50,51]

6.3.4 MICROEMULSION POLYMERIZATION

Transparent or translucent dispersions of homopolymer and copolymer nanoparticles (size of 10–100 nm) can be synthesized by oil/water (o/w) atomized microemulsion process as shown in Fig. 6.3a.[58,59]

In this method, atomized reaction is used in which monomer (purified and distilled under reduced pressure before use) is sprayed through the nozzle of atomizer by reciprocating compressor at controlled temperature and pressure with constant rate. The baffles are mounted at the top of the reactor to bounce back the monomer stream from outgoing air. The exhaust is then led through the distillation column for the recovery of monomer. After complete addition of the monomer, the polymerization reaction was maintained for 1 h, which can be stopped by cooling it to room temperature. A transparent or translucent dispersion was formed, which indicates the formation of microemulsion of nPS nanoparticles.[58] The above phenomenon indicates that as the amount of initiator increased, the numbers of free radicals also increased. These free radicals had enough time to react with the sprayed monomer mists well protected by micelle. The maximum number of free radicals gets larger surface area to react with large number of well-protected single monomer droplets, resulting in more number of smaller particles in the reaction system. Hence, growth of the particles can be initiated at a number of sites in semi-batch mode. In this way, coagulation of two or more particles is inhibited to obtain particles in nanoform as shown in Fig. 6.3b.[58] Acrylate latex nanocomposites can also be prepared by emulsion polymerization of MMA and butyl acrylate by stabilizing Prussian blue using mixture of Tween 80 in deionized water containing APS.[70]

Transparent or translucent dispersion of spherical nPS (10–100 nm),[23,59,60] and nPMMA (20–50 nm)[26,27] can be easily synthesized by oil/water (o/w atomized) microemulsion process. The isolated nPS particles are generally used to prepare blend with PP.[59,60] Nanocomposites with core–shell structure, that is, nanoCaCO$_3$/PS[23,25] and nanoCaCO$_3$/nPMMA[26,27] and PMMA/biosurfactants (rhamnolipid/surfactin/trehalose lipid)[69] were synthesized by atomized microemulsion technique.

6.3.5 IN-SITU POLYMERIZATION

PNCs based on poly(styrene–butylacrylate–acrylic acid) of (PSBA)/BaSO$_4$ were synthesized *in-situ* copolymerization under ultrasound treatment of

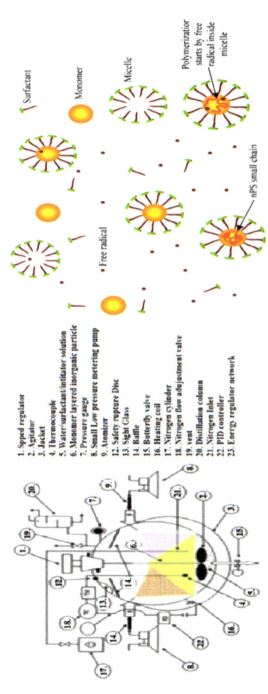

FIGURE 6.3 (a) Preparation of PNC by microemulsion method[23,58] and (b) mechanism of microemulsion polymerization[58] (Reprinted with permission from Mishra, S.; Chatterjee, A. Novel Synthesis of Polymer and Copolymer Nanoparticles by Atomized Micro Emulsion Technique and its Characterization. *Polym. Adv. Technol.* **2011**, *22*, 1593–1601. © 2011 John Wiley.)

20 min.[40] PANI/γ-Fe_2O_3 nanocomposites were prepared by *in-situ* chemical polymerization route under ultrasound environment.[65]

6.3.6 ULTRASOUND-ASSISTED POLYMERIZATION

Nanoclay-based PNCs have been undertaken by using PAA in aqueous medium and ultrasound environment.[46] ER/nano$CaCO_3$ was synthesized using ultrasound-assisted polymerization technique with controlled reaction parameters.[21,24] PS:PLA and PS:PLA:OMMT nanocomposites were prepared under controlled ultrasonic cavitation technique for 30 min.[53,55] PNCs-filled conductive nanofillers can be prepared by melt mixing process.[64,66,67] Dispersion of CNTs in polymers (PEO) was also carried out using a probe sonicator for 15 min, keeping the beaker immersed in cold water (Fig. 6.4).[64] PNCs of ER/n$BaSO_4$ were successfully synthesized using solution polymerization technique with controlled reaction parameters under ultrasound cavitation technique to remove the agglomeration of nanoparticles in the resin.[41]

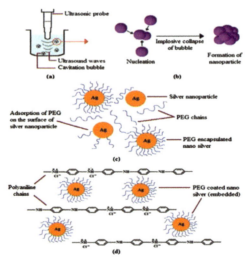

FIGURE 6.4 Pictorial representation of (a) cavitation, (b) nuclei growth due to bubble collapse, (c) encapsulation of silver nanoparticles by PEG, and (d) incorporation of silver nanoparticles between PANi chains[64] (Reprinted from Mishra, S.; Shimpi, N. G.; Sen, T. The Effect of PEG Encapsulated Silver Nanoparticles on the Thermal and Electrical Property of Sonochemically Synthesized Polyaniline/Silver Nanocomposite. *J. Polym. Res.* **2013**, *20*, 49. With kind permission from Springer Science and Business Media.)

6.3.7 SOLUTION CASTING

Solution casting or drawdown coating can be used to make nanocomposite films. These films were prepared by depositing a known mass of blended suspension onto an aluminum pan and air drying in an oven at either 40 or 120°C.[56]

6.4 PROPERTIES OF PNCS

Advances in the fundamental understanding of role of the filler-polymer interface/interphase have driven growth in PNCs applications. Consequently, achieving enhanced properties of PNCs and performance depends critically on maximizing dispersion of nanoparticles in the polymer matrix. The improvement of properties of PNCs basically depends on structure, surface area of nanoparticles, and interfacial bonding with polymer matrix. The improvement in properties of PNC basically depends on the crystalline structure and the degree of the crystallinity, so the melt behavior of the thermoplastic nanocomposites is technologically important for industrial-processing conditions.

6.4.1 MECHANICAL PROPERTIES

An improvement in mechanical properties of PNCs can be observed by proper intercalation, reinforcement of polymer matrix with different nanomaterials. Mechanical properties of inorganic nanomaterial-filled PNCs were found to be improved as compared to virgin polymer. Nanoparticles transfer the heat uniformly thoroughly the matrix during cross-linking. Also, uniform dispersion of nanofiller prevents catastrophic failure of the specimen by arresting the crack growth and to increase the extensibility of rubber-based PNCs. Improvement in mechanical properties of rubber-based PNCs is due to greater degree of cross-linkage during vulcanization of rubber by uniform transfer of heat in to the rubber matrix. It is also due to the very fine size of the particles, which produced more interfacial bonding along with good dispersion and homogeneity of bonding. Table 6.1 shows mechanical properties of various PNCs filled with nano-$CaCO_3$.

The addition of other inorganic nanofillers like $Mg(OH)_2$,[6–8] nano-$CaCO_3$,[9–28,49] $CaSO_4$,[29–36] and $Ca_3(PO_4)_2$,[29,32,37–39] $BaSO_4$,[40,41] $Al(OH)_3$,[42] and

$BaCO_3$[43] is also capable for improvement in mechanical properties of PNCs due to higher mobilization of the matrix molecules, higher nucleation, and the intercalation of polymer chains in the nanolayers galleries.

TABLE 6.1 Mechanical Properties of n-$CaCO_3$-filled PNCs.

Polymer or rubber used	Tensile strength	Hardness	% Elongation break	Young's modulus	Moduli at elongation
PBR[9,16,20]	1.03 MPa (8 wt%)	65 (12 wt%)	261% (8 wt%)	3.06 MPa (8 wt%)	1.41 MPa (8 wt%)
ER[10,21,24]	9 MPa (10 wt%)	–	–	1400 MPa (10 wt%)	–
SBR[11,12,14,49]	2.58 MPa (0.4 wt%)	73 (8 wt%)	663% (0.4 wt%)	1300 MPa (10 wt%)	1.2 MPa (0.4 wt%)
EPDM[17]	2.4 MPa (10 wt%)	–	625% (10 wt%)	–	–
PVC[18]	79.5 MPa (9 wt%)	72 (9 wt%)	22% (9 wt%)	–	–
PA[19]	9 MPa (4 wt%)	98 (4 wt%)	8% (4 wt%)	67 (4 wt%)	–
SR[22]	8.5 MPa (10 wt%)	–	780% (10 wt%)	–	–
PS[25]	34 MPa (1 wt%)	–	50% (0.25 wt%)	–	–

Parenthesis indicates loading of nanofiller with minimum size.

The improvement in mechanical properties of rubber-based PNCs might be due to the over curing of the matrix at the mold surface as a result of higher temperature for longer period of time. This increment is also observed in compressive strength of smaller nanoparticles due to their role as nucleating agents, which can create greater numbers of reduced closed cells, and also it can dissipate compressive load. Nanoclay-based PNCs were found to be improved due to reduced viscosity during processing.

6.4.2 SURFACE MORPHOLOGICAL PROPERTIES

For rubber, inorganic nanoparticles are strongly fixed by electrostatic forces, so it is necessary to add a hydrophobic layer with coupling agent to reduce the agglomeration problem. This phenomenon is also observed from the SEM (Fig. 6.5a and b), in which uniform dispersion and the agglomerated structure are clearly observed in rubber matrix for 8 and 12 wt% loading of $CaCO_3$. SEM of pure PBR and commercial $CaCO_3$-filled PBR are also given in Fig. 6.5c and Fig. 6.5d, respectively. TEM of pure nano-$CaCO_3$ is also observed in Fig. 6.5e, which looks like a nano-flower structure.[16]

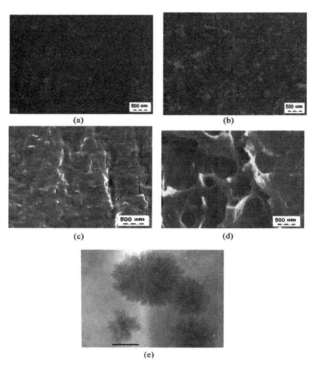

FIGURE 6.5 (a) SEM micrographs of 8 wt% nano-$CaCO_3$ filled in PBR (9 nm), (b) SEM micrographs of 12 wt% nano-$CaCO_3$ filled in PBR (9 nm), (c) SEM micrographs of pristine PBR, (d) SEM micrographs of 8 wt% CaCO3 filled in PBR (commercial), and (e) TEM image of nanosize $CaCO_3$ (9 nm)[16].

Inorganic nanoparticles are strongly fixed by electrostatic forces in rubber, so it is necessary to add a hydrophobic layer with coupling agent to reduce the agglomeration problem. This phenomenon gives the clear idea of agglomeration and uniform dispersion of $CaCO_3$ nanoparticles in to the rubber matrix at 8 and 10 wt% loading of fillers in SBR.[49] This phenomenon can also be observed by SEM (Fig. 6.6) of nano-$CaSO_4$/SBR nanocomposites, which gives a clear idea of the agglomeration and uniform dispersion of the nanoparticles in the rubber matrix at 10 and 12 wt% loadings of the filler.[31]

FIGURE 6.6 (a) SEM of 10 wt% nano-CaSO$_4$/SBR, (b) SEM of 12 wt% nano-CaSO$_4$/SBR, and (c) TEM of CaSO$_4$ nanoparticles[31] (Reprinted with permission from Mishra, S.; Shimpi, N. G. Effect of the Variation in the Weight Percentage of the Loading and the Reduction in the Nanosizes of CaSO4 on the Mechanical and Thermal Properties of Styrene–Butadiene Rubber. *J. Appl. Polym. Sci.* **2007,** *104,* 2018–2026. © 2007, John Wiley.)

The synthesized CSNPs were characterized by transmission electron microscopy that indicated the particle size <100 nm (Fig. 6.7a). The microemulsion reaction phenomena can give rise to the formation of CSNPs as well as free nano-CaCO$_3$ and PMMA nanoparticles.

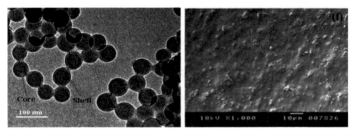

FIGURE 6.7 (a) Nano-CaCO$_3$/PMMA core–shell nanoparticles (CSNP) (b) 1 wt% CSNP reinforced PP composites[26,27]. (a: Reprinted from Chatterjee, A.; Mishra, S. Novel Synthesis with an Atomized Micro Emulsion Technique and Characterization of Nano-calcium Carbonate (CaCO3)/Poly(Methyl Methacrylate) Core–Shell Nanoparticles. *Particles* **2013,** *11,* 760–767. © 2013, with permission from Elsevier; b: Reprinted from Chatterjee, A.; Mishra, S. Rheological, Thermal and Mechanical Properties of Nano-Calcium Carbonate (CaCO3)/Poly(Methyl Methacrylate) (PMMA) Core–Shell Nanoparticles Reinforced Polypropylene (PP) Composites. *Macromol. Res.* **2013,** *21*(5), 474–483. © 2013, with kind permission from Spring Science and Business Media.)

So, it is necessary to measure the percent grafting of CSNPs and therefore, CSNPs have to be separated from free nano-CaCO$_3$ and PMMA nanoparticles. These CSNPs can be reinforced in to PP matrix. It is clearly observed that PMMA shells were completely merged with PP matrix, and nano-CaCO$_3$ particles were integrated with PP matrices by the grafted polymer chains.[27]

Figure 6.8 presents the TEM micrographs of pure PANI/PEG-Ag (0 and 2 wt%) nanocomposites and silver nanoparticles. Plate-like, nanosized silver with dimension of about 50 nm × 35 nm is observed in all the nanocomposite systems.[64]

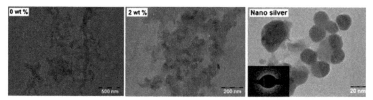

FIGURE 6.8 PEG encapsulated silver-nanoparticle-based PANI/Ag nanocomposites[64] (reproduced by permission). (Reprinted from Mishra, S.; Shimpi, N. G.; Sen, T. The Effect of PEG Encapsulated Silver Nanoparticles on the Thermal and Electrical Property of Sonochemically Synthesized Polyaniline/Silver Nanocomposite. *J. Polym. Res.* **2013,** *20,* 49. © 2012 with the kind permission of Springer Science and Business Media.)

Fig. 6.9a–d presents the FE-SEM micrographs of γ-Fe_2O_3 nanoparticles and PANi/γ-Fe_2O_3 nanocomposites (1, 2, and 3 wt%). Uniform nanospheres of γ-Fe_2O_3 with a diameter of ~22 nm were observed, as shown in Fig. 6.9a. This shows the effectiveness of ultrasound in producing non-aggregated nanoparticles even in the absence of an encapsulating agent. Fig. 6.9(b–d) shows the nanospheres of γ-Fe_2O_3 were uniformly dispersed in the PANi matrix. Nanofibrillar morphology of PANi with a diameter of about 100, 96, and 70 nm (±5 nm) can be seen for 1, 2, and 3 wt% of PANi/-Fe_2O_3 nanocomposites, respectively.[65]

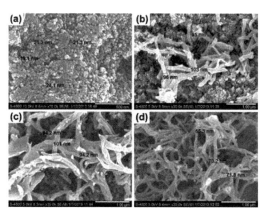

FIGURE 6.9 FE-SEM micrographs of (a) γ-Fe_2O_3 nanoparticles (500 nm) and PANi/γ-Fe_2O_3 nanocomposites at 1 wt% (b), 2 wt% (c), and 3 wt% (d) (1 μm)[65] (Reprinted from Sen, T.; Shimpi, N. G.; Mishra, S.; Sharma, R. Polyaniline/γ-Fe2O3 Nanocomposite for Room Temperature. *LPG Sens.: Sens. Act. B* **2014,** *190,* 120–126. © 2014, with permission from Elsevier.)

6.4.3 FLAME-RETARDANT PROPERTIES

The relationship between the nanofiller concentration and the rate of burning for the PNCs represents the flame-retarding property of PNCs. The time required to burn the nanofilled PNCs was found to be reduced than the time required for polymer. This reduction in flammability was due to the endothermic nature of nanofiller dispersed in the polymer matrix. The nanofiller formed an effective layer on the surface by its uniform dispersion. Thus, the absorption of energy by nanoparticles was uniform (endothermic), and the evolution of flue gases is hampered. Table 6.2 shows flame retarding values of various nano-$CaCO_3$-based PNCs.

TABLE 6.2 Flame Retarding Values of Various $CaCO_3$-based PNCs.

Polymer or rubber used	Flammability (s/mm)
PBR[9,16,20]	2.32
ER[10,21,24]	30
SBR[11,12,14,49]	1.97
EPDM[17]	1.44
PA[19]	4
SR[22]	32

All values are at minimum loading of nanofiller.

6.4.4 THERMAL PROPERTIES

Thermal behavior of PNC can give idea about crystalline nature and % weight loss with respect to temperature. The temperature of melting, glass transition, onset, off set, and heat of fusion, of the various PNCs, can be determined by differential scanning calorimetry and thermogravimetric analysis. The melting temperatures of commodity polymer PP are reported to be increasing as amount of inorganic nanofiller loading increases. While change in enthalpy was observed to be decreasing as compared to the virgin PP, the decrement in degree of crystallization was reported with addition of inorganic nanofiller. Moreover, it is worth mentioning that the reduction in particle size reduces the degree of crystallinity, which is the reverse of cooling rate. Furthermore, it can be said that the higher interaction of filler with matrix facilitates uniform absorption of energy by nanoparticles and thereby faster cooling. Also, crystallization temperature was found to be decreasing on increasing the wt% of loading and varying the particle size of nanofillers in PP-based PNC. Table 6.3 shows thermal properties of n-CaCO$_3$ filled PNCs.

TABLE 6.3 Thermal properties of n-CaCO$_3$-filled PNCs.

Polymer or rubber used	T_m (°C)	T_g (°C)	Degradation temperature (°C)	Enthalpy
PBR[9,16,20]	–	−67	491	−23.98 J/mol
ER[10,21,24]	400	232	345	5–10 kJ/mol
SBR[11,12,14,49]	–	−55	460	−23.98 J/mol
PP[13]	167	–	–	82.88 J/g
HIPS[15]	200	–	180	5.92 J/g
EPDM[17]	–	–	457	–
PVC[18]	450	87	275	–
PA[19]	–	–	571	–
SR[22]	590	–	529	–
HIPS[25]	465	−53	432	–
PS/PP[23]	550	125	390	–
PMMA[26]		136	369	–
PMMA/PP[26,27]	172	–	435	–

It can be said that the incorporation of inorganic nanofiller in rubber matrix with reduced size can show better thermal stability as compared to commercial filled one. This enhancement in thermal stability was due to uniform dispersion of nanofiller throughout the matrix. It is clear that nano-inorganic filler provided better thermal stability; moreover, the reduction in nanosize and increase in amount of nano-inorganic filler improve the thermal stability of rubber-based PNCs. This was due to higher increment in surface area of nanoparticles.[8–17,20,22,38,42,43,49–52]

6.4.5 SWELLING PROPERTIES

Swelling index (SI) of inorganic nanofiller-based PNCs is reported to be increasing which may be due to the greater crosslinking of polymer matrix, as the uniform dispersion of nanofiller brought the chains closer and kept them intact with nanoparticles. The SI of rubber-based PNCs is reported decreasing as % loading of nanofiller increases. This is due to greater cross-linking of rubber, as the uniform dispersion of nanofiller brings the chains closer and keeps them intact with nanoparticles. Swelling in elastomeric nanocomposites depends on elastomer crosslinking density and solvent used.

6.4.6 PHYSICAL PROPERTIES

Specific gravity of inorganic nanofiller-based PNCs and rubber matrices generally increases by decreasing the size of nanofiller, which is due to greater and uniform dispersion in polymer matrix, which brings chains of matrix closer to reduce the free volume to the greater extent in crosslinking of chains.[11,15,16,49,31,32] Another reason may be the formation of greater number of closed cells at nanoscale that can provide greater facilitation for the expansion of foam, while the formation of closed cells with bigger size may reduce foam expansion. Thus, the smaller size nanoparticles act as nano-spacers more efficiently than bigger size nanoparticles, which result in increment in relative density in comparison to smaller size.[38]

6.4.7 RHEOLOGICAL PROPERTIES

It was reported that for increasing the size of inorganic nanofiller, the initial torque required to incorporate the filler into the matrix is more than that of the smaller nanosize particles. It is also noted that the torque, during the mixing of the reduced nanosize particles, lasted for a longer time compared to that of the bigger nanosize and virgin polymer which might have been due to the dispersion of smaller nanoparticles that closely bound the polymer matrix and, hence, created a restriction to flow with less variation in torque during processing. It was also reported that shear viscosity of commodity PNCs decreased with increasing weight percentage of nanofiller and it showed shear thinning behavior of PNCs. Shear thickening can be observed for PNC due to needle-like structure of nanofiller.[29] The effective decrease in the viscosity with increasing nanoclay content might be due to the lubrication effect of nanoclay during processing of the polymer, and the rheological data indicated that the behavior of the melts was highly non-Newtonian due to the lubrication effect of OMMT.

The pseudo-plastic nature of the polyamide/OMMT blends is observed, indicating that the apparent viscosity decreased as shear rate increased.[45]

6.4.8 ELECTRICAL PROPERTIES

The electrical conductivity of pure PANi and PANi/Ag nanocomposite is shown in Fig. 6.10. The electrical conductivity is measured as a function of temperature. Conductivity of PANi/Ag nanocomposites with nano-silver content of 1 wt% and above is slightly higher than the rest. An increasing trend in electrical conductivity is observed when the PANi/Ag nanocomposites are heated from room temperature to 150°C. This confirms the semiconducting nature of the PANi/Ag nanocomposites. PANi/Ag nanocomposites with 0.5–0.75 wt% nano-silver content show a marginal increase in conductivity with temperature.[64]

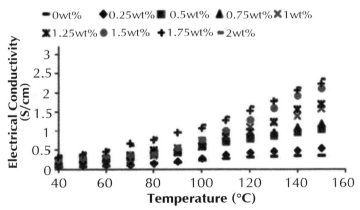

FIGURE 6.10 Variation in electrical conductivity of pure PANi (0 wt%) and PANi/Ag nanocomposites (0.25–2 wt%) with temperature[64] (Reprinted from Mishra, S.; Shimpi, N. G.; Sen, T. The Effect of PEG Encapsulated Silver Nanoparticles on the Thermal and Electrical Property of Sonochemically Synthesized Polyaniline/Silver Nanocomposite. *J. Polym. Res.* **2013**, *20*, 49. © 2012 with the kind permission of Springer Science and Business Media.)

Scanning spreading resistance microscopy is a very efficient mode of AFM for materials characterization in terms of local resistance. A bias voltage is applied to a conducting probe in contact with the sample surface, and the resulting current flow through the sample is measured as a function of the probe with simultaneous surface topography measurements.

Assuming that there is a constant contact between the probe and the surface, for a given bias voltage, the magnitude of the measured current flow is proportional to local resistance of the sample under investigation. Fig. 6.11 shows the image for current signal of pure PANi (Fig. 6.11a) and PANi/Ag nanocomposites at 2 wt% nano-silver content (Fig. 6.11b). The lighter areas in the signal image correspond to higher conductivity.[64]

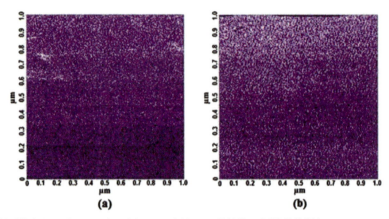

FIGURE 6.11 Current signal image of (a) pure PANi and (b) PANi/Ag nanocomposites at 2 wt% nano-silver content[64] (Reprinted from Mishra, S.; Shimpi, N. G.; Sen, T. The Effect of PEG Encapsulated Silver Nanoparticles on the Thermal and Electrical Property of Sonochemically Synthesized Polyaniline/Silver Nanocomposite. *J. Polym. Res.* **2013**, *20*, 49. © 2012 with the kind permission of Springer Science and Business Media.)

6.5 APPLICATIONS OF PNC

PNCs are increasingly utilized in automotive, aerospace, energy, and flame retardant applications. PNCs may offer significant performance advantages over traditional polymer composites, including (i) enhanced physical properties, (ii) ability to tailor material properties for new applications, (iii) improved performance/weight ratio achieved by reduction of filler loadings from 15–40 vol% to as little as 1–5%, and (iv) improved processing performance. Fig. 6.12 shows various applications of PNCs.

Commercial developments of PNCs involve mainly two types of nano-objects: oxide-based particles and clays (some examples of hybrid materials based on NPs/clays PNCs are illustrated in Fig. 6.12). This could be easily explained by the commercial availability and the relatively low cost of such nano-objects.[72] It is important to recognize that PNCs research is extremely broad, encompassing areas such as electronics and computing, data storage, communications, aerospace and sporting materials, health and medicine, energy, environmental, transportation, and national defense applications.

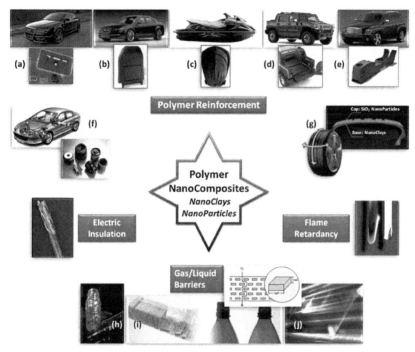

FIGURE 6.12 Application of PNCs[72] (Reproduced from Sanchez, C.; Belleville, P.; Popall, M.; Nicole, L. Applications of Advanced Hybrid Organic–Inorganic Nanomaterials: From Laboratory to Market. *Chem. Soc. Rev.* **2011**, *40*, 696–753. with permission of The Royal Society of Chemistry.)

Recently, nanocomposites have been introduced in structural applications, such as automotive parts, gas barrier films, scratch-resistant coating, and flame-retardant cables. Filament-winding techniques can be used to manufacture nanocomposite parts for various applications including commercial aircraft structures for Boeing, Airbus, as well as many products in the industrial markets. Conductive filler-based PNCs are found to be useful in varieties of applications. Carbon nanofibers have been used for different aerospace applications like aircraft engine anti-icing, fire retardant coatings (forming a char layer over combustible composites), lightning strike protection, solid rocket motor nozzles, conductive aerospace adhesives, thermo-oxidative resistant structures, missile/airframe structure. Many potential applications have been proposed for CNTs, including conductive and high-strength composites, energy storage and energy conversion devices, sensors, field emission displays and radiation sources, hydrogen storage media and nanometer-sized semiconductor devices, probes, and interconnect. Following are the broad areas where PNCs are useful in modern life.[1–72]

6.5.1 BARRIER PACKAGING AND SPORTS GOODS

Rubber- and nanosilica-filled PNCs can be specifically shaped and used as golf balls. A number of PNCs based on rubbers, such as BR, SBR, PBR, and EPDM, have been used commercially for barrier applications for many gases such as CO_2, O_2, N_2, and chemicals such as toluene, HNO_3, H_2SO_4, HCl, etc. Due to excellent solvent barrier properties, PNCs have been utilized in chemical protective and surgical gloves in order to protect against chemical warfare agents and for avoiding contamination from medicine.

6.5.2 ENERGY STORAGE DEVICES AND SENSORS

Typical membranes are made of organic polymers containing acidic functions such as carboxylic, sulfonic or phosphonic groups which dissociate when solvated with water, allowing H_3O^+ (hydrated proton) transport. Fuel cells act as electrochemical devices, which convert chemical energy of carbon, hydrogen, and oxygen directly and efficiently into useful electrical energy with heat and water as the only byproducts. Due to incorporation of nanomaterials, their efficiency increases considerably. Various conducting and semiconducting metal oxides' nanomaterial-based PNCs have been found by their application in sensing of gases like LPG, ammonia, organic vapors, etc.

6.5.3 TRANSPARENT MATERIALS

Clay-incorporated PNCs have been proved as transparent materials with both toughness and hardness without sacrificing light transmission characteristics. Owing to this reason and improved optical properties of these transparent materials, they have been commercialized in contact lens and optical glass applications.

6.5.4 LOW-FLAMMABLE PRODUCTS

The nanoclays are capable to reduce the flammability of polymeric materials which is incredible. The improvement of flame resistance by incorporation of clay can be useful in applications of cable wire jacket, car seat, packaging films, textile cloths, surface coatings for many steel products, and paints; one of the higher end applications is rocket ablative materials' core manufacturing. PU/clay-based PNCs also exhibit superior flame retardancy, and they have already been commercialized in automobile seats manufacturing.

6.5.5 AUTOMOBILE SECTOR

Application of thermoset/clay nanocomposites for automobile sectors is another big milestone. The ability of nanoclay incorporation to reduce solvent transmission through polymers such as specialty elastomers, polyimides, PU, etc. has been demonstrated. A study reveals the significant reduction in fuel transmission through PA/nanoclay-based PNCs.

6.5.6 COATINGS

Coatings are important for modifying properties of surfaces. Several strategies have been tried by researchers for improving surface properties of products. One of the well-versed developments is nanoclay-based PNC coatings. Nanoclay-incorporated thermoset PNC coatings exhibit superior properties such as super hydrophobicity, improved wettability, excellent resistance for chemicals, corrosion resistance, improved weather resistance, better abrasion resistance, improved barrier properties and resistance to impact, scratch, etc.

6.6 SUMMARY

PNCs with substantial improvements in the characteristic properties can be obtained by the addition of small amounts of nanofillers. This improvement can be observed due to even dispersion of nanofillers in polymer matrix, which was endothermic in nature and intercalated and nucleated the polymer chains. Exfoliation imparted by nanosize materials is also helpful to improve the characteristic properties of PNCs.

Among many highly hyped technological products, PNCs are those which have lived up to the expectations. PNCs exhibit superior properties, such as mechanical, barrier, optical, etc. as compared to micro- or macro-composites. Owing to this, PNCs have shown ubiquitous presence in various fields of applications. PNCs for various applications could be synthesized by proper selection of matrix, nanoreinforcements, and synthesis methods. Many products based on PNCs have been commercialized. This review has tried to highlight various types of PNCs, their preparation methods, unique properties, and various technological applications with some specific examples of commercialized products.

KEYWORDS

- **nanotechnology**
- **polymer nanocomposites**
- **preparation**
- **properties**
- **applications**

REFERENCES

1. Patil, Y. P.; Gajre, B.; Dusane, D.; Chavan, S.; Mishra, S. Effect of Maleic Anhydride Treatment on Steam and Water Absorption of Wood Polymer Composites Prepared from Wheat Straw, Cane Bagasse, and Teak Wood Sawdust Using Novolac as Matrix. *J. Appl. Polym. Sci.* **2000,** *77,* 2963–2967.
2. Mishra, S.; Talele, N. R. Filler Effect of Potato Starch and Urea on Degradation of Linear Low Density Polyethylene Composites. *Polym. Plast. Technol. Eng.* **2002,** *41*(2), 361–381.

3. Mishra, S.; Patil, Y. P. Kinetics of Esterification and its Effect on Mechanical Properties of Cane Bagasse Pith Filled Melamine Formaldehyde Composites. *Nat. Polym. Comp.* **2002**, *IV*, 1–6.
4. Mishra, S.; Patil, Y. P. Compatibilizing Effect of Different Anhydrides on Cane Bagasse Pith and Melamine–Formaldehyde–Resin Composites. *J. Appl. Polym. Sci.* **2003**, *88*, 1768–1774.
5. Mishra, S.; Naik, J. B.; Patil, Y. P. Studies on Swelling Properties of Wood/Polymer Composites Based on Agro-waste and Novolac. *Adv. Polym. Technol.* **2004**, *23*(3), 1–5.
6. Mishra, S.; Sonawane, S. H.; Singh, R. P.; Bendale, A.; Patil, K. Effect of Nano-$Mg(OH)_2$ on the Mechanical and Flame-retarding Properties of Polypropylene Composites. *J. Appl. Polym. Sci.* **2004**, *94*, 116–122.
7. Sonawane, S. S.; Mishra, S.; Shimpi, N. G.; Rathod, A. P.; Wasewar, K. L. Comparative Study of the Mechanical and Thermal Properties of Polyamide-66 Filled with Commercial and Nano-$Mg(OH)_2$ Particles. *Polym. Plast. Technol. Eng.* **2010**, *49*(5), 474–480.
8. Shimpi, N. G.; Sonawane, H. A.; Mali, A. D.; Mishra, S. Effect of $Mg(OH)_2$ Nanoparticles on Thermal, Mechanical and Morphological Properties of Millable Polyurethane Elastomer. *J. Reinf. Plast. Comp.* **2013**, *32*(13), 935–946.
9. Mishra, S.; Sonawane, S. H.; Badgujar, N.; Gurav, K.; Patil, D. Comparative Study of the Mechanical and Flame-retarding Properties of Polybutadiene Rubber Filled with Nanoparticles and Fly Ash. *J. Appl. Polym. Sci.* **2005**, *96*, 6–9.
10. Mishra, S.; Sonawane, S.; Chitodkar, V. Comparative Study on Improvement in Mechanical and Flame Retarding Properties of Epoxy-$CaCO_3$ Nano and Commercial Composites. *Polym. Plast. Technol. Eng.* **2005**, *44*, 463–473.
11. Mishra, S.; Shimpi, N. G. Comparison of Nano-$CaCO_3$ and Fly Ash Filled Styrene Butadiene Rubber on Mechanical and Thermal Properties. *J. Sci. Ind. Res.* **2005**, *64*, 744–751.
12. Mishra, S.; Shimpi, N. G. Mechanical and Flame-retarding Properties of Styrene–Butadiene Rubber Filled with Nano-$CaCO_3$ as a Filler and Linseed Oil as an Extender. *J. Appl. Polym. Sci.* **2005**, *98*, 2563–2571.
13. Mishra, S.; Sonawane, S. H.; Singh, R. P. Studies on Characterization of Nano $CaCO_3$ Prepared by the In Situ Deposition Technique and its Application in PP Nano $CaCO_3$ Composites. *J. Polym. Sci. B: Polym. Phys.* **2005**, *43*, 107–113.
14. Mishra, S.; Shimpi, N. G.; Verma, J. Effect of Nano-$CaCO_3$ on Thermal Properties of Styrene–Butadiene Rubber (SBR), In: *Rubber Chemistry*; Munich, Germany, 2006; pp 1–8.
15. Mishra, S.; Mukherji, A. Studies of Thermal Conductivity of Nano $CaCO_3$/HIPS Composites by Unsteady State Technique and Simulation with Nielsen's Model. *Polym. Plast. Technol. Eng.* **2007**, *46*, 239–244.
16. Mishra, S.; Shimpi, N. G. Studies on Mechanical, Thermal, and Flame Retarding Properties of Polybutadiene Rubber (PBR) Nanocomposites. *Polym. Plast. Technol. Eng.* **2008**, *47*, 72–81.
17. Mishra, S.; Patil, U. D.; Shimpi, N. G. Synthesis of Mineral Nanofiller Using Solution Spray Method and its Influence on Mechanical and Thermal Properties of EPDM Nanocomposites. *Polym. Plast. Technol. Eng.* **2009**, *48*(100), 1078–1083.

18. Shimpi, N. G.; Verma, J.; Mishra, S. Dispersion of Nano $CaCO_3$ on PVC and its Influence on Mechanical and Thermal Properties. *J. Comp. Mater.* **2010**, *44*, 211.
19. Sonawane, S. S.; Mishra, S.; Shimpi, N. G. Effect of Nano-$CaCO_3$ on Mechanical and Thermal Properties of Polyamide Nanocomposites. *Polym. Plast. Technol. Eng.* **2010**, *49*(1), 38–44.
20. Shimpi, N. G.; Mishra, S. Synthesis of Nanoparticles and its Effect on Properties of Elastomeric Nanocomposites. *J. Nanopart. Res.* **2010**, *12*, 2093–2099.
21. Shimpi, N. G.; Kakade, R. U.; Sonawane, S. S.; Mali, A. D.; Mishra, S. Influence of Nano-Inorganic Particles on Properties of Epoxy Nanocomposites. *Polym. Plast. Technol. Eng.* **2011**, *50*, 758–761.
22. Mishra, S.; Shimpi, N. G.; Mali, A. D. Influence of Stearic Acid Treated Nano-$CaCO_3$ on the Properties of Silicone Nanocomposites. *J. Polym. Res.* **2011**, *18*(6), 1715–1724.
23. Mishra, S.; Chatterjee, A.; Singh, R. P. Novel Synthesis of Nano-calcium Carbonate ($CaCO_3$)/Polystyrene (PS) Core–Shell Nanoparticles by Atomized Microemulsion Technique and its Effect on Properties of Polypropylene (PP) Composites. *Polym. Adv. Technol.* **2011**, *22*, 2571–2582.
24. Shimpi, N. G.; Mishra, S. Sonochemical Synthesis of Mineral Nanoparticles and its Applications in epoxy Nanocomposites. *Polym. Plast. Technol. Eng.* **2012**, *51*(2), 111–115.
25. Chatterjee, A.; Mishra, S. Nano-Calcium Carbonate ($CaCO_3$)/Polystyrene (PS) Core–Shell Nanoparticle: It's Effect on Physical and Mechanical Properties of High Impact Polystyrene (HIPS). *J. Polym. Res.* **2013**, *20*, 249.
26. Chatterjee, A.; Mishra, S. Novel Synthesis with an Atomized Micro Emulsion Technique and Characterization of Nano-calcium Carbonate ($CaCO_3$)/Poly(Methyl Methacrylate) Core–Shell Nanoparticles. *Particles* **2013**, *11*, 760–767.
27. Chatterjee, A.; Mishra, S. Rheological, Thermal and Mechanical Properties of Nano-Calcium Carbonate ($CaCO_3$)/Poly(Methyl Methacrylate) (PMMA) Core–Shell Nanoparticles Reinforced Polypropylene (PP) Composites. *Macromol. Res.* **2013**, *21*(5), 474–483.
28. Shimpi, N.; Mali, A.; Hansora, D. P.; Mishra, S. Synthesis and Surface Modification of Calcium Carbonate Nanoparticles Using Ultrasound Cavitation Technique. *Nanosci. Nanoeng.* **2015**, *3*(1), 8–12.
29. Mishra, S.; Sonawane, S.; Mukherji, A.; Mruthyunjaya, H. C. Effect of Nanosize $CaSO_4$ and $Ca_3(PO_4)_2$ Particles on the Rheological Behavior of Polypropylene and its Simulation with a Mathematical Model. *J. Appl. Polym. Sci.* **2006**, *100*, 4190–4196.
30. Mishra, S.; Mukherji, A.; Sharma, D. K. Nonisothermal Crystallization Modeling and Simulation for Polypropylene/Nano-$CaSO_4$ Composites with Variations in Nanosizes and Wt% of Loading. *J. Appl. Polym. Sci.* **2007**, *45*(11), 1191–1198.
31. Mishra, S.; Shimpi, N. G. Effect of the Variation in the Weight Percentage of the Loading and the Reduction in the Nanosizes of $CaSO_4$ on the Mechanical and Thermal Properties of Styrene–Butadiene Rubber. *J. Appl. Polym. Sci.* **2007**, *104*, 2018–2026.
32. Mishra, S.; Mukherji, A. Phase Characterization and Mechanical and Flame-retarding Properties of Nano-$CaSO_4$/Polypropylene and Nano-$Ca_3(PO_4)_2$/Polypropylene Composites. *J. Appl. Polym. Sci.* **2007**, *103*, 670–680.

33. Sonawane, S. S.; Mishra, S.; Shimpi, N. G. Polyamide Nanocomposites: Investigation of Mechanical, Thermal and Morphological Characteristics. *Polym. Plast. Technol. Eng.* **2009**, *48*, 1055–1061.
34. Shimpi, N. G.; Verma, J.; Mishra, S. Preparation, Characterization and Properties of Poly(Vinyl Chloride)/$CaSO_4$ Nanocomposites. *Polym. Plast. Technol. Eng.* **2009**, *48*(10), 997–1001.
35. Kundu, D.; Hazra, C.; Chatterjee, A.; Chaudhari, A.; Mishra, S. Biopolymer and Biosurfactant-Graft-Calcium Sulfate/Polystyrene Nanocomposites: Thermophysical, Mechanical and Biodegradation Studies. *Polym. Degrad. Stab.* **2014**, *107*, 37–52.
36. Vaia, R. A.; Maguire, J. F. Polymer Nanocomposites with Prescribed Morphology: Going Beyond Nanoparticle-filled Polymers. *Chem. Mater.* **2007**, *19*, 2736–2751.
37. Mishra, S.; Mukherji, A.; Sonawane, S. H.; Sharma, D. K. Nonisothermal Crystallization Modeling and Simulation of Polypropylene/Nano $Ca_3(PO_4)_2$ Composites with Variation in Wt.% of Loading and Reduction in Nanosize. *J. Appl. Polym. Sci.* **2006**, *45*(5), 641–651.
38. Mukherji, A.; Mishra, S. Effect of sizes of Nano-$Ca_3(PO_4)_2$ on Mechanical and Thermal Properties of Polyurethane Foam Composites. *J. Appl. Polym. Sci.* **2007**, *46*, 675–681.
39. Mishra, S.; Sonawane, S. S.; Shimpi, N. G. Effect of Commercial and Nano-$Ca_3(PO_4)_2$ on Mechanical and Thermal Properties of Polyamide Composites. *Polym. Plast. Technol. Eng.* **2009**, *48*(3), 265–271.
40. Kulkarni, R. D.; Ghosh, N.; Patil, U. D.; Mishra, S. In Situ Synthesis of Poly(Styrene–Butylacrylate–Acrylic Acid) Latex/Barium Sulfate Nanocomposite and Evaluation of their Film Properties. *Polym. Comp.* **2013**, *34*, 1–12.
41. Shimpi, N. G.; Mishra, S. Ultrasonic-assisted Synthesis of Nano-$BaSO_4$ and its Effect on Thermal and Crosslinking Density of Epoxy Nanocomposites. *J. Reinf. Plast. Comp.* **2013**, *32*(13), 947–954.
42. Shimpi, N. G.; Sonawane, H. A.; Mali, A. D.; Mishra, S. Effect of $nAl(OH)_3$ on Thermal, Mechanical and Morphological Properties of Millable Polyurethane (MPU) Rubber. *Polym. Bull.* **2014**, *71*, 515–531.
43. Shimpi, N. G.; Mali, A. D.; Sonawane, H. A.; Mishra, S. Effect of $nBaCO_3$ on Mechanical, Thermal and Morphological Properties of Isotactic PP-EPDM Blend. *Polym. Bull.* **2014**, *71*(8), 2067–2080.
44. Winey, K. I.; Vaia, R. A. Polymer Nanocomposites. *MRS Bull.* **2007**, *32*, 314–322.
45. Mishra, S.; Sonawane, S. S.; Shimpi, N. G. Influence of Organo-montmorillonite on Mechanical and Rheological Properties of Polyamide Nanocomposites. *Appl. Clay Sci.* **2009**, *46*, 222–225.
46. Sonawane, S. H.; Chaudhari, P. L.; Ghodke, S. A.; Parande, M. G.; Bhandari, V. M.; Mishra, S.; Kulkarni, R. D. Ultrasound Assisted Synthesis of Polyacrylic Acid–Nanoclay Nanocomposite and its Application in Sonosorption Studies of Malachite Green Dye. *Ultrason. Sonochem.* **2009**, *16*, 351–355.
47. Shimpi, N. G.; Mishra, S. Influence of Surface Modification of Montmorillonite on Properties of PVC Nanocomposites. *J. Comp. Mater.* **2011**, *45*(23), 2447–2453.
48. Shimpi, N. G.; Mishra, S. Studies on Effect of Improved d-Spacing of Montmorillonite on Properties of Poly(Vinyl Chloride) Nanocomposites. *J. Appl. Polym. Sci.* **2011**, *119*, 148–154.

49. Mishra, S.; Shimpi, N. G.; Patil, U. D.; Effect of Nano-CaCO$_3$ on Thermal Properties of Styrene Butadiene Rubber (SBR). *J. Polym. Res.* **2007**, *14*, 449–459.
50. Mishra, S.; Shimpi, N. G.; Mali, A. D. Surface Modification of Montmorillonite (MMT) Using Column Chromatography Technique and its Application in Silicone Rubber Nanocomposites. *Macromol. Res.* **2012**, *20*(1), 44–50.
51. Mishra, S.; Shimpi, N. G.; Mali, A. D. Effect of Surface Modified Montmorillonite on Photo-oxidative Degradation of Silicone Rubber Composites. *Macromol. Res.* **2013**, *21*(5), 466–473.
52. Mali, A. D.; Shimpi, N. G.; Mishra, S. Thermal, Mechanical and Morphological Properties of Surface-modified Montmorillonite-Reinforced Viton Rubber Nanocomposites. *Polym. Int.* **2014**, *63*, 338–346.
53. Shimpi, N. G.; Borane, M.; Mishra, S. Kadam, M. Biodegradation of Polystyrene (PS)–Poly(Lactic Acid) (PLA) Nanocomposites Using *Pseudomonas aeruginosa*. *Macromol. Res.* **2012**, *20*(2), 181–187.
54. Paul, D. R.; Robeson, L. M. Polymer Nanotechnology: Nanocomposites. *Polymer* **2008**, *49*, 3187–3204.
55. Shimpi, N. G.; Borane, M.; Mishra, S. Preparation, Characterization, and Biodegradation of PS:PLA and PS:PLA:OMMT Nanocomposites Using *Aspergillus niger*. *Polym. Comp.* **2014**, *35*, 263–272.
56. Shori, S. K. Surface Modification of Nanoplatelets in Polymer Nanocomposites, PhD Theses, University of South Carolina: Columbia, 2014.
57. Ghodke, S. A.; Sonawane, S. H.; Bhanvase, B. A.; Mishra, S.; Joshi, K. S. Studies on Fragrance Delivery from Inorganic Nanocontainers: Encapsulation, Release and Modeling Studies. *J. Inst. Eng. India Ser. E* **2015**, *96*, 45.
58. Mishra, S.; Chatterjee, A. Novel Synthesis of Polymer and Copolymer Nanoparticles by Atomized Micro Emulsion Technique and its Characterization. *Polym. Adv. Technol.* **2011**, *22*, 1593–1601.
59. Mishra, S.; Chatterjee, A. Effect of Nano-polystyrene (nPS) on Thermal, Rheological, and Mechanical Properties of Polypropylene (PP). *Polym. Adv. Technol.* **2011**, *22*, 1547–1554.
60. Mishra, S.; Chatterjee, A. Particle Size, Morphology and Thermal Properties of Polystyrene Nanoparticles in Micro Emulsion Process. *Polym. Plast. Technol. Eng.* **2010**, *49*(8), 791–795.
61. Chatterjee, A.; Mishra, S.; Novel Synthesis of Crystalline Polystyrene Nanoparticles (nPS) by Monomer Atomization in Micro-emulsion and their Effect on Thermal, Rheological, and Mechanical Properties of Polypropylene (PP). *Macromol. Res.* **2012**, *20*(8), 780–788.
62. Mishra, S.; Chatterjee, A.; Rana, V. K. Polymer Nanoparticles: Their Effect on Rheological, Thermal, and Mechanical Properties of Linear Low-density Polyethylene (LLDPE). *Polym. Adv. Technol.* **2011**, *22*, 1802–1811.
63. Ratna, D.; Jagtap, S. B.; Rathor, R.; Kushwaha, R. K.; Shimpi, N. G.; Mishra, S. A Comparative Studies on Dispersion of Multiwall Carbon Nanotubes in Poly(Ethylene Oxide) Matrix Using Dicarboxylic Acid and Amino Acid Based Modifiers. *Polym. Comp.* **2013**, *34*, 1004–1011.

64. Mishra, S.; Shimpi, N. G.; Sen, T. The Effect of PEG Encapsulated Silver Nanoparticles on the Thermal and Electrical Property of Sonochemically Synthesized Polyaniline/Silver Nanocomposite. *J. Polym. Res.* **2013**, *20*, 49.
65. Sen, T.; Shimpi, N. G.; Mishra, S.; Sharma, R. Polyaniline/γ-Fe_2O_3 Nanocomposite for Room Temperature. *LPG Sens.: Sens. Act. B* **2014**, *190*, 120–126.
66. Potschke, P.; Bhattacharyya, A. R.; Alig, I. Dudkin, S. M.; Leonhardt, A.; Taschner, C.; Ritschel, M.; Roth, S.; Hornbostel, B.; Cech, J. Dispersion of Carbon Nanotubes into Thermoplastic Polymers Using Melt Mixing. *Am. Inst. Phys.* **2004**, 478–482.
67. Potschke, P.; Bhattacharyya, A. R.; Janke, A.; Pegel, S.; Leonhardt, A.; Taschner, C.; Ritschel, M.; Roth, S.; Hornbostel, B.; Cech, J. Melt Mixing as Method to Disperse Carbon Nanotubes into Thermoplastic Polymers. *Full Nanotube Carbon Nanostruct.* **2005**, *13*, 211–224.
68. Rana, V. K.; Akhtar, S.; Chatterjee, S.; Mishra, S.; Singh, R. P.; Ha, C. S. Chitosan and Chitosan-*co*-Poly(Caprolactone) Grafted Multiwalled Carbon Nanotube Transducers for Vapor Sensing. *J. Nanosci. Nanotechnol.* **2014**, *14*, 2425–2435.
69. Hazra, C.; Kundu, D.; Chatterjee, A.; Chaudhari, A.; Mishra, S. Poly(Methyl Methacrylate) (core)–Biosurfactant (Shell) Nanoparticles: Size Controlled Sub-100 nm Synthesis, Characterization, Antibacterial Activity, Cytotoxicity and Sustained Drug Release Behavior. *Colloid Surf. A: Physicochem. Eng. Aspects* **2014**, *449*, 96–113.
70. Kulkarni, R. D.; Ghosh, N.; Patil, U. D.; Mishra, S. Surfactant Assisted Solution Spray Synthesis of Stabilized Prussian Blue and Iron Oxide for Preparation of Nanolatex Composites. *J. Vac. Sci. Technol. B* **2009**, *27*(3), 1478–1483.
71. Purohit, K.; Khitoliya, P.; Purohit, R. Recent Advances in Nanotechnology. *Inter. J. Sci. Eng. Res.* **2012**, *3*(11), 1–11.
72. Sanchez, C.; Belleville, P.; Popall, M.; Nicole, L. Applications of Advanced Hybrid Organic–Inorganic Nanomaterials: From Laboratory to Market. *Chem. Soc. Rev.* **2011**, *40*, 696–753.

CHAPTER 7

SYNTHESIS AND CHARACTERIZATION OF SnO_2/POLYANILINE AND AL-DOPED SnO_2/POLYANILINE COMPOSITE NANOFIBER-BASED SENSORS FOR HYDROGEN GAS SENSING

HEMLATA J. SHARMA[*] and SUBHASH B. KONDAWAR

Department of Physics, Polymer Nanotech Laboratory, Rashtrasant Tukadoji Maharaj Nagpur University, Nagpur 440033, India

[*]Corresponding author. E-mail: hemlatasharma208@gmail.com

CONTENTS

Abstract 124
7.1 Introduction 124
7.2 Experimental 126
7.3 Results and Discussion 128
7.4 Conclusion 134
Acknowledgment 135
Keywords 135
References 135

ABSTRACT

In this chapter, tin oxide/polyaniline (SnO_2/PANI) composites and aluminum (Al)-doped SnO_2/PANI composites nanofiber-based gas sensors for hydrogen-gas-sensing application are presented. PANI in powder form was prepared by chemical oxidative polymerization of aniline using ammonium persulfate in acidic medium at 0–5°C. SnO_2/PANI and Al-doped SnO_2/PANI composites nanofibers were synthesized by electrospinning technique and subsequent calcinations. These composite nanofibers have been characterized by X-ray diffraction, ultraviolet–visible spectroscopy and scanning electron microscopy. The response of these sensors for hydrogen gas was evaluated by monitoring the change in electrical resistance at room temperature. It was observed that Al-doped SnO_2/PANI composite nanofiber-based sensors show a higher response as compared to SnO_2/PANI composite nanofibers sensor. On exposure to hydrogen gas, it was also observed that composite nanofibers showed high sensitivity in temperature range of 45–50°C with relatively faster response/recovery behavior compared to pure SnO_2 and Al-doped SnO_2 nanofibers reported earlier.

7.1 INTRODUCTION

In air-quality control, hydrogen gas sensor is very important as hydrogen gas is colorless, odorless, and extremely flammable gas due to which there have been significant efforts to enhance the sensitivity of hydrogen gas sensors to be operated at low temperature. To ensure the safety of hydrogen, efficient and safety hydrogen sensors are still demanded.[1] In recent years, there has been significant progress in one-dimensional (1D) nanostructures due to their unique physical and chemical properties. Compared to the other three dimensions, 1D nanostructure are highly suitable for moving charges in integrated nanoscale systems due to their low dimension structure and high aspect ratio, which could efficiently transport electrical carriers along one controllable direction.[2,3] Nanofibers of semiconducting metal oxides have been successfully fabricated and widely utilized for gas sensors due to their sensing properties based on the surface reaction between the metal oxides and adsorbed gas species on exposure to specific gas.[4] Nanofibers of pure and doped SnO_2 have been exposed with high-sensing characteristics, but the high-operating temperature (200–400°C) of these sensors may be inadequate for measuring high gas concentrations due to the danger of explosions.[5–9] The conducting polymers have improved the gas-sensing properties especially in lowering the operating temperature to around room temperature. In addition to this, the ability to incorporate specific binding sites into conducting polymers promises the improvement of selectivity and sensitivity of material. Among the various conducting polymers, polyaniline (PANI) has been investigated as a potential material for gas-sensing applications, due to its controllable electrical conductivity, environmental stability, and interesting redox chemistry (or electroactivity). It is the unique type of conducting polymer in which the charge delocalization can, in principle, offer multiple active sites on its backbone for the adsorption and desorption of gas analyte.[10] However, PANI is not as sensitive as metal oxides toward gas species, and its poor solubility in organic solvents limits its applications. Therefore, there has been increasing interest of the researchers for the preparation of nanocomposites based on PANI as it has been successfully utilized for the preparation of nanocomposites.[11,12]

In this chapter, a high-efficiency hydrogen sensor based on SnO_2/PANI and Al-doped SnO_2/PANI composite nanofibers operated at low temperature has been demonstrated via electrospinning technique and calcination procedure. Electrospinning seems to be the simplest and most versatile technique capable of generating 1D nanostructures. Compared to the commercial mechanical spinning process for generating microfibers, electrospinning mainly makes use of the electrostatic repulsions between surface charges to reduce the diameter of a viscoelastic jet or a glassy filament. One of the most important advantages of the electrospinning is that it is relatively easy and not expensive to produce the large numbers of different kinds of nanofibers. Other advantages of the electrospinning technique are the liability to control the fiber diameters, the high surface-to-volume ratio, high aspect ratio, and pore size as nonwoven fabrics. Excellent hydrogen-sensing properties such as high sensitivity, fast response-recovery behavior, and good selectivity have been obtained at very low temperature compared with that of the pure or doped metal oxides.

7.2 EXPERIMENTAL

7.2.1 MATERIALS AND METHODS

Tin chloride ($SnCl_2 \cdot 2H_2O$, purity 99%), aluminum nitrate ($Al(NO_3)_3 \cdot 9H_2O$, purity 99%), aniline (purity 98.5%), ammonium peroxydisulphate (purity 99%), camphor sulfonic acid (CSA, purity 99%), and polyvinyl pyrrolidone (PVP, M_w = 1300,000, purity 99%) were purchased from Sigma-Aldrich. Aniline was purified under reduced pressure prior to use for synthesis. All other chemicals were used as received without any further purification.

Scanning electron microscopy (SEM) was done by using SEM model—Carl Zeiss EVO-18. Ultraviolet–visible (UV–vis) absorption spectra were recorded on Shimadzu UV-1800 spectrophotometer. X-ray diffraction (XRD) patterns were obtained on Philips PW1710 automatic X-ray diffractometer. The gas-sensing behavior was studied by using laboratory built up sensing apparatus by measuring change in the resistance of sensing film with temperature toward pure air and hydrogen gas exposure. The resistance variation was measured by Keithley 2000 Multimeter and temperature was controlled by Temperature Controlled VI Characterization System.

7.2.2 PREPARATION OF SnO_2/PANI AND AL-DOPED SnO_2/PANI COMPOSITE NANOFIBERS

Sol–gel method was used for the synthesis of SnO_2 nanoparticles. In a typical method, 0.1 M $SnCl_2 \cdot 2H_2O$ was added in 1 M starch solution and the mixture was stirred for half an hour. Then 0.2 M ammonia was added drop wise in the solution under constant stirring. The stirring was continued for further 2 h, and then the solution was allowed to settle for overnight. Supernatant liquid was then discarded carefully and the remaining solution was centrifuged for 10 min and then filtered. The precipitate of SnO_2 was washed completely using double distilled water to remove by-product, and the excessive starch those were bound with the nanoparticles. The product was dried at 80°C for overnight. Then powder was sintered at 600°C for 10 h and nanocrystalline SnO_2 was obtained. In the similar way, nanoparticles of Al-doped SnO_2 were prepared by adding 1 wt% $Al(NO_3)_3 \cdot 9H_2O$ with $SnCl_2 \cdot 2H_2O$ and the same procedure was followed.

Synthesis and Characterization of SnO_2/Polyaniline

For the preparation of SnO_2/PANI composite nanofibers, 0.4 g of as-synthesized SnO_2 nanoparticles and 0.4 g of CSA-doped PANI as-prepared from chemical oxidative polymerization were dissolved in 10 ml DMF under vigorous stirring for 30 min at room temperature. Subsequently, 1.0 g of PVP was added and stirred for further 45 min so as to form a desired viscous solution. Then the solution was loaded into a glass syringe having a stainless steel needle of an orifice of 0.5 mm and electrospun by using ESPIN-NANO modified electrospinning apparatus (Fig. 7.1) at an applied electric field of 17 kV and flow rate of 0.4 ml/h. Nanofibers were collected on aluminum foil wrapped on rotating collector which was grounded and fixed at a distance of 20 cm from needle. Nanofibers collected on foil were dried to remove the organic constituents of PVP. In the similar way, Al-doped SnO_2/PANI composite nanofibers have been prepared by using 0.4 g of as-synthesized Al-doped SnO_2 nanoparticles instead of pure SnO_2 and the same procedure was followed keeping all parameters of electrospinning fixed as that for SnO_2/PANI composite nanofibers.

FIGURE 7.1 Electrospinning apparatus used for the fabrication of nanofibers.

7.3 RESULTS AND DISCUSSION

7.3.1 SCANNING ELECTRON MICROSCOPY

SEM images of SnO_2/PANI and Al-doped SnO_2/PANI composite nanofibers are shown in Figure 7.2(a) and (b), respectively. From SEM micrographs, the average diameter of as-synthesized Al-doped SnO_2/PANI composite nanofibers was found to be changed as compared to that of SnO_2/PANI composite nanofibers, which may be due to change in viscosity and surface tension of the solution when 1 wt% aluminum nitrate was mixed with tin chloride which was electrospun at the same condition as that of SnO_2/PANI composite. While comparing the porosity from their SEM images, Al-doped SnO_2/PANI composite showed more porous structure than that of SnO_2/PANI composite nanofibers due to which Al-doped SnO_2/PANI composite nanofibers have shown better sensing properties at low operating temperature.

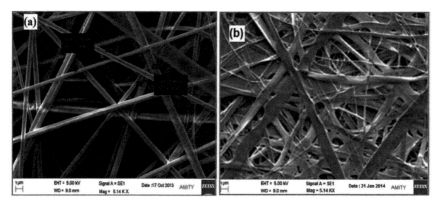

FIGURE 7.2 SEM images of (a) SnO_2/PANI and (b) Al-doped SnO_2/PANI composite nanofibers.

7.3.2 UV–VIS SPECTROSCOPY

UV–vis spectra of SnO_2/PANI and Al-doped SnO_2/PANI composite nanofibers are shown in Figure 7.3. Both the composite nanofibers showed two characteristic bands as compared to a band only in UV region for pure SnO_2 and Al-doped SnO_2. The band in SnO_2/PANI at 407 nm in visible region corresponding to inter ring charge transfer ratio of benzenoid to quinoid moieties showing polaron–π* transition has been shifted to 404 nm in case of Al-doped SnO_2/PANI. The band at 326 nm in UV region corresponding to π–π* transition of benzenoid ring has been shifted to 324 nm in case of Al-doped SnO_2/PANI indicating the existence of Al-doped SnO_2 in PANI matrix.[13–15]

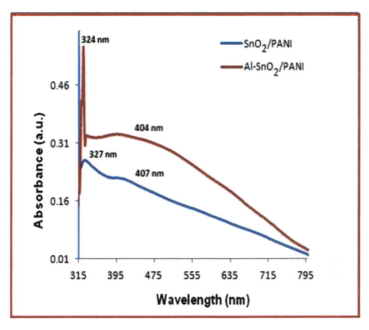

FIGURE 7.3 UV–vis spectra of SnO_2/PANI and Al-doped SnO_2/PANI composite nanofibers.

7.3.3 X-RAY DIFFRACTION

XRD patterns of SnO_2/PANI and Al-doped SnO_2/PANI composite nanofibers are shown in Figure 7.4. All the strong diffraction peaks of SnO_2 present in composite can be perfectly indexed as the tetragonal rutile structure for SnO_2 (ICDD file 41-1445). In the SnO_2/PANI composite nanofibers, most of the peaks are found to be broadened due the polycrystalline effect of PANI as compared to those of SnO_2.[4] The broad peak due to PANI around 26° has been found to be merged with that of SnO_2 at 26.66°. In addition, the reduced intensity of the peaks was observed compared with the XRD pattern of pure SnO_2. The main dominant peaks of SnO_2 were identified at $2\theta = 26.66°$, $34.18°$, $52.3°$, $61.34°$, $64.4°$, and $65.54°$ corresponding to (1 1 0), (1 0 1), (2 1 1), (1 1 2), (3 0 1), and (3 0 2). The crystallization behavior of SnO_2 particles was found to be not much affected by equal mass PANI deposition on the surface of SnO_2 particles in the composite. While comparing composite nanofibers, it has been observed that by doping Al in SnO_2, the peaks in XRD of Al-doped SnO_2/PANI composite nanofibers broadened and decreased in counts indicating the change in fiber diameter due to solution viscosity and surface tension as observed in SEM images.

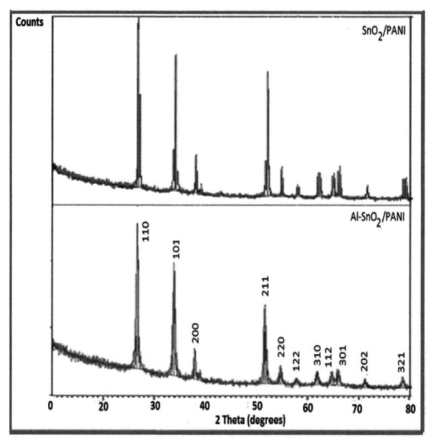

FIGURE 7.4 XRD patterns of SnO_2/PANI and Al-doped SnO_2/PANI composite nanofibers.

7.3.4 HYDROGEN GAS SENSING

In order to systematically investigate the gas-sensing properties of SnO_2/PANI and Al-doped SnO_2/PANI composite nanofibers, the study of gas-sensing response with time and sensitivity with temperature was carried out toward 1000 ppm of H_2 gas. The sensitivity of SnO_2/PANI composite nanofibers to 1000 ppm of H_2 gas at different temperature is shown in Figure 7.5. SnO_2/PANI composite nanofibers have shown better sensitivity and response to H_2 gas at lower operating temperature (~50°C), whereas pristine SnO_2 nanofibers are insensitive at such low temperature and sensitive only at more than 200°C.[4] Similarly, Al-doped SnO_2/PANI composite nanofibers have shown appreciable sensitivity even for 1000 of H_2 gas. The sensitivity of Al-doped SnO_2/PANI composite nanofibers was found to be increased and reached its maximum around 48°C for 1000 ppm of H_2 gas than that of SnO_2/PANI composite nanofibers. The response of sensor was monitored in terms of the normalized resistance calculated by response = R_0/R_g and the sensitivity factor was monitored in terms of the % sensitivity calculated by % sensitivity = $\Delta R/R_0$, where ΔR is the variation

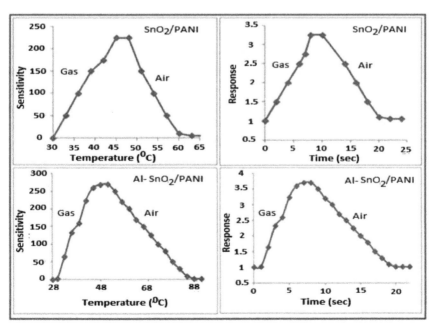

FIGURE 7.5 Sensitivity and response of SnO_2/PANI and Al-doped SnO_2/PANI composite nanofibers.

in resistance of composite films from baseline after exposure to H_2 gas, R_g is the resistance of the sensor in presence of H_2 gas, and R_0 is the initial baseline resistance of the sensor. The suitability of composite nanofibers as H_2 gas sensor was also investigated in terms of "response time (R_p)" and "recovery time (R_C)." The response of the composite nanofibers was observed with respect to time of expose to hydrogen gas and air as shown in Figure 7.5. The values of R_p and R_C were found to be ~3 and ~4 s for SnO_2/PANI and ~2 and ~2 s for Al-doped SnO_2/PANI, respectively. Over a long period of hydrogen exposure, it was observed that composite film sensor exhibited a good stability and repeatability with consistent pattern when exposed to the H_2 gas at 50°C.

7.3.5 MECHANISM OF CHANGE IN RESISTANCE FOR HYDROGEN GAS

On exposure to H_2 gas, the composite nanofibers resistance was found to be increased by more than an order of magnitude from its original value, indicating that the electrical resistance of composite is a sensitive parameter in the presence of hydrogen gas. In presence of SnO_2 crystallites, the PANI matrix gets a modified structure electronically. In composite, SnO_2 crystallites being an n-type surrounded by p-type PANI molecules make a p–n junction like formation locally. The n-type nature of SnO_2 crystallites annihilate the holes of PANI molecules near its boundary making a depletion region, which in turn makes the overall PANI matrix electrically more insulating in nature.[12] On exposing the composite film with hydrogen, the H_2 molecules reach into the depletion region and act as a dielectric between the PANI and SnO_2 border. The depletion region field polarizes the hydrogen molecules, which in turn provide a positive charge to PANI molecules and become mobile on its transfer to the central N atom of PANI molecule. This process creates some free holes on PANI molecules, which make the composite film relatively more conducting electrically. Once the process of polarizing the hydrogen molecules by p–n heterojunction like formation is saturated, this mechanism cannot generate additional holes in the composite film and therefore no additional changes in film resistance even the sensor is exposed hydrogen gas illustrated in Figure 7.6. Al doping in SnO_2 further increases the conductivity of Al–SnO_2/PANI composite nanofibers which showed highly sensitive at low temperature than SnO_2/PANI composite nanofibers.

There are several reasons which should be considered to explain the enhanced sensing performances based on the addition of Al dopant in pure SnO_2. It is well-known n-type semiconducting metal oxide and is exactly stoichiometry they cannot chemisorbs oxygen. The oxygen vacancies play a critical role in determining the sensing performances. To restore their stoichiometry, n-type semiconducting metal oxide, oxygen molecules will absorb on their surfaces and generate chemisorbed oxygen species resulting in high resistance. When reductive target is introduced at close to room temperature, the reductive target will react with oxygen species on the outer surface of n-type semiconducting metal oxide and increase the electron concentrations. In this experiment, both the partial substitution of Sn^{4+} cations with lower valence. Al^{3+} cations, at low concentrations of Al element and the different ions radius of Al^{3+} cations and Sn^{4+} cations, will generate more oxygen vacancies through the SnO_2 crystals.[4] Thus, more oxygen species will form, resulting in higher sensing performances. Thus external heterojunction will form between the Al_2O_3 and SnO_2 within the composite fibers. While those Al_2O_3 nanoclusters can act as the catalytic sites ("spill over" effect) for redox processes and oxygen dissociation, resulting in enhanced sensing performance comparing with the pristine SnO_2 nanofibers.[16] However, in contrast to the Al–SnO_2 metal solid solution, those Al_2O_3 nanoclusters will reduce not only the surface areas of the SnO_2 nanofibers but also the amount of oxygen vacancies within the SnO_2 nanofibers, which diminishes the SnO_2 crystal ability to chemisorbs hydrogen as proven in Figure 7.5.

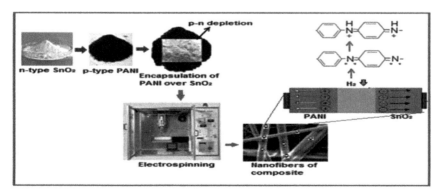

FIGURE 7.6 Mechanism of change in resistance for hydrogen gas.

The response of sensor was monitored in terms of the normalized resistance calculated by response = R_0/R_g and the sensitivity factor was monitored in terms of the % sensitivity calculated by % sensitivity = $\Delta R/R_0$, where ΔR is the variation in resistance of composite films from baseline after exposure to H_2 gas, R_g is the resistance of the sensor in presence of H_2 gas, and R_0 is the initial baseline resistance of the sensor. As per the definitions of "response time (R_P)" and "recovery time (R_C)" for gas sensing, the values of R_P and R_C for SnO_2/PANI were estimated to be ~3 and ~4 s and for Al-doped SnO_2/PANI ~2 and ~2 s, respectively. The comparison of response and sensitivity behavior of both composites is shown in Figure 7.7.

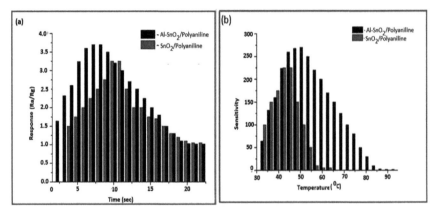

FIGURE 7.7 Comparison of gas-sensing parameters toward 1000 ppm of H_2 gas (a) response and (b) sensitivity.

7.4 CONCLUSION

SnO_2/PANI and Al-doped SnO_2/PANI composite nanofibers were successfully fabricated using electrospinning technique. UV–vis analysis and XRD patterns showed the existence of Al-doped SnO_2 nanoparticles in PANI matrix. SEM images revealed the formation of fibers of average diameter in nanoscale regime. Both the composite nanofibers showed highly sensitive to H_2 gas even at room temperature. Al-doped SnO_2/PANI composite nanofibers exhibited the higher sensitivity for hydrogen gas at low temperature with faster response and recovery as compared to that of SnO_2/PANI composite nanofibers.

ACKNOWLEDGMENT

The authors acknowledge Department of Science & Technology, New Delhi (India) for financial assistance under INSPIRE Fellowship 2013, Sanction Order No. & Date: DST/INSPIRE Fellowship/2013/92, dated May 17, 2013. Registration No.: [IF130149].

KEYWORDS

- nanofibers
- aluminum-doped tin dioxide
- PANI
- electrospinning
- hydrogen sensing

REFERENCES

1. Hung, C. W.; Lin, H. L.; Chen, H. I.; Tsai, Y. Y.; Lai, P. H.; Fu, S. I.; Chuang, H. M.; Liu, W. C. Comprehensive Study of a Pd–GaAs High Electron Mobility Transistor (HEMT)-based Hydrogen Sensor. *Sens. Actuat., B: Chem.* **2007,** *122*(1), 81–88.
2. Lu, X.; Zhang, W.; Wang, C.; Wen, T. C.; Wei, Y. One-dimensional Conducting Polymer Nanocomposites: Synthesis, Properties and Applications. *Prog. Polym. Sci.* **2011,** *36*(5), 671–712.
3. Long, Y. Z.; Li, M. M.; Gu, C.; Wan, M.; Duvail, J. L.; Liu, Z.; Fan, Z. Recent Advances in Synthesis, Physical Properties and Applications of Conducting Polymer Nanotubes and Nanofibers. *Prog. Polym. Sci.* **2011,** *36*(10), 1415–1442.
4. Xu, X.; Sun, J.; Zhang, H.; Wang, Z.; Dong, B.; Jiang, T.; Wang, W.; Li, Z.; Wang, C. Effects of Al Doping on SnO_2 Nanofibers in Hydrogen Sensor. *Sens. Actuat., B: Chem.* **2011,** *160*(1), 858–863.
5. Berry, L.; Brunet, J. Oxygen Influence on the Interaction Mechanisms of Ozone on SnO_2 Sensors. *Sens. Actuat., B: Chem.* **2008,** *129*(1), 450–458.
6. Hieu, V. N.; Kim, H. R.; Ju, B. K.; Lee, J. H. Enhanced Performance of SnO_2 Nanowires Ethanol Sensor by Functionalizing with La_2O_3. *Sens. Actuat., B: Chem.* **2008,** *133*(1), 228–234.
7. Kuang, Q.; Lao, C.; Wang, Z. L.; Xie, Z.; Zheng, L. High-sensitivity Humidity Sensor Based on a Single SnO_2 Nanowire. *J. Am. Chem. Soc.* **2007,** *129*(19), 6070–6071.

8. Zhang, Y.; He, X.; Li, J.; Miao, Z.; Huang, F. Fabrication and Ethanol-sensing Properties of Micro Gas Sensor Based on Electrospun SnO_2 Nanofibers. *Sens. Actuat., B: Chem.* **2008,** *132*(1), 67–73.
9. Kolmakov, A.; Klenov, D. O.; Lilach, Y.; Stemmer, S.; Moskovits, M. Enhanced Gas Sensing by Individual SnO_2 Nanowires and Nanobelts Functionalized with Pd Catalyst Particles. *Nano Lett.* **2005,** *5*(4), 667–673.
10. Kargirwar, S. R.; Thakare, S. R.; Choudhary, M. D.; Kondawar, S. B.; Dhakate, S. R. Morphology and Electrical Conductivity of Self Doping Polyaniline Synthesized via Self-assembly Process. *Adv. Mater. Lett.* **2011,** *2*(6), 397–401.
11. Kondawar, S. B.; Thakare, S. R.; Bompilwar, S.; Khati, V. Nanostructure Titania Reinforced Conducting Polymer Composites. *Int. J. Mod. Phys. B* **2009,** *23*(15), 3297–3302.
12. Deshpande, N. G.; Gudage, Y. G.; Sharma, R.; Vyas, J. C.; Kim, J. B.; Lee, Y. P. Studies on Tin Oxide-intercalated Polyaniline Nanocomposite for Ammonia Gas Sensing Applications. *Sens. Actuat., B: Chem.* **2009,** *138*(1), 76–84.
13. MacDiarmid, A. G.; Epstein, A. J. The Concept of Secondary Doping as Applied to Polyaniline. *Synth. Met.* **1994,** *65*(2), 103–116.
14. Jiang, H.; Geng, Y.; Li, J.; Wang, F. Organic Acid Doped Polyaniline Derivatives. *Synth. Met.* **1997,** *84*(1), 125–126.
15. Kim, B. J.; Oh, S. G.; Han, M. G.; Im, S. S. Synthesis and Characterization of Polyaniline Nanoparticles in SDS Micellar Solutions. *Synth. Met.* **2001,** *122*(2), 297–304.
16. Dong, K. Y.; Choi, J. K.; Hwang, I. S.; Lee, J. W.; Kang, B. H.; Ham, D. J.; Lee, J. H.; Ju, B. K. Enhanced H_2S Sensing Characteristics of Pt Doped SnO_2 Nanofibers Sensors with Micro Heater. *Sens. Actuat., B: Chem.* **2011,** *157*(1), 154–161.

CHAPTER 8

NANOCOMPOSITES FOR FOOD PACKAGING APPLICATIONS

BADAL DEWANGAN[1,*] and UMESH MARATHE[2]

[1]Department of Polymer Technology, Laxminarayan Institute of Technology, Nagpur 440001, Maharashtra, India

[2]Center for Polymer Science and Engineering, Indian Institute of Technology, New Delhi 110016, India

*Corresponding author.

CONTENTS

Abstract	138
8.1 Introduction	138
8.2 Nanocomposite	140
8.3 Preparation of Nanocomposites	144
8.4 Characterization of Polymer Nanocomposites	145
8.5 Properties of Polymer Nanocomposites	146
8.6 Recent Applications of Polymer Nanocomposite	150
8.7 Correlation of Food Packaging and Nanocomposite	151
8.8 Polymer Nanocomposites for Food Packaging	151
8.9 Role of Biopolymer Nanocomposites in Food Packaging	159
8.10 Scope for Research in Food Packaging Based on Polymer Nanocomposites	161
8.11 Summary	161
Keywords	162
References	162

ABSTRACT

The chapter focuses on the packaging application of polymer nanocomposites. It comprise various polymer matrices, nanoparticles as reinforcing phase, preparation methods along with improved properties for instance mechanical, thermal and barrier properties etc. It also comprises advance packaging methods such as active packaging and intelligent packaging and their working principles.

8.1 INTRODUCTION

Food packaging is one of the imperative aspects of packaging industry. In food packaging, assorted types of material are utilized, and they are competent to have properties, such as mechanical strength and barrier property that enables the package to protect and enhance the shelf life of foodstuff. Consequently, with the intention of fulfilling the necessities of the food packaging, basic packaging materials were used, for example, paper, plastic, metal, glass, materials from hybrid mixture or combination of two materials (composites). Apart from this material, plastics or polymeric material produced by utilizing petroleum sources are incredibly popular from last few decades; this is because of their property to provide simplicity in their processing as well as their lower cost than an alternative material.[1] Pristine polymer materials possess the ability to provide assistance in order to produce food packaging up to certain critical point, that is, after or at certain level of advancement, it cannot be improved because of limited properties of pure polymers. Above statement was the motive to develop or introduce novel type of material or more extensively new class of material with advanced properties known as composites. In last two or three decades evolvement of composites were significant. They are fabricated from two or more materials of diverse chemical natures and physical structures. It is primed by incorporation of ample variety of reinforcing material in polymer materials, such as particles (e.g., calcium carbonate), fibers (e.g., glass fibers), or plate-shaped particles (e.g., mica). These composite materials possess major disadvantages for instance increment in weight, loss in optical properties, brittleness, etc.[2–6] On the other hand, there are developments of another new class of composite material with the aim of conquering above-mentioned disadvantages by replacing reinforcing material by inorganic nanosize materials; the entire system known as polymer nanocomposites. Development of polymer nanocomposite is in the boom phase to accomplish highest properties in order to advance food packaging in a unique approach. This refers to the evolvement of the Smart Packaging materials, which are not only able to bestow the basic requirements of the packaging material but also gives additional effort to enhance the food shelf life, protection from bacterial attack on the food, etc. These types of smart materials are categorized in some categories in this review. It is the material, in which the reinforcement material is into nanosize. By this substitution, there is a drastic amendment in the properties of the material. Amid all the potential nanocomposite precursor, those based on clay and layered silicates have been mostly studied as well as developed due to fact that there is an availability of information concerning the chemistry of clay or layered silicate as well as easy availability of these materials.[2,7] The function of nanostructure in the polymer nanocomposites is to improve mechanical properties, gas permeability, etc. Besides the above-mentioned functions, some of the nanostructures are able to provide extra features, for example, biosensing, antimicrobial properties, gas detection in side of package, etc.;[8] therefore, due to this reason, polymer nanocomposites are widely favorite among the researchers, industries, or academic institutions for research work. As mentioned above, it possesses two constituents: polymer and inorganic reinforcing material. Nanocomposite is not only analogous but also is similar to conventional composite, but different in the size of the reinforcing, that is, it is in nanoscale in case of nanocomposites. To achieve the supreme properties via nanocomposite, it is necessary for the polymer to have exfoliated or well-dispersed structure of nanomaterial/nanostructure. The effect of plastic or polymeric material on the environment is quite an anxiety to human beings because of its sky-scraping mass production. It can hamper the environment as well as human health. So with the intention to solve the problem, innumerable researches are going on or done regarding polymer nanocomposite material with assistance of biopolymers.[1,9] In addition to the application of polymer nanocomposite in the packaging field, many other applications were developed, for example, automotive, e.g., timing-belt cover, engine cover, barrier fuel line, furniture, bottles, heavy duty electronics enclosure, multilayer container, barrier films, paper-coating applications, furniture, etc. There was a boom seen in the research of polymer nanocomposite, when Toyota group disclosed their work on the same areas. This chapter comprises preparation of the nanocomposites, their constituent material, their types, fabrication method, applicability of nanocomposites in the field of food packaging, advantages in the sense of high properties which are serving food safety, transportation, self-life, smart packaging, etc.

8.2 NANOCOMPOSITE

Nanocomposite is a composite material filled with nanosize reinforcing material, for example, clay or layered silicate into polymer material. The property of nanocomposites extends on the basis of the structure of nanocomposite (intercalated or exfoliated), aspect ratio of reinforcement. Small particle size of the inorganic clay or particles lends to great larger surface area. Nanocomposite may be intercalated or exfoliated structured. Intercalated structure comes in occurrence, when a polymer single chain is intercalated in between two layers of layered clay or silicate. Distance between two adjacent layers of the layered silicate is 20–30 Å in intercalated nanocomposite. So, resultant will be separation of layered structure of the layered structure by polymeric chain with limited extend.[2,5,10–13] On the other hand, exfoliated or disintegrated structure can be obtained. In exfoliated type, interlayer distance in layered silicate is much more than intercalated one. A distance of 80–100 Å can occur in exfoliated structure.[2,5,10,13] This higher in-between distance gives confirmation of well dispersion of the reinforcing phase. As dispersion is high, automatically properties become higher. Exfoliated nanocomposites have major attention because of their enhanced properties. The mechanical properties are much better than the intercalated structure of exfoliated one. Exfoliated nanocomposite provides uncomplicated structure to transfer the load or stress to the reinforcement and it allows enhancement in mechanical properties.[2,14] Thus, preparation of polymer nanocomposite with exfoliated structure is not an easy task because of the presence of van der Waal's force which has become a hurdle in exfoliation. With the intention of accomplishing such structure, it is necessary to modify the inorganic-layered clay or silicate by means of chemical modification. There are methods to prepare nanocomposites:

1. In situ intercalative polymerization.
2. In situ template synthesis.
3. Melt intercalation.
4. Intercalation of polymer from solution.

Prior to the introduction of the layered silicate into polymer, it is necessary to treat it with some solvent to get good structure.[2] There is another approach in front of researchers that uses super critical CO_2 gas to get exfoliated structure.[15] Polymer nanocomposites consist of the two constituents: polymer matrix and inorganic nanoscale reinforcement or layered silicate.

8.2.1 POLYMER MATRIX

There are ample variety of materials which were investigated under this subtitle. It is due to polymer matrix's potential to give improved properties. These polymers include poly(ethyleneterephthalate), poly(1-caprolactone), unsaturated polyesters, polyurethanes, polystyrene, styrene copolymers, polypropylene, polyethylene, polyethylene oligomers or copolymers, poly(vinyl alcohol), poly(acrylamide), poly(methyl methacrylate), methyl methacrylate copolymers, other polyacrylates, poly(acrylic acid), poly(ethylene oxide), ethylene oxide copolymers, poly(ethylene imine), epoxidized natural rubber, polybutadiene, butadiene copolymers, poly(dimethylsiloxane), polypyrrole, poly(vinylpyrrolidone), poly(vinyl pyrrolidinone), poly(vinyl pyridine), poly(vinyl pyridinium) salts and poly(*N*-vinyl carbazole), polyaniline, poly(*p*-phenylene vinylene), polyimides, poly(amic acid), liquid–crystalline polymers, epoxy polymers, phenolic resins, DNA, and met-hemoglobin.[15,16]

8.2.2 LAYERED SILICATE

The role of layered clay, that is, layered silicate, is especially crucial in preparation of the clay nanocomposites, since properties of clay nanocomposites extensively depend on the properties of clay that is used to reinforce composite structure of clay as well as morphology of clay inside the polymer. Due to this number of reasons, the structure of clay or nanoclay is widely studied by researchers. Structure of layered silicate is shown in Figure 8.1. Generally, while preparing clay nanocomposites, use of range of smectite clay minerals is vastly drastic. The clay utilized to prepare nanocomposites is prepared synthetically or it may be natural, made up of the layers which are separated by very small gap (in unit nanosize), which is filled with the exchangeable cations. The layered clay is made up of the

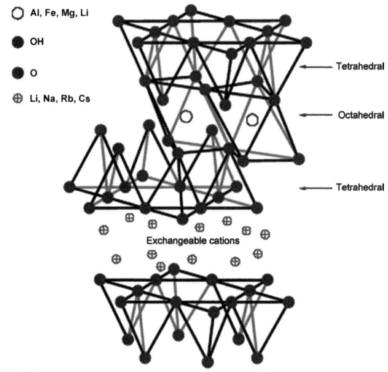

FIGURE 8.1 The structure of a 2:1 layered silicate.[17] (Reprinted from Beyer, G. Nanocomposites: A New Class of Flame Retardants for Polymers. *Plast. Addit. Compound* **2002**, *4*(10), 22–27. © 2002. with permission from Elsevier.)

tetrahedral and octahedral sheets. This is all about very general discussion around clay, but it is a necessary focus on the material that is widely used by various researchers namely phillosilicates. It consists of one octahedral sheet of aluminum, which is sandwiched in between tetrahedral sheet of silica.[8] Both of the sheets (tetrahedral and octahedral sheets) are fused with each other. The basic structure of layered clay/silicate is also categorized by number of sheets in each layer such as 1:1 kaolinite or 2:1 montmorillonite (MMT). When the Al ion of the octahedral sheet is replaced by divalent Mg cation, then the resultant structure is called MMT.[2] Interlayer distance between the adjacent clay layers is 1 nm as well as lateral dimension is approximately 200 nm. These clay layers are stacked to each other and held together through van der Waal forces and separated from each other by 1-nm gaps called galleries. These galleries are usually filled with exchangeable cations. Role of these cations are to balance the negative charge formed by clay/silicate layer. Generally, alkali and alkaline-earth cations like Na^+ and K^+ are present in galleries. An analog to polymer blends the compatibility of in-between clay, and polymer is very necessary to get best of properties as well as exfoliated nanocomposites. In order to obtain higher properties, it is necessary to modify clay by chemical means. This modification consists of replacement of exchangeable cations located at galleries by another cations like with quarternary alkyl ammonium or alkyl phosphonium cations. Due to this modification, the hydrophobic nature of clay becomes organophilic. Because of this modification, clay has become more compatible with polymers as well as it plays an important role to get exfoliated structure of nanocomposites.

8.2.3 MODIFICATION OF LAYERED SILICATE

Modification of layered silicate plays extremely crucial role in the fabrication of polymer nanocomposites. It is nothing but making improvement in the dispersion properties by means of certain chemical treatments. It predominantly consists of replacement of exchangeable cations, which are placed in between two adjacent layers in layered silicate. Originally layered silicate possesses hydrophilic nature, so it is compatible or possesses good dispersion properties with the hydrophilic polymers like poly(vinyl alcohol). So, in order to make it compatible with other polymer, its modification was carried out by exchanging cations placed in between two layers of silicate.[18] Therefore, to obtain exchange of pristine cations by modification cations in galleries, water swelling of the silicate is prime, and for this reason, alkali cations are preferred. Hydrate formation of monovalent gallery cation assist as a driving force for water swelling. The alkali cations are not structural and can be easily replaced by other positively charged atoms or molecules; thus, they are called exchangeable cations.[2,18,19] The improvement in the wetting and dispersion property of the layered silicate can be achieved by organic cations by lowering the surface energy of silicate surface. There is another advantage of exchanging cations; these cations possess long-organic chain with positively charged ends; these ends are tethered with surface of the layers of silicates and lead to increased gallery height of layered structure. The organic modification of surface of layered silicate both way increases in basal space as well as it acts as compatibilizer in between organic silicate and polymer material.[12,20]

8.3 PREPARATION OF NANOCOMPOSITES

Fabrication or preparation of nanocomposite is carried out by four methods by which one can prepare nanocomposite; thus, the methods are enlisted as well as explained in the following sections.

8.3.1 IN SITU INTERCALATIVE POLYMERIZATION

In in situ polymerization technique, layered silicate or nanoparticles are swollen in the monomer; in swollen state, monomer migrates in between layers of silicate. Subsequently, polymerization is carried out "in situ," by which, there is increase in the layer basal space because swollen silicates possess the monomer in between the adjacent layer. This gap is modified with catalyst and initiators that initiate the polymerization in between the silicate layer. Due to polymerization, there is increase in basal space. Polymerization reaction can be initiated by heat or radiation.[10,17,21]

8.3.2 IN SITU TEMPLATE SYNTHESIS

In case of other three methods, one should have to add inorganic material as reinforcement externally, but this method is quite different. In this method, clay minerals are synthesized in polymer matrix while preparing nanocomposite. In this case, there is a use of aqueous solution or gel containing polymer and silicate-building blocks. While the process is going on, the polymer aids the nucleation and growth of the inorganic crystal and gets trapped within the layer as they grow. Along with this, presence of disadvantages is not surprising. First, for the synthesis of clay, the required high temperature can easily degrade polymer material. Another difficulty is regarding aggregation tendency of the growing silicate layers. The one advantageous point is that it can provide dispersed-layered silicate in one-step method.[2,4,10,22]

8.3.3 MELT INTERCALATION

This technique possesses commercial enticement to prepare polymer nanocomposite. It is the result of its simplicity and economy. In this case, polymer matrix or material and layered clay are simply blended in molten state (polymer). It is able to give exfoliated structure in case when the polymer and clay or layered silicates both are compatible with each other. In this process, polymer chains simply crawl into basal space.[2,10,17,21]

8.3.4 INTERCALATION OF POLYMER FROM SOLUTION

In case of the intercalation of polymer from solution technique, polymer replaces the prior intercalated solvent. Such replacement requires negative variation in Gibbs free energy. It is obtained through the reduced entropy because the confinement of the polymer is compensated by an increase due to the desorption of intercalated solvent molecules.[2,23–28] This technique has been mostly used with water-soluble polymer, such as PEO, PVE, PVP, and PPA.[2,4,29] This method may involve the copious use of organic solvent which is usually hazardous to environment and costlier.[30]

8.4 CHARACTERIZATION OF POLYMER NANOCOMPOSITES

The characterization of polymer nanocomposites can be done by using equipment with high accuracy. The two most widespread methods of structure characterization of polymer nanocomposites are X-ray diffraction (XRD) and Transmission Electron Microscopy (TEM). XRD analysis is easy to perform and gives graphical illustration as a result. But there are several things that must be kept in mind when analyzing polymer nanocomposite samples in this tactic[31]—prime is the low sensitivity of XRD and can depend on several parameters that may not be optimized. XRD is not quantitative, and it is normally used to detect the absence of a peak. But, XRD cannot be used as definite proof of an exfoliated structure, because of it lower sensitivity. XRD can be a helpful method to present a rapid analysis of the nanocomposite. TEM is extremely helpful and effective in providing a straight way of determining nanocomposite morphology. It can also yield the quantitative data along with visualization from TEM regard to the extent of exfoliation and/or intercalation. TEM gives clear picture of morphology of nanocomposite. Dynamics of the exfoliated nanocomposite, morphology, and surface chemistry can be characterized by solid-state nuclear magnetic resonance spectroscopy.[32] Some authors were attempted with Fourier transform infrared spectroscopy to make clear the structure of the nanocomposites,[33] but it is not that much reliable technique. Differential scanning calorimetry provides information about intercalation. Restrictions on polymer chain mobility increase its glass transition temperature (T_g). A similar increase is anticipated to occur in a nanocomposite due to elevation of the energy threshold needed for the transition. This effect is promptly detected by DSC. The resulting peak of inorganic material gives evidence of characterization of polymer nanocomposite.[2,22] These are several techniques used to characterize polymer nanocomposites.

8.5 PROPERTIES OF POLYMER NANOCOMPOSITES

8.5.1 MECHANICAL PROPERTIES

A reason behind development of whole class of composites is to obtain higher properties, rather than pristine polymer material. The higher properties of composites material are in the sense mechanical properties. In case of nanocomposites, addition of nanoclay or other nanosize reinforcement affect mechanical properties of the nanocomposites very efficiently. Increment in such mechanical properties comprises increases in the tensile strength, Young's modulus, flexural strength, creep resistance, dimensional stability, etc.[2] As an example, Toyota group did develop Nylon 6 and clay nanocomposites with improved mechanical properties, which were utilized in automobile industries. Mechanism behind the improvement in mechanical properties of nanocomposite is similar as conventional composites. Properties of nanocomposites depend upon the aspect ratio, structure of nanocomposites, and affinity of nano-reinforcement with polymer matrix. Higher aspect ratio and exfoliated structure plays vital role in order to get higher tensile strength, Young's modulus, and flexural strength. Formation of the microvoids in nanocomposite gives variation in brittleness. In 2004, Zhang and Yang reported that incorporation of the nanoclay improves the poor creep resistance and dimensional stability.[35] As surface area of nano-reinforcement is higher, greater are the mechanical properties. Generally, layered silicate or nanoclay possesses higher surface area of about 800 m^2/g. Therefore, such higher surface area gives mechanical properties a drastic path. The mechanism behind reinforcing nature of the nanoclay or nano-reinforcement is similar to the mechanism of conventionally utilized reinforcement. Some authors reported that lesser loading of nano-reinforcement is more effective rather than higher loading in weight %. Higher surface area and aspect ratio are responsible for significant improvement in the modulus with very little amount of nanofiller loading.[36]

8.5.2 BARRIER PROPERTY

Barrier property is one of the important properties of polymer material or composites that can play crucial role in final applications. Barrier property of nanocomposites is excellent compared to virgin polymer materials. Improvement in barrier property of nanocomposites is due to exfoliated structure of polymer nanocomposites. Exfoliated structure gives higher barrier properties than partially exfoliated or intercalated structure. Barrier property or permeability of material depends upon an orientation of clay layers or particles in polymer matrix. Mechanism of improvement in the barrier property can be explained by concept of tortuous paths. Gas or vapor easily passes through virgin polymer material due to straightforward pathway or with lower hindrance in their path, but in case of polymer material reinforced with nanoclays or nanofillers, pathway of gases is quite difficult. Gases or vapors cannot easily pass through nanocomposites, and it is because of tortuous path developed by nano-reinforcement. Gas or vapor should take jiggled path in order to pass the barrier created by the nano-fillers or nano-clay layers.[2,6,37] In 1995, Messersmith and Giannelis reported that permeability of liquid and gases in polymer nanocomposites reduced significantly.[38]

8.5.3 THERMAL PROPERTIES

8.5.3.1 THERMAL STABILITY

Thermal stability of nanocomposite depends upon thermal stability of nano-reinforcement or organically modified nanoclay, as reported by Chang and Sur in 2003.[39] Increase in thermal stabilities of poly(lactic acid) polymer nanocomposite is with 5% of clay content. Paul reported that as clay content advances toward higher loading than 5%, thermal stability of nanocomposite decreases.[40] If polymer material is exposed to the high temperature, it degrades without residue, but polymer nanocomposites leave some residue after degradation.[41] In polymer nanocomposites, organomodified clay or nanofiller acts as a shield to polymer material, in order to protect it from action of oxygen, and due to this phenomenon, thermal stability of whole nanocomposite increases drastically. Some of the authors argued that the thermal stability of the polymer decreases after addition of the nanoclay or nanosized filler.[37]

8.5.3.2 COEFFICIENT OF THERMAL EXPANSION

Variation in coefficient of thermal expansion of polymers is directly affected to dimensional stabilities as well as dimensional changes while molding. When composite or nanocomposite is subjected to the temperature change, the matrix attempts to extend or contract. Reinforcement present in composite opposes these dimensional changes by creating hindrance; this hindrance comes in occurrence because the reinforcing material creates opposing stress to expansion and contraction. Due to mechanism mentioned above, the coefficient of thermal expansion of composite material reduces effectively. Akin phenomena were observed in case of nanocomposite of Nylon 6 and MMT. Coefficient of thermal expansion Nylon 6/MMT is reduced due to incorporation of nano-reinforcement as reported by Yoon, Fornes, and Paul.[42]

8.5.3.3 HEAT DEFLECTION TEMPERATURE

It was reported by many researchers that significant improvement in the heat deflection temperature (HDT) cannot be achieved by incorporating convention filler. It is possible to improve HDT by adding nanoclays, nanoparticles, or more commonly nano-reinforcement in polymers rather than conventional fillers.[30,37] Improvement in HDT was reported by Kojima in 1993 by 90°C, in case of Nylon 6 and organically modified clay or layered silicate nanocomposites.[43]

8.5.4 RHEOLOGICAL PROPERTIES

Rheological properties of polymer nanocomposites are crucially important from their processing viewpoint. Rheological properties of nanocomposite are dependent upon structure, particle size, shape, and surface characteristics of reinforcing phase. Cho and Paul reported that viscosity of polymer nanocomposite at low shear rate increases with filler loading. As shear rate increases, shear thinning behavior occurs.[44] Behavior analogs to solid were observed because of the percolation of layered silicate or filler. Occurrence of solid-like behavior at low-volume fraction is on account of anisotropy.[37] Value of absolute viscosity of poly(amide) 6 nanocomposite is quite lower than that of pristine poly(amide) 6 or conventional composite of it. Low value of the absolute viscosity increases the melt processability over a wide range of processing conditions. A possible reason behind this behavior is nothing but slip between poly(amide) 6 matrix and exfoliated clay layers.[2]

8.5.5 OPTICAL PROPERTIES

Optical properties of conventional composite or composite having their reinforcing phase in macrosize show poor optical properties. Polymer nanocomposite gives excellent optical properties like good optical clarity, excellent transmittance, etc. Cause behind this improvement is nothing but nano-size of reinforcing phase. Generally, used MMT mineral clay possesses the layer size in nanoscale, that is, 1 nm. One of the important negative aspects was reported is the loss in intensity in the UV region. This loss in intensity in the UV region is because of the scattering by the MMT particles.[7]

8.5.6 FLAME RETARDANCE

Flame retardancy is a property of a material that is considered widely while designing various applications. Polymer nanocomposite shows potential in flame retardancy in exclusive approach. Number of researchers reported performance of polymer nanocomposites in tests carried out in laboratories. Nanocomposite possesses number of advantages in the sense of flame retardancy over conventional polymer composite or compounded polymer material with flame retardance additives. The quantity of flame retardance additives is quite higher, which motivates increase in density, discoloration, loss in optical properties, etc. So, it is concluded that the nanocomposites possess abilities to become substitute for conventional polymers as well as composite materials.[2]

8.6 RECENT APPLICATIONS OF POLYMER NANOCOMPOSITE

Along with advanced food packaging, nanocomposites are used in various applications. Polymer nanocomposites can be utilized to prepare high-temperature lubricating coating application.[47] Nanocomposite can be applicable for high-temperature application that was reported by Provenzano, Holtz.[48] Cellulose nanocomposites with nanofibers can be utilized for medical applications.[49] Nanostructured conducting polymers/nanocomposites can be utilized to sensor applications; this was reported by Rajesh, Ahuja, and Kumar.[50] Polymer and biopolymer–clay nanocomposites show attractive properties for electrochemical and electroanalytical applications, so they can utilized for such applications.[51] A novel glucose biosensor was developed by using the chitosan–polypyrrole nanocomposites that was reported by Fang, Ni, Zhang, Mao, Huang, and Shen. This can be utilized for active packaging.[52] Chitosan–clay nanocomposites are used as electrochemical sensors for the potentiometric determination of anionic species.[53] Polymer clay nanocomposite can be used for catalytic degradation, adsorptive removal, and detection of contaminants; it is quite helpful in environmental view.[54] Sepiolite-based nanocomposites make this nanoparticle the most attractive material for tissue engineering and environmental industrial applications.[55] Nanocomposites of gold and poly(3-hexylthiophene) containing fullerene moieties are used in application in solar cell.[56] One of the earliest application or nanocomposite developed by Toyota group for automobile application, the first commercial product of clay-based polymer nanocomposites was a timing-belt cover made from PA6 nanocomposites by Toyota Motors in the early of 1990s. This timing-belt cover exhibited good rigidity, excellent thermal stability, and no wrap. It also saved weight by up to 25%.[57] Meanwhile, Ube America was attempted to prepare nanocomposite barriers for automotive fuel systems, using up to 5% nanoclay in PA6 and PA6/66 blends.[58] There were various applications of different nanocomposites, tabulated in review by Pavlidoua and Papaspyrides.[2] Lithium electrode application of $ZnO-ZnFe_2O_4$ nanocomposites was reported.[59] The applicability of biopolymer nanocomposite to membrane application was reported in 2011.[60]

This is a broad overview of scope of polymer nanocomposite and their research-based future application along with earliest applications.

8.7 CORRELATION OF FOOD PACKAGING AND NANOCOMPOSITE

From last century or the last four to five decades approximately, the usage of plastic material was prodigious. It is a result of its quality of leading in diverse application than any other packaging material. When one can see plastic material in the sense of environment friendly aspect, it gets futile to prove its versatility. Plastic is not environment friendly in nature and hence it is futile to prove its versatility. So, in order to conquer such problem, evolution of biopolymer is vital and it is a drastic step. A major disadvantage to biopolymer is that it doesn't have properties requisite for food packaging or any other types of packaging. So, in order to trounce such problem, an introduction of the bio-composite material was already done. In bio-composites, properties are enhanced by incorporating reinforcements, that is, filler materials. Hence, this filler enhances the properties of biopolymer.[8,31] Fillers not only give enhanced properties but also additionally give increase in brittleness, weight, and loss in optical properties.[2] Consequently, with the intention of accomplishing sound properties without any losses in the prime properties of the material, it is necessary to add reinforcement in the nanoscale and this is known as nanocomposite. Nanocomposite possesses the potential to substitute the conservatively used packaging material. Hence, this relation in between nanocomposites and food packaging is based on the disadvantages of materials that are used conventionally. It means that food packaging industry needs material with versatile properties that can be superior and sincere to meet its food packaging application and it is possible via merely nanocomposites.

8.8 POLYMER NANOCOMPOSITES FOR FOOD PACKAGING

According to food packaging, polymer nanocomposite can be categorized in three different categories and that are (1) improved polymer nanocomposite, (2) active polymer nanocomposites, and (3) smart/intelligent polymer nanocomposites.

These categories are made by the criteria by how efficiently nanocomposites are serving in food packaging application so that the descriptive explanation is given on each category in the following sections.

8.8.1 IMPROVED POLYMER NANOCOMPOSITES FOR FOOD PACKAGING

After polymer material is incorporated with clay nanoparticles, it is then identified as improved polymer nanocomposite. It is proficient to fulfill basic requirements, like barrier property, mechanical strength, and ensures safety of food for food packaging.[32] In case of improved polymer nanocomposites, it was reported by Adame and Beall in 2009 that structure of clay in polymer gives only way to gas or water to travel in tortuous path, so this improves the gas and water barrier property.[33] Along with this, many studies researchers reported the potential of clay nanoparticles to reduce oxygen and water–vapor permeability.[22,45,46] A unique attempt was reported by Christopher Thellen, Sarah Schirmer in 2009 that coextrusion of two-layer film consists a MMT-layered silicate/poly(*m*-xylylene adipimide) nanocomposite as the oxygen barrier layer and low-density polyethylene (LDPE) as moisture-resistant layer[61]. It was reported that improved oxygen barrier properties of the modified polyamide and Nylon 6 clay films were explained in terms of the reduced free-volume properties and demarcated structures of an overlapped clay mineral layers.[62] Improved nanocomposites not only featured with evolved barrier properties but also there is significant enhancement in the mechanical property. Discussion on the mechanical properties was done by Pavlidou and Papaspyrides with remarkable depth in their review.[2] Nanocomposite prepared by using MMT modified with citric acid gives higher Young's modulus than unmodified MMT as investigated by Majdzadeh-Ardakani et al. in 2010.[63] Organic clay and LDPE nanocomposite are inept to give intercalated or exfoliated structure. But if ethylene-*co*-vinyl acetate (EVA) copolymer is added in it as compatibilizer, it gives properties of nanocomposites. Elastic modulus of the prepared material is also higher.[64] It was reported in 2012 by Hong and Rhim that nanocomposite film prepared by using LLDPE and two types of clay gives better smooth, homogeneous, and flexible, but less transparent than the neat pristine, LLDPE film. Very slight intercalated nanostructures were formed with both organoclays.[65] Another unique attempt was done by Yang, Wu, Saito, and Isogai that nanocomposite of cellulose and MMT was prepared, by using cellulose/LiOH/urea solutions. Result shows somewhat intercalated structure.[66] Regenerated cellulose/MMT nanocomposite by using ionic liquid, 1-butyl-3-methylimidazolium chloride via solution casting method gives tensile strength and Young's modulus of regenerated cellulose films improved by 12%. Regenerated cellulose/MMT nanocomposite films exhibited improved gas barrier properties and water absorption resistance compared to regenerated cellulose.[67] Nanocomposite films were prepared by using corn starch, polyvinyl alcohol, nanosized PMMA-*co*-acrylamide particles, and additives, that is, glycerol, xylitol, and citric acid. The results show that the mechanical properties and water resistance were improved up to 70–400% by the addition of nanosized PMMA-*co*-acrylamide.[68] Swelling parameter, diffusion coefficient rate, and oxygen transmission rate of cellulose microfiber/EVA composite were studied and resulted that it gives increased crystallinity of composite, investigated by Sonia and Dasan.[69] Coating of polymer nanocomposite over virgin polymer film can give very good result in the sense of permeability and barrier property. Such novel attempt was tried by coating corn zein nanocomposite on polypropylene. Results are positive; reduce in the oxygen permeability was nearly four times, while water vapor permeability was reduced by 30%.[70]

8.8.2 ACTIVE POLYMER NANOCOMPOSITES FOR FOOD PACKAGING

Active polymer nanocomposite for food packaging is nothing but the dynamic participation of packaging material to enhance the food shelf life. Active packaging mainly concerns about antimicrobial property. Active packaging was developed to provide shield to food inside the packet from microbes. In addition, it is under development to serve various other functions under the title of active packaging for example, oxygen scavengers, ethylene removers, and carbon dioxide absorbers/emitters. In case of preparation of active packaging metal nanoparticles or metal oxide, nanoparticles are preferred over inorganic layered silicates. It is because metal or metal oxides such as silver, gold, zinc oxide, and magnesium oxide have antimicrobial property. Ample of research was done beneath this title, so there are plenty materials available that are able to give an antimicrobial property. Bruna, Peñaloza, Guarda, Rodríguez, and Galotto prepared nanocomposite by melt-mixing using extruder; copper ($MtCu^{2+}$)-modified MMT melt intercalated. An antibacterial effect of the polymer nanocomposite increases with the proportion of $MtCu^{2+}$ added, obtaining a 94% reduction in microbial attack, when 4% of $MtCu^{2+}$ was added to the polymer.[71] Polymer nanocomposite containing polyethylene and silver nanoparticles was produced via in situ polymerization, which ultimately gives 99.99% efficiency against bacteria.[72] The cellulose–silver nanocomposites possess a high antimicrobial activity against the model microbes *Escherichia coli* (Gram-negative) and *Staphylococcus aureus* (Gram-positive), reported in 2011 by Li, Jia, Ma, Zhang, Liu, Sun.[73] Chitosan and MMT nanocomposite were prepared by using ion exchange reaction; result showed synergistic effect in the antimicrobial activity against *E. coli* and *S. aureus*.[74] In an ordinary case, MMT is modified by using Na^+ cation, when priorly present Na^+ cations were replaced by silver cations such as MMT shows better antimicrobial properties.[75] ZnO–nanorod sago starch films gives excellent antimicrobial activity against *S. aureus*. This result made ZnO–nanorod and starch nanocomposites' readiness to use it as an active packaging material as reported by Nafchi et al.[80] Another attempt by preparing nanocomposite of hydroxypropyl methylcellulose and silver nanoparticles was done by de Moura et al. and results shown by nanocomposite were interesting; silver nanoparticles with particle size of 41 nm show greater antibacterial property than silver nanoparticles having size of 100 nm.[77] Another literature revealed some disadvantages of use of silver nanoparticles for instance easy aggregation, uncontrollable release of silver ions, and potential cytotoxicity effectively affect its practical use.[78] It was reported in 2012 by Busolo et al. that there is significant potential for the use of this novel oxygen scavenger additive to constitute active packaging of value in the shelf-life extension of oxygen-sensitive food products.[83] The polystyrene was grafted with two different monomers: acrylic and maleic acids in presence of MMT using potassium persulfate as initiator under nitrogen atmosphere. Results show good antimicrobial properties against *K. pneumonia, E. coli*, and Gram-positive bacterium (*S. lutea*), in addition to the yeast fungus (*C. albicans*).[80] Natural rubber/rutile-TiO2 nanocomposites with different *n*-TiO2 contents were prepared. The *n*-TiO2 with median particle size of 73 nm was prepared by ultrasonication. Thus, result shows that nanocomposite gives improved mechanical properties and it possesses high UV-protection properties as well as antibacterial property.[81] Nanocomposite LDPE films containing silver nanoparticles (AgNPs) were prepared by melt mixing in a twin-screw extruder and used to pack and store fresh orange juice at 4°C. The juice was stored for 56 days.[82] Methanol-diluted ZnO nanoparticles were dispersed in solution of polystyrene in toluene. Afterward, this whole solution was casted and placed in vacuum oven at room temperature to evaporate solvent, and after that, it was heated at 80°C to remove the trapped solvent in between polystyrene. This prepared film of ZnO nanoparticle–polystyrene nanocomposite shows no antimicrobial property. These were reported by Jin et al.[87] In 2010, it was reported that the application of nano-ZnO to LDPE packages prolonged the microbial stability of orange juice.[84] Chitosan nanoparticles have higher antimicrobial property and it could be readily dispersed in biopolymer to give biodegradable food packaging.[85] It was reported that in 2011, the ions Na^+ were replaced by Ag^+ ions in order to develop nanocomposite of Ag-MMT and ager with good microbial property.[86] Effectiveness of an antimicrobial packaging system consisting of agar and Ag-MMT nanoparticles on cheese stability was evaluated by Incoronato et al.[90] In order to get the antimicrobial property, issue regarding migration of Ag^+ ion was reported by Lloret et al.[92] Nanocomposite was prepared by modified MMT (MMT–Cu_2^+) and LDPE by melt mixing in extruder shows that the antibacterial effect of the nanocomposite increases with the concentration of MMT–Cu_2^+.[89] It was investigated that radiolytic method is more suitable to produce Ag nanoparticles without use of any solvents.[90] If hybrid Ag–TiO2 nanoparticles are activated by UV light, then they exhibit stronger antibacterial activity than UV combined only with Ag or TiO2 nanoparticles, as was examined by Li et al.[95] More detailed discussion about antimicrobial nanostructure and properties was reported by de Azeredo et al. in 2013.[92]

8.8.2.1 TESTING OF ACTIVE OR ANTIMICROBIAL PACKAGING MATERIAL

To characterize antimicrobial behavior of polymer nanocomposites, there are four typical food microorganisms including two Gram-positive bacteria–*Listeria monocytogenes* ATCC-19111 and *Staphylococcus aureus* ATCC-14458, and two Gram-negative bacteria–*Salmonella typhimurium* ATCC-14028 and *Escherichia coli* O157:H7 ATCC-11775, which are utilized to test the antimicrobial activity of polymer nanocomposite films using a viable cell count method. Film sample can be prepared by cutting film into square pieces (10 cm × 10 cm) and placed in individual germ-free flasks. The Gram-positive and Gram-negative bacteria were separately incubated in two separate broths at 37 and 30°C, respectively, under aerobic conditions for 16 h. Each 100 mL of the prepared inoculums with the 1/10 diluted broth is aseptically added to the flasks containing the test films.[93]

8.8.3 SMART/INTELLIGENT POLYMER NANOCOMPOSITES

Intelligent or smart packaging can be achieved with nanocomposite materials. Intelligent packaging consists of packaging material which can be proficient to sense condition of food inside the packaging, freshness of food, toxicity, oxygen level, microbial activities, time, and temperature. For sensing application, it is essential to incorporate materials that are able to sense such parameters. Hence, development of intelligent packaging for food packaging can give information to its customers about how much fresh product they are going to buy. The oxygen indicator was used to sense oxygen, prepared from a combination of electrochrome, titanium dioxide, and EDTA. It can be used in intelligent or smart packaging applications.[93] Nanocrystalline SnO_2 was used as a photosensitiser in a colorimetric O_2 indicator. It can be utilized to sense level of oxygen.[94] Metal oxides were also very popular because of their sensitivity and stability as sensors.[95] pH value can be sensed by incorporating silicate-based pH indicators.[96] H_2S-sensitive indicators are based on a visually detectable color change of agarose immobilized myoglobin; when the freshness of food is goes on decreasing, it librates H_2S. This H_2S is sensed by myoglobin-based indicators. Such type of indicator was reported by Hurme et al.[102] A colorimetric-mixed pH dye-based indicator is used as carbon dioxide indicator. It was reported by Nopwinyuwong et al.[103] It has potential for the development of intelligent packaging. In the case of packaging of fishes, changes in the ammonia content of the fish could be monitored with the NH_4^+-ISE. The indication gives an idea about freshness of fish, as reported in 2012 by Heising et al.[104] Chitosan-based carbon dioxide sensor can be utilized in smart packaging, as reported by Jung et al.[105] A time-monitoring sensor based on the oxidation of leuco-methylene blue to methylene blue is developed. The sensor changes its color from yellow to green in the presence of oxygen; therefore, it may be utilized in intelligent food packaging application, as investigated by Marek et al.[106] Kuswandi, Jayus, Restyana, Abdullah, Heng, and Ahmad examined that a novel colorimetric method based on polyaniline film could be used to develop a smart packaging. In case of fish packaging, such polyaniline film acts as chemical sensor for real-time monitoring of the microbial breakdown products in the headspace of packaged fish. This on-package indicator contains polyaniline film, which responds through visible color change to a variation of basic volatile amines (specifically known as total volatile basic nitrogen) released during fish spoilage period.[102] A smart temperature indicator packaging material was developed based on a natural and heat-sensitive pigment (anthocyanin) and was able to serve smart food packaging.[103] Enzyme-based time and temperature indicator was developed by using laccase.[104] Colorimetric detection of melamine in raw milk using gold nanoparticles and crown-ether-modified thiols with a limit of detection of 6 ppb was reported.[106] Ozdemir et al. reported that gold nanoparticles and glucose-sensitive enzymes can be used to measure glucose concentrations in commercial beverages.[50] In a broad sense, it can be observed that nanosensors are the sensors which sense various parameters such as time and temperature, humidity, and detection of gases. Detection of microbes is possible due to the unique chemical and electrooptical properties at nanoscale sizes known as nanosensor. Brief discussion regarding sensors and indicator is explained in further part of same section.

8.8.3.1 SENSORS

A sensor is a device that can be used to detect, quantify, as well as locate any physical or chemical properties. Development in chemical sensors and biosensors is in boom phase in recent years. Sensors are important aspects of today's intelligent packaging material and are utilized in food packaging applications. Sensors are used in an extremely drastic nature to detect spoilage of food, gas in side of food packages, etc. Sensors can be classified in categories like gas sensors, biosensors, and florescence-based oxygen sensors.

8.8.3.1.1 Gas Sensors

Gas sensors are devices, which detect gas by means of their inherent properties or by using external devices. It was reported that the majority of carbon dioxide sensors, developed in recent years, were utilized in biomedical application and food packaging application.[108,109] In last century, a number of optical oxygen sensors pertain to gas sensors were reported. Optical oxygen sensor comprises solid metal material and operates on the principles of luminescence quenching. Changes caused which are to be sensed were caused by direct contact with analyte, that is, oxygen.[108,110] Optochemical sensors were developed to sense hydrogen sulfide, amines, and carbon dioxide. They detect deterioration of product by sensing gases.[108]

8.8.3.1.2 Florescence-based Oxygen Sensor

Detection of oxygen via florescence-based sensor was reported in 1996 by Reiniger, Kolle, Trettnak, and Gruber. It is one of the economical ways to develop a sensor by using luminescent dyes. Fluorescent or phosphorescent dyes were used as key component of these florescence-based sensors. These dyes are generally utilized as sensors by combing them with polymer matrix. Oxygen in package was measured by measuring luminescence change in a sensor device.[111]

8.8.3.1.3 Biosensors

Biosensors come under intelligent packaging but till date, there is no strong commercialization occurred. Thus, it is a very sensitive class of sensors used to detect various contaminants in food packaging. Biosensors mainly comprises enzymes, antigens, microbes, hormones, and nucleic acids.[108,112]

8.8.3.2 INDICATORS

Indicators are the materials that indicate characteristics of food like freshness, integrity of package as well as time and temperature. Indicator do not comprise transducer and receptor, which made indicators different than sensors. Indicator indicates the changes in object by changing its own characteristics or color.[108]

8.8.3.2.1 Freshness Indicators

Freshness indicator is key tool to determine the spoilage of food, chemical changes in food stuff, and progress in microbial growth in package.

Consequently, this key function made freshness indicator one of the essential elements of intelligent food packaging. As comparable to sensor, it does not possess any receptor or transducer. In last two centuries, various researchers reported that the basis on which freshness indicators developed was provided by chemical detection of spoilage of food and chemical changes occur in storage of food stuff. Chemical changes in organic acid, like *n*-butyrate, L-lactic acid, D-lactate, and acetic acid, offer potential to indicator metabolites. Numbers of freshness indicators developed were based on indicator color changes in response to microbial metabolites that occur during spoilages of food. In the last decade, COX Technologies, USA, launched Fresh Tag color change indicator labels. This label was utilized to determine the freshness of seafood.[108]

8.8.3.2.2 Time and Temperature Indicator

Time and temperature indicator indicates full or partial temperature history of food product. The response of time and temperature indicators is generally in the form of change in color or mechanical deformation. Time and temperature indicator is classified into two types: partial-temperature history indicators and full-temperature history indicators. A partial-temperature history indicator does not respond until it is exposed to temperature that causes it to change in food quality. In case of full-temperature history indicators, temperature history regarding food product is indicated utterly. Diffusion, enzymatic, and polymer-based systems are used to commercialize time and temperature indicators.[108]

8.9 ROLE OF BIOPOLYMER NANOCOMPOSITES IN FOOD PACKAGING

Current environmental issues of synthetic polymer packaging are the main driving force to develop the biopolymer packaging. Biopolymer packaging materials are ready to substitute non-eco-friendly synthetic plastic packaging with excellent biodegradable plastics.[113–123] Ahead of biopolymer, there is bio-nanocomposites, a class of composite material developed with more effective and enhanced barrier, mechanical and thermal properties to serve food packaging applications. It consists of biopolymer matrix reinforced with filler or particles having their size at nanoscale.[2,116–122] A reason behind the changeover of biopolymer to bio-nanocomposite is poor mechanical property, barrier property, challenges in processing and cost, etc. of biopolymer.[123,124] Bio-nanocomposites are generally prepared by using layered silicate or inorganic clay minerals and variety of matrix chosen from the ample number of biopolymers. After preparation of the bio-nanocomposites, characterization of its biodegradability antimicrobial property, mechanical property, and thermal stabilities will be carried out. Biopolymers used to prepare bio-nanocomposites are enlisted as follows: polylactic acid, polyglycolic acid, polybutylene succinate, cellulose and its derivatives, starch, ager, polyhydroxybutyvate-co-hydroxvalevate, and poly-co-lactum.[1] There are a number of examples of bio-nanocomposite with improved mechanical and barrier properties. Huang reported 450% increment in tensile strength of corn starch/MMT nanocomposites.[125] Another example of soya protein/MMT nanocomposite possesses enhanced tensile strength from 8.77 to 15.43 MPa at 16% loading of MMT clay.[126] It was reported that ager/unmodified MMT nanocomposite film possesses lower water–vapor transmission rate than ager/organically modified MMT nanocomposite film.[127] Similar result regarding water–vapor transmission rate of nanocomposite consist of biopolymer, starch, whey protein isolates, soy protein isolates, wheat gluten, and poly(caprolactum) was reported.[128–138] Cloisite 30B is a mineral clay that is able to give reduction on water–vapor transmission rate when nanocomposite formed of Cloisite 30B and chitosan-based nanocomposite.[131] Water vapor transmission rate was examined by Park with nanocomposites like cellulose acetate/clay nanocomposite with different triethyl citrate plasticizers.[139] Along with mechanical and barrier properties, bio-nanocomposite are featured with property that differentiate bio-nanocomposites from polymer nanocomposite that is nothing but biodegradation property. Ample numbers of attempts were made to prepare biodegradable nanocomposites by using poly(lactic acid) matrix, and it was reported by Nieddu. It can be able to give enhanced biodegradation of poly(lactic acid)-based nanocomposites prepared with different types nanoclays.[141] Rhim reported that organoclay, especially Cloisite 30B, gives strong antimicrobial activity against food-poisoning bacteria.[1,131] Silver nanoparticles are effective antimicrobial agent used to prepare polymer nanocomposites with better antimicrobial property. Silver nanoparticles have been loaded in biopolymers in order to get strong antimicrobial property against bacteria.[141,142–144]

8.10 SCOPE FOR RESEARCH IN FOOD PACKAGING BASED ON POLYMER NANOCOMPOSITES

In case of food packaging, intensive research is required on active food packaging and smart/intelligent food packaging rather than the improved food packaging. For this reason, these developing materials have potential to change the face of packaging material used today. It is vital to do the studies as well as research on nanosensor materials in order to develop intelligent packaging material for diverse packaging applications. Antimicrobial property given by nanosilver particle to nanocomposite is great, but along with it, there arises a problem of toxicity of nanosilver particle; so the research should be directed in such way that problem of toxicity of nanosilver is avoided or reduced. There is also scope to develop material having antimicrobial property which can replace conventional metal oxides and silver, etc. Abundant numbers of researches are done on the polymer nanocomposites, but there are still a number of challenges in preparation, structure, and properties of polymer nanocomposites. Challenges, for example, how changes in polymer crystalline structure induced by the clay affect overall composite properties? How does one can tailor organoclay chemistry to achieve high degrees of exfoliation reproducibility for a given polymer system? How process parameters and fabrication affect composite properties? Thus, in order to overcome such questions, thorough study on the highlighted problem is very necessary.[2,145]

8.11 SUMMARY

Today, there are much critical issue regarding environment, shelf life, and safety of food due to use of conventionally packaging material, that is, pristine plastics, metals, and papers. To avoid such issues, polymer nanocomposite will be a smart choice. Polymer nanocomposites are able to provide passage to overcome above-mentioned problem by forming improved/active/smart nanocomposites for food packaging, with or without combining biopolymers. Polymer nanocomposite is the marvelous product from the combination of polymer science and nanotechnology. Nanocomposite is a composite material of polymer and nanoscale reinforcement. Nanocomposite can be prepared from different methods that are greatly emphasized in the review reported by Pavlidou and Papaspyrides. It was accentualy reported by a number of reviewers that melt intercalation and in situ polymerization are mostly preferred for commercial as well as for research work. Layered silicate or nanoparticles is a crucial aspect of a nanocomposite, because it defines the properties via its structure, that is, whether it is intercalated or exfoliated. Chemical modification of nanoparticles is pivotal treatment that is carried out on layered silicate, succors to achieve exfoliated nanocomposite which gives higher properties. Characterization of polymer nanocomposite is quite essential because it gives an idea about its structure, morphology, and whether it is exfoliated or intercalated. Correlation in between food packaging and nanocomposites perceptively unfolds the need of polymer nanocomposite for packaging application, especially food packaging by illustrating its advantages. As mentioned in Section 8.8.2, advanced food packaging is disunited in three diverse categories as per their advancement. Improved polymer nanocomposite possesses good mechanical and barrier properties, active consist of antimicrobial property, and intelligent one senses the condition of food, gases, and temperature of packet or food which is quite assisting to improve shelf life of food. Along with sensors, this subtitle comprises indicators, like freshness indicator, time, and temperature indicator. Along with synthetic polymer nanocomposite, importance of biopolymer nanocomposites in packaging field is crucially explained.

KEYWORDS

- **nanocomposites**
- **layered silicate**
- **food packaging**
- **indicators**
- **biosensors**

REFERENCES

1. Rhim, J.-W.; Park, H.-M.; Ha, C.-S. Bio-nanocomposites for Food Packaging Applications. *Prog. Polym. Sci.* **2013**, 38(10–11), 1629–1652.
2. Pavlidou, S.; Papaspyrides, C. D. A Review on Polymer-layered Silicate Nanocomposites. *Prog. Polym. Sci.* **2008**, *33*, 1119–1198.

3. Fischer, H. Polymer Nanocomposites: From Fundamental Research to Specific Applications. *Mater. Sci. Eng. C* **2003,** *23,* 763–772.
4. Lagaly, G. Introduction: From Clay Mineral–Polymer Interactions to Clay Mineral–Polymer Nanocomposites. *Appl. Clay Sci.* **1999,** *15,* 1–9.
5. Giannelis, E. P. Polymer Layered Silicate Nanocomposites. *Adv. Mater.* **1996,** *8,* 29–35.
6. Varlot, K.; Reynaud, E.; Kloppfer, M. H.; Vigier, G.; Varlet, J.; Clay-reinforced Polyamide: Preferential Orientation of the Montmorillonite Sheets and the Polyamide Crystalline Lamellae. *J. Polym. Sci. Polym. Phys.* **2001,** *39,* 1360–1370.
7. Gorrasi, G.; Tortora, M.; Vittoria, V.; Galli, G.; Chiellini, E. Transport and Mechanical Properties of Blends of Poly(caprolactone) and a Modified Montmorillonite–Poly(Caprolactone) Nanocomposite. *J. Polym. Sci. Polym. Phys.* **2002,** *40,* 1118–1124.
8. de Azeredo, H. M. C. Nanocomposites for Food Packaging Applications. *Food Res. Int.* **2009,** *42* 1240–1253.
9. Nafchi, A. M.; Nassiri, R.; Sheibani, S.; Ariffin, F.; Karim, A. A. Preparation and Characterization of Bionanocomposite Films Filled with Nanorod-rich Zinc Oxide. *Carbohydr. Polym.* **2013,** *96,* 233–239.
10. Alexandre, M.; Dubois, P. Polymer-layered Silicate Nanocomposites: Preparation, Properties and Uses of a New Class of Materials. *Mater. Sci. Eng. Res.* **2000,** *28,* 1–63.
11. Chin, I.-J.; Thurn-Albrecht, T.; Kim, H.-C.; Russell, T. P.; Wang, J. On Exfoliation of Montmorillonite in Epoxy. *Polymer* **2001,** *42,* 5947–5952.
12. Kim, C.-M.; Lee, D.-H.; Hoffmann, B.; Kressler, J.; Stoppelmann, G. Influence of Nanofillers on the Deformation Process in Layered Silicate/Polyamide 12 Nanocomposites. *Polymer* **2001,** *42,* 1095–1100.
13. Dennis, H. R.; Hunter, D. L.; Chang, D.; Kim, S.; White, J. L.; Cho, J. W. Effect of Melt Processing Conditions on the Extent of Exfoliation in Organoclay-based Nanocomposites. *Polymer* **2001,** *42,* 9513–9522.
14. Wu, S. H.; Wang, F. Y.; Ma, C. C. M.; Chang, W. C.; Kuo, C. T.; Kuan, H. C.; et al. Mechanical, Thermal and Morphological Properties of Glass Fiber and Carbon Fiber Reinforced Polyamide 6 and Polyamide 6/clay Nanocomposites. *Mater. Lett.* **2001,** *49,* 327–333.
15. Nguyen, Q. T. Process for Improving the Exfoliation and Dispersion of Nanoclay Particles into Polymer Matrices Using Supercritical Carbon Dioxide, Ph.D. Dissertation, 2007.
16. Sur, G. S.; Sun, H. L.; Lyu, S. G.; Mark, J. E. Synthesis, Structure, Mechanical Properties, and Thermal Stability of Some Polysulfone/Organoclay Nanocomposites. *Polymer* **2001,** *42,* 9783.
17. Beyer, G. Nanocomposites: A New Class of Flame Retardants for Polymers. *Plast. Addit. Compound* **2002,** *4*(10), 22–27.
18. Manias, E.; Touny, A.; Wu, L.; Strawhecker, K.; Lu, B.; Chung, T. C.; Polypropylene/Montmorillonite Nanocomposites. Review of the Synthetic Routes and Materials Properties. *Chem. Mater.* **2001,** *13,* 3516–3523.
19. Xie, W.; Gao, Z.; Liu, K.; Pan, W. P.; Vaia, R.; Hunter, D.; et al. Thermal Characterization of Organically Modified Montmorillonite. *Thermochim. Acta* **2001,** *367/368,* 339–350.

20. Kornmann, X.; Lindberg, H.; Berglund, L. A. Synthesis of Epoxy–Clay Nanocomposites: Influence of the Nature of the Clay on Structure. *Polymer* **2001**, *42*, 1303–1310.
21. Solomon, M. J.; Almusallam, A. S.; Seefeldt, K. F.; Somwangthanaroj, A.; Varadan, P. Rheology of Polypropylene/Clay Hybrid Materials. *Macromolecules* **2001**, *34*, 1864–1872.
22. Zanetti, M.; Lomakin, S.; Camino, G. Polymer Layered Silicate Nanocomposites. *Macromol. Mater. Eng.* **2000**, *279*, 1–9.
23. Vaia, R. A.; Giannelis, E. P. Lattice of Polymer Melt Intercalation in Organically Modified Layered Silicates. *Macromolecules* **1997**, *30*, 7990–7999.
24. Arada, P.; Ruiz-Hitzky, E. Polymer–Salt Intercalation Complexes in Layer Silicates. *Adv. Mater.* **1990**, *2*, 545–547.
25. Arada, P.; Ruiz-Hitzky, E. Poly(Ethylene Oxide)–Silicate Intercalation Materials. *Chem. Mater.* **1992**, *4*, 1395–1403.
26. Tunney, J. J.; Detellier, C.; Poly(Ethylene Glycol)–Kaolinite Intercalates. *Chem Mater.* **1996**, *8*, 927–935.
27. Fischer, H. R.; Gielgens, L. H.; Koster, T. P. M. Nanocomposites from Polymers and Layered Minerals. *Acta Polym.* **1999**, *50*, 122–126.
28. Theng, B. K. G. *Formation and Properties of Clay–Polymer Complexes.* Elsevier: Amsterdam, 1979.
29. Greenland, D. J. Adsorption of Poly(Vinyl Alcohols) by Montmorillonite. *J. Colloid Sci.* **1963**, 647–664.
30. Ray, S. S.; Bousima, M. Biodegradable Polymers and their Layered Silicate Nanocomposites: In Greening the 21st Century Materials World. *Prog. Mater. Sci.* **2005**, *50*, 962–1079.
31. Morgan, A. B.; Gilman, J. W. Characterization of Polymer-layered Silicate (Clay) Nanocomposites by Transmission Electron Microscopy and X-ray Diffraction: A Comparative Study. *J. Appl. Polym. Sci.* **2003**, *87*, 1329–1338.
32. VanderHart, D. L.; Asano, A.; Gilman, J. W. NMR Measurements Related to Clay-dispersion Quality and Organic-modifier Stability in Nylon-6/clay Nanocomposites. *Macromolecules* **2001**, *34*, 3819–3822.
33. Wu, H. D.; Tseng, C. R.; Chang, F. C. Chain Conformation and Crystallization Behavior of the Syndiotactic Polystyrene Nanocomposites Studied Using Fourier Transform Infrared Analysis. *Macromolecules* **2000**, *31*, 2992–2999.
34. Tharanathan, R. N. Biodegradable Films and Composite Coatings: Past, Present and Future. *Trends Food Sci. Technol.* **2003**, *14*(3), 71–78.
35. Zhang, Z.; Yang, J. L. Creep Resistant Polymeric Nanocomposites. *Polymer* **2004**, *45*, 3481–3485.
36. Shia, D.; Hui, C. Y.; Burnside, S. D.; Giannelis, E. P. An Interface Model for the Prediction of Young's Modulus of Layered Silicate–Elastomer Nanocomposites. *Polym. Compos.* **1998**, *19*, 608–617.
37. Ray, S. S.; Okamoto, M. Polymer-layered Silicate Nanocomposite: A Review from Preparation to Processing. *Prog. Polym. Sci.* **2003**, *28*, 1539–1641.
38. Messersmith, P. B.; Giannelis, E. P. Synthesis and Barrier Properties of poly(ε-caprolactone)-layered Silicate Nanocomposites. *J. Polym. Sci. Polym. Chem.* **1995**, *33*, 1047–1057.

39. Chang, J.-H.; Uk-An, Y.; Sur, G. S. Poly(Lactic Acid) Nanocomposites with Various Organoclays. I. Thermomechanical Properties, Morphology, and Gas Permeability. *J. Polym. Sci. Polym. Phys.* **2003**, *41*, 94–103.
40. Paul, M-A.; Alexandre, M.; Degee, P.; Henrist, C.; Rulmont, A.; Dubois, P. New Nanocomposite Materials Based on Plasticized Poly(L-Lactide) and Organo-modified Montmorillonites: Thermal and Morphological Study. *Polymer* **2003**, *44*, 443–450.
41. Vyazovkin, S.; Dranka, I.; Fan, X.; Advincula, R. Kinetics of the Thermal and Thermo-oxidative Degradation of a Polystyrene–Clay Nanocomposite. *Macromol. Rapid Commun.* **2004**, *25*, 498–503.
42. Yoon, P. J.; Fornes, T. D.; Paul, D. R. Thermal Expansion Behavior of Nylon 6 Nanocomposites. *Polymer* 2002, *43*, 6727–6741.
43. Kojima, Y.; Usuki, A.; Kawasumi, M.; Okada, A.; Kurauchi, T.; Kamigaito, O. Synthesis of Nylon-6 Hybrid by Montmorillonite Intercalated with Caprolactam. *J. Polym. Sci. Polym. Chem.* **1993**, *31*, 983–986.
44. Cho, J. W.; Paul, D. R. Nylon 6 Nanocomposites by Melt Compounding. *Polymer* **2001**, *42*, 1083–1094.
45. Silvestre, C.; Duraccio, D.; Cimmino, S. Food Packaging Based on Polymer Nanomaterials. *Prog. Polym. Sci.* **2011**, *36*, 1766–1782.
46. Adame, D.; Beall, G. W. Direct Measurement of the Constrained Polymer Region in Polyamide/Clay Nanocomposites and the Implications for Gas Diffusion. *Appl. Clay Sci.* **2009**, *42*, 545–552.
47. Kim, K.; Kim, E.; Lee, S. J. New Enzymatic Time–Temperature Integrator (TTI) that Uses Laccase. *J. Food Eng.* **2012**, *113*(1), 118–123.
48. Duncan, T. V. Applications of Nanotechnology in Food Packaging and Food Safety: Barrier Materials, Antimicrobials and Sensors. *J. Colloid Interface Sci.* **2011**, *363*, 1–24.
49. Kuang, H.; Chen, W.; Yan, W.; Xu, L.; Zhu, Y.; Liu, L.; Chu, H.; Peng, C.; Wang, L.; Kotov, N. A.; Xu, C. Crown Ether Assembly of Gold Nanoparticles: Melamine Sensor. *Biosens. Bioelectron.* **2011**, *26*, 2032–2037.
50. Ozdemir, C.; Yeni, F.; Odaci, D.; Timur, S. Electrochemical Glucose Biosensing by Pyranose Oxidase Immobilized in Gold Nanoparticle–Polyaniline/AgCl/Gelatin Nanocomposite Matrix. *Food Chem.* **2010**, *119*, 380.
51. Fornes, T. D.; Yoon, P. J.; Keskkula, H.; Paul, D. R. Nylon 6 Nanocomposites: The Effect of Matrix Molecular Weight. *Polymer* **2001**, *42*, 9929–9940.
52. Choudhury, N. R.; Kannan, A. G.; Dutta, N. Novel Nanocomposites and Hybrids for High-temperature Lubricating Coating Applications. *Tribology of Polymeric Nanocomposites Friction and Wear of Bulk Materials and Coatings*, 2nd ed.; 2013; pp 717–778.
53. Provenzano, V.; Holtz, R. L. Nanocomposites for High Temperature Applications. *Mater. Sci. Eng.* **1995**, *204*(1–2), 125–134.
54. Cherian, B. M.; Leão, A. L.; Ferreira de Souza, S.; Costa, L. M. M.; de Olyveira, G. M.; Kottaisamy, M.; Nagarajan, E. R.; Thomas, S. Cellulose Nanocomposites with Nanofibres Isolated from Pineapple Leaf Fibers for Medical Applications. *Carbohydr. Polym.* **2011**, *86*(4), 1790–1798.
55. Rajesh; Ahuja, T.; Kumar, D. Recent Progress in the Development of Nano-structured Conducting Polymers/Nanocomposites for Sensor Applications. *Sens. Actuators, B: Chem.* **2009**, *136*(1), 275–286.

56. Aranda, P.; Darder, M.; Fernández-Saavedra, R.; López-Blanco, M.; Ruiz-Hitzky, E. Relevance of Polymer– and Biopolymer–Clay Nanocomposites in Electrochemical and Electroanalytical Applications. *Thin Solid Films* **2006**, *495*(1–2), 104–112.
57. Fang, Y.; Ni, Y.; Zhang, G.; Mao, C.; Huang, X.; Shen, J. Biocompatibility of CS–PPy Nanocomposites and their Application to Glucose Biosensor. *Bioelectrochemistry* **2012**, *88*, 1–7.
58. Darder, M.; Colilla, M.; Ruiz-Hitzky, E. Chitosan–Clay Nanocomposites: Application as Electrochemical Sensors. *Appl. Clay Sci.* **2005**, *28*(1–4), 199–208.
59. Zhao, X.; Lv, L.; Pan, B.; Zhang, W.; Zhang, S.; Zhang, Q. Polymer-supported Nanocomposites for Environmental Application: A Review. *Chem. Eng. J.* **2011**, *170*(2–3), 381–394.
60. Fukushima, K.; Wu, M.-H.; Bocchini, S.; Rasyida, A.; Yang, M.-C. PBAT Based Nanocomposites for Medical and Industrial Applications. *Mater. Sci. Eng., C* **2012**, *32*(6, 1), 1331–1351.
61. Bharadwaj, R. K.; Mehrabi, A. R.; Hamilton, C.; Trujillo, C.; Murga, M.; Fan, R.; Chavira, A.; et al.; Structure–Property Relationships in Cross-linked Polyester–Clay Nanocomposites. *Polymer* **2002**, *43*(13), 3699–3705.
62. Cabedo, L.; Gimenez, E.; Lagaron, J. M.; Gavara, R.; Saura, J. J.; Development of EVOH–Kaolinite Nanocomposites. *Polymer* **2004**, *45*(15), 5233–5238.
63. Cava, D.; Gimenez, E.; Gavara, R.; Lagaron, J. M. Comparative Performance and Barrier Properties of Biodegradable Thermoplastics and Nanobiocomposites versus PET for Food Packaging Applications. *J. Plast. Film Sheet* **2006**, *22*, 265–274.
64. Lotti, C.; Isaac, C. S.; Branciforti, M. C.; Alves, R. M. V.; Liberman, S.; Bretas, R. E. S. Rheological, Mechanical and Transport Properties of Blown Films of High Density Polyethylene Nanocomposites. *Eur. Polym. J.* **2008**, *44*, 1346–1357.
65. Thellen, C.; Schirmer, S.; Ratto, J. A.; Finnigan, B.; Schmidt, D. Co-extrusion of Multilayer Poly(*m*-Xylene Adipimide) Nanocomposite Films for High Oxygen Barrier Packaging Applications. *J. Membr. Sci.* **2009**, *340*(1–2), 45–51.
66. Yeh, J.-T.; Chang, C.-J.; Tsai, F.-C.; Chen, K.-N.; Huang, K.-S. Oxygen Barrier and Blending Properties of Blown Films of Blends of Modified Polyamide and Polyamide-6 Clay Mineral Nanocomposites. *Appl. Clay Sci.* **2009**, *45*(1–2), 1–7.
67. Majzdadeh-Ardakani, K.; Navarchian, A. H.; Sadeghi, F. Optimization of Mechanical Properties of Thermoplastic Starch/Clay Nanocomposites. *Carbohydr. Polym.* 2010, *79*(3), 547–554.
68. Ali Dadfar, S. M.; Alemzadeh, I.; Reza Dadfar, S. M.; Vosoughi, M. Studies on the Oxygen Barrier and Mechanical Properties of Low Density Polyethylene/Organoclay Nanocomposite Films in the Presence of Ethylene Vinyl Acetate Copolymer as a New Type of Compatibilizer. *Mater. Des.* **2011**, *32*(4), 1806–1813.
69. Hong, S.-I.; Rhim, J.-W. Preparation and Properties of Melt-intercalated Linear Low Density Polyethylene/Clay Nanocomposite Films Prepared by Blow Extrusion. *Lebensm. Wiss. Technol.* **2012**, *48*(1), 43–51.
70. Yang, Q.; Wu, C.-N.; Saito, T.; Isogai, A. Cellulose–Clay Layered Nanocomposite Films Fabricated from Aqueous Cellulose/LiOH/Urea Solution. *Carbohydr. Polym.* **2014**, *100*, 179–184.

71. Mahmoudian, S.; Wahit, M. U.; Ismail, A. F.; Yussuf, A. A. Preparation of Regenerated Cellulose/Montmorillonite Nanocomposite Films via Ionic Liquids. *Carbohydr. Polym.* **2012**, *88*(4), 1251–1257.
72. Yoon, S.-D.; Park, M.-H.; Byun, H.-S. Mechanical and Water Barrier Properties of Starch/PVA Composite Films by Adding Nano-sized Poly(Methyl Methacrylate-*co*-Acrylamide) Particles. *Carbohydr. Polym.* **2012**, *87*(1), 676–686.
73. Sonia, A.; Dasan, P. Celluloses Microfibers (CMF)/Poly(Ethylene-*co*-Vinyl Acetate) (EVA) Composites for Food Packaging Applications: A Study Based on Barrier and Biodegradation Behavior. *J. Food Eng.* **2013**, *118*(1), 78–89.
74. Ozcalik, O.; Tihminlioglu, F. Barrier Properties of Corn Zein Nanocomposite Coated Polypropylene Films for Food Packaging Applications. *J. Food Eng.* **2013**, *114*(4), 505–513.
75. Bruna, J. E.; Peñaloza, A.; Guarda, A.; Rodríguez, F.; Galotto, M. J. Development of MtCu^{2+}/LDPE Nanocomposites with Antimicrobial Activity for Potential Use in Food Packaging. *Appl. Clay Sci.* **2012**, *58*, 79–87.
76. Zapata, P. A.; Tamayo, L.; Páez, M.; Cerda, E.; Azócar, I.; Rabagliati, F. M. Nanocomposites Based on Polyethylene and Nanosilver Particles Produced by Metallocenic "In Situ" Polymerization: Synthesis, Characterization, and Antimicrobial Behavior. *Eur. Polym. J.* **2011**, *47*(8), 1541–1549.
77. Li, S.-M.; Jia, N.; Ma, M.-G.; Zhang, Z.; Liu, Q.-H.; Sun, R.-C. Cellulose–Silver Nanocomposites: Microwave-assisted Synthesis, Characterization, their Thermal Stability, and Antimicrobial Property. *Carbohydr. Polym.* **2011**, *86*(2), 441–447.
78. Han, Y.-S.; Lee, S.-H.; Choi, K. H. Preparation and Characterization of Chitosan–Clay Nanocomposites with Antimicrobial Activity. *J. Phys. Chem. Solids* **2010**, *71*(4), 464–467.
79. Costa, C.; Conte, A.; Buonocore, G. G.; Del Nobile, M. A. Antimicrobial Silver-montmorillonite Nanoparticles to Prolong the Shelf Life of Fresh Fruit Salad. *Int. J. Food Microbiol.* **2011**, *148*(3), 164–167.
80. Nafchi, A. M.; Alias, A. K.; Mahmud, S.; Robal, M. Antimicrobial, Rheological, and Physicochemical Properties of Sago Starch Films Filled with Nanorod-rich Zinc Oxide. *J. Food Eng.* **2012**, *113*(4), 511–519.
81. de Moura, M. R.; Mattoso, L. H. C.; Zucolotto, V. Development of Cellulose-based Bactericidal Nanocomposites Containing Silver Nanoparticles and their Use as Active Food Packaging. *J. Food Eng.* **2012**, *109*(3), 520–524.
82. Guo, L.; Yuan, W.; Lu, Z.; Li, C. M. Polymer/Nanosilver Composite Coatings for Antibacterial Applications. *Colloids Surf. A: Physicochem. Eng. Asp.* **2013**, *439*, 69–83.
83. Busolo, M. A.; Lagaron, J. M. Oxygen Scavenging Polyolefin Nanocomposite Films Containing an Iron Modified Kaolinite of Interest in Active Food Packaging Applications. *Innov. Food Sci. Emerg. Technol.* **2012**, *16*, 211–217.
84. Haroun, A. A.; Ahmed, E. F.; El-Halawany, N. R.; Taie, H. A. A. Antimicrobial and Antioxidant Properties of Novel Synthesized Nanocomposites Based on Polystyrene Packaging Material Waste. *IRBM* **2013**, *34*(3), 206–213.
85. Seentrakoon, B.; Junhasavasdikul, B.; Chavasiri, W. Enhanced UV-protection and Antibacterial Properties of Natural Rubber/Rutile-TiO$_2$ Nanocomposites. *Polym. Degrad. Stab.* **2013**, *98*(2), 566–578.

86. Emamifar, A.; Kadivar, M.; Shahedi, M.; Soleimanian-Zad, S.; Effect of Nanocomposite Packaging Containing Ag and ZnO on Inactivation of *Lactobacillus plantarum* in Orange Juice. *Food Control* **2011**, *22*, 408–413.
87. Jin, T.; Sun, D.; Su, J. Y.; Zhang, H.; Sue, H. Antimicrobial Efficacy of Zinc Oxide Quantum Dots against *L. monocytogenes*, *S. enteritidis*, and *E. coli* O157:H7. *J. Food Sci.* **2009**, *1*, M46–M52.
88. Emamifar, A.; Kadivar, M.; Shahedi, M.; Soleimanian-Zad, S. Evaluation of Nanocomposite Packaging Containing Ag and ZnO on Shelf Life of Fresh Orange Juice. *Innov. Food Sci. Emerg. Technol.* **2010**, *11*, 742–748.
89. Watthanaphanit, A.; Supaphol, P.; Tamura, H.; Tokura, S.; Rujiravanit, R. Wet-spun Alginate/Chitosan Whiskers Nanocomposite Fibers: Preparation, Characterization and release Characteristic of the Whiskers. *Carbohydr. Polym.* **2010**, *79*, 738–746.
90. Incoronato, A. L.; Buonocore, G. G.; Conte, A.; Lavorgna, M.; Del Nobile, M. A. Active Systems Based on Silver Montmorillonite Nanoparticles Embedded into Bio-based Polymermatrices for Packaging Applications. *J. Food Protect.* **2010**, *73*(12), 225–262.
91. Incoronato, A. L.; Conte, A.; Buonocore, G. G.; Del Nobile, M. A. Agar Hydrogel with Silver Nanoparticles to Prolong Shelf Life of Fior di Latte Cheese. *J Dairy Sci.* **2011**, *94*(4), 1697–1704.
92. Lloret, E.; Picouet, P.; Fernandez, A. Matrix Effects on the Antimicrobial Capacity of Silver Based Nanocomposite Absorbing Materials. *Lebensm. Wiss. Technol.* **2012**, *49*(2), 333–338.
93. Bruna, J. E.; Penaloza, A.; Guarda, A.; Rodrıguez, F.; Galotto, M. J. Development of MtCu^{2+}/LDPE Nanocomposites with Antimicrobial Activity for Potential Use in Food Packaging. *Appl. Clay Sci.* **2012**, *58*, 79–87.
94. Djurdjevic, Z.; Radosavljevic, A.; Šiljegović, M.; Bibić, N.; Mišković-Stanković, V. B.; Kačarević-Popović, Z.; Structural and Optical Characteristics of Silver/Poly(*N*-vinyl-2-Pyrrolidone) Nanosystems Synthesized by g-Irradiation. *Radiat. Phys. Chem.* **2012**, *81*, 1720–1728.
95. Li, M.; Noriega-Trevino, M. E.; Nino-Martinez, N.; Marambio-Jones, C.; Wang, J.; Damoiseaux, R.; et al. Synergistic Bactericidal Activity of AgeTiO$_2$ nanoparticles in both Light and Dark Conditions. *Environ. Sci. Technol.* **2011**, *45*(20), 8989–8995.
96. de Azeredo, H. M. C. Antimicrobial Nanostructures in Food Packaging. *Trends Food Sci. Technol.* **2013**, *30*, 56–69.
97. Rhim, J.-W.; Hong, S.-I.; Ha, C.-S. Tensile, Water Vapor Barrier and Antimicrobial Properties of PLA/Nanoclay Composite Films. *Lebensm. Wiss. Technol.* **2009**, *42*, 612–617.
98. Roberts, L.; Lines, R.; Reddy, S.; Hay, J. Investigation of Polyviologens as Oxygen Indicators in Food Packaging. *Sens Actuators, B: Chem.* **2011**, *152*(1), 63–67.
99. Mills, A.; Hazafy, D. Nanocrystalline SnO$_2$-based, UVB-activated, Colourimetric Oxygen Indicator. *Sens. Actuators, B: Chem.* **2009**, *136*(2), 344–349.
100. Setkus, A. Heterogeneous Reaction Rate Based Description of the Response Kinetics in Metal Oxide Gas Sensors. *Sens. Actuators, B: Chem.* **2002**, *87*, 346–357.
101. Jurmanovi, S.; Kordi, S.; Steinberg, M. D.; Steinberg, I. M. Organically Modified Silicate Thin Films Doped with Colourimetric pH Indicators Methyl Red and Bromocresol Green as pH Responsive Sol–Gel Hybrid Materials. *Thin Solid Films* **2010**, *518*, 2234–2240.

102. Hurme, M. S. E.; Latva-Kala, K.; Luoma, T.; Alakomi, H.-L.; Ahvenainen, R. Myoglobin-based Indicators for the Evaluation of Freshness of Unmarinated Broiler Cuts. *Innov. Food Sci. Emerg. Technol.* **2002,** *3*(3), 279–288.
103. Nopwinyuwong, A.; Trevanich, S.; Suppakul, P. Development of a Novel Colorimetric Indicator Label for Monitoring Freshness of Intermediate-moisture Dessert Spoilage. *Talanta* **2010,** *81*(3), 1126–1132.
104. Heising, J. K.; Dekker, M.; Bartels, P. V.; van Boekel, M. A. J. S. A Non-destructive Ammonium Detection Method as Indicator for Freshness for Packed Fish: Application on Cod. *J. Food Eng.* **2012,** *110*(2), 254–261.
105. Jung, J.; Puligundla, P.; Ko, S. Proof-of-concept Study of Chitosan-based Carbon Dioxide Indicator for Food Packaging Applications. *Food Chem.* **2012,** *135*(4), 2170–2174.
106. Marek, P.; Velasco-Veléz, J. J.; Haas, T.; Doll, T.; Sadowski, G. Time-monitoring Sensor Based on Oxygen Diffusion in an Indicator/Polymer Matrix. *Sens. Actuators, B: Chem.* **2013,** *178*, 254–262.
107. Kuswandi, B.; Jayus; Restyana, A.; Abdullah, A.; Heng, L. Y.; Ahmad, M. A Novel Colorimetric Food Package Label for Fish Spoilage Based on Polyaniline Film. *Food Control* **2012,** *25*(1), 184–189.
108. Kerry, J. P.; O'Grady, M. N.; Hogan, S. A. Past, Current and Potential Utilisation of Active and Intelligent Packaging Systems for Meat and Muscle-based Products: A Review. *Meat Sci.* **2006,** *74*, 113–130.
109. Kerry, J. P.; Papkovsky, D. B. Development and Use of Nondestructive, Continuous Assessment, Chemical Oxygen Sensors in Packs Containing Oxygen Sensitive Foodstuffs. *Res. Adv. Food Sci.* **2002,** *3*, 121–140.
110. Trettnak, W.; Gruber, W.; Reininger, F.; Klimant, I. Recent Progress in Optical Sensor Instrumentation. *Sens. Actuators, B: Chem.* **1995,** *29*, 219–225.
111. Wolfbeis, O. S. Fibre Optic Chemical Sensors and Biosensors. CRC Press: Boca Raton, FL, 1991.
112. Alocilja, E. C.; Radke, S. M. Market analysis of biosensors for food safety. *Biosens. Bioelectron.* **2003,** *18*, 841–846.
113. Han, J. H. Antimicrobial Food Packaging. *Food Technol.* **2000,** *54*(3), 56–65.
114. Imran, H.; Revol-Junelles, A. M.; Martyn, A.; Tehrany, E. A.; Jacquot, M.; Lin-der, M.; Desobry, S. Active Food Packaging Evolution: Transformation from Micro- to Nanotechnology. *Crit. Rev. Food Sci.* **2010,** *50*, 799–821.
115. Clarinval, A. M.; Halleux, J. Classification of Biodegradable Polymers. In: *Biodegradable Polymers for Industrial Applications*; Smith, R. Ed.; Woodhead Publishing Ltd.: Cambridge, UK, 2005; pp 3–31.
116. Sorrentino, A.; Gorrasi, G.; Vittoria, V. Potential Perspectives of Bio-nanocomposites for Food Packaging Applications. *Trends Food Sci. Technol.* **2007,** *18*, 84–95.
117. Pandey, J. K.; Kumar, A. P.; Misra, M.; Mohanty, A. K.; Drzal, L. T.; Singh, R. P. Recent Advances in Biodegradable Nanocomposites. *J. Nanosci. Nanotechnol.* **2005,** *5*, 497–526.
118. Rhim, J. W.; Ng, P. K. Natural Biopolymer-based Nanocomposite Films for Packaging Applications. *Crit. Rev. Food Sci.* **2007,** *47*, 411–433.
119. Yang, K. K.; Wang, X. L.; Wang, Y. Z. Progress in Nanocomposite of Biodegradable Polymer. *J. Ind. Eng. Chem.* **2007,** *13*, 485–500.

120. Peterse, K.; Nielsen, P. V.; Bertelsen, G.; Lawther, M.; Olsen, M. B.; Nilsson, N. H.; Mortensen, G.; Potential of Biobased Materials for Food Packaging. *Trends Food Sci. Technol.* **1999**, *10*, 52–68.
121. Akbari, Z.; Ghomashchi, T.; Moghadam, S. Improvement in Food Packaging Industry with Biobased Nanocomposites. *Int. J. Food Eng.* **2007**, *3*(4), 3/1–24.
122. Arora, A.; Padua, G. W. Review: Nanocomposite in Food Packaging. *J Food Sci.* **2010**, *75*(1), R43–R49.
123. Scott, G. 'Green' Polymer Degradation and Stability. *Polymer* **2000**, *68*, 1–7.
124. Trznadel, M. Biodegradable Polymer Materials. *Int. Polym. Sci. Technol.* 1995, *22*, 58–65.
125. Huang, M.; Yu, J.; Ma, X. High Mechanical Performance MMT–Urea and Formamide-plasticized Thermoplastic Cornstarch Biodegradable Nanocomposites. *Carbohydr. Polym.* **2005**, *52*, 1–7.
126. Tang, X. G.; Kumar, P.; Alavi, S.; Sandeep, K. P. Recent Advances in Biopolymers and Biopolymer-based Nanocomposites for Food Packaging Materials. *Crit. Rev. Food Sci.* **2012**, *52*, 426–442.
127. Rhim, J. W. Effect of Clay Contents on Mechanical and Water Vapor Barrier Properties of Agar-based Nanocomposite Films. *Carbohydr. Polym.* **2011**, *86*, 691–699.
128. Rhim, J. W.; Lee, S. B.; Hong, S. I. Preparation and Characterization of Agar/Clay Nanocomposite Films: The Effect of Clay Type. *J. Food Sci.* **2011**, *76*(3), N40–N48.
129. Choudalakis, G.; Gotsis, A. D. Permeability of Polymer/Clay Nanocomposites: A Review. *Eur. Polym. J.* **2009**, *45*, 967–984.
130. Yano, K.; Usuki, A.; Okad, A. Synthesis and Properties of Polyimide–Clayhybrid Films. *J. Polym. Sci. A: Polym. Chem.* **1997**, *35*, 2289–2294.
131. Rhim, J. W.; Hong, S. I.; Park, H. M.; Ng, P. K. W. Preparation and Characterization of Chitosan-based Nanocomposite Films with Antimicrobial Activity. *J. Agric. Food. Chem.* **2006**, *54*, 5814–5822.
132. Rhim, J. W.; Hong, S. I.; Ha, C. S. Tensile, Water Barrier and Antimicrobial Properties of PLA/Nanoclay Composite Films. *Lebensm. Wiss. Technol.* **2009**, *42*, 612–617.
133. Tang, X.; Alavi, S.; Herald, T. J. Barrier and Mechanical Properties of Starch–Clay Nanocomposite Films. *Cereal Chem.* **2008**, *85*, 433–439.
134. Sothornvit, R.; Rhim, J. W.; Hong, S. I. Effect of Nano-clay Type on the Physical and Antimicrobial Properties of Whey Protein Isolate/Clay Composite Film. *J. Food Eng.* **2009**, *91*, 468–473.
135. Sothornvit, R.; Hong, S. I.; An, D. J.; Rhim, J. W. Effect of Clay Content on the Physical and Antimicrobial Properties of Whey Proteinisolate/Organo-clay Composite Films. *Lebensm. Wiss. Technol.* **2010**, *43*, 279–284.
136. Kumar, P.; Sandeep, K. P.; Alavi, S.; Truong, V. D.; Gorga, R. E. Preparation and Characterization of Bio-nanocomposite Films Based on Soy Protein Isolate and Montmorillonite Using Melt Extrusion. *J. Food Eng.* **2010**, *100*, 480–489.
137. Kumar, P.; Sandeep, K. P.; Alavi, S.; Truong, V. D.; Gorga, R. E. Effect of Type and Content of Modified Montmorillonite on the Structure and Properties of Bio-nanocomposite Films Based on Soy Protein Isolate and Montmorillonite. *J. Food Sci.* **2010**, *75*(5), 46–56.

138. Tunc, S.; Angellier, H.; Cahyana, Y.; Chalier, P.; Gontard, N.; Gastaldi, E. Functional Properties of Wheat Gluten/Montmorillonite Nanocomposite Films Processed by Casting. *J. Mater. Sci.* **2007,** *289*, 159–168.
139. Park, H. M.; Mohanty, A.; Misra, M.; Drzal, L. T. Green Nanocomposites from Cellulose Acetate Bioplastic and Clay: Effect of Eco-friendly Triethyl Citrate Plasticizer. *Biomacromolecules* **2004,** *5*, 2281–2288.
140. Nieddu, E.; Mazzucco, L.; Gentile, P.; Benko, T.; Balbo, V.; Mandrile, R.; Ciardelli, G. Preparation and Biodegradation of Clay Composite of PLA. *React. Funct. Polym.* **2009,** *69*, 371–379.
141. Bi, L.; Yang, L.; Narsimhan, G.; Bhunia, A. K.; Yao, Y. Designing Carbohydrate Nanoparticles for Prolonged Efficacy of Antimicrobial Peptide. *J. Controlled Release* **2011,** *150*, 150–156.
142. Yoksan, R.; Chirachanchai, S. Silver Nanoparticle-loaded Chitosan-starch Based Films: Fabrication and Evaluation of Tensile, Barrier and Antimicrobial Properties. *Mater. Sci. Eng., C* **2010,** *30*, 891–897.
143. Vimala, K.; Mohan, Y. M.; Sivudu, K. S.; Varaprasad, K.; Ravindra, S.; Reddy, N. N.; Padma, Y.; Sreedhar, B.; Mohana Raju, K. Fabrication of Porous Chitosan Films Impregnated with Silver Nanoparticles: A Facile Approach for Superior Antibacterial Application. *Colloids Surf., B* **2010,** *76*, 248–258.
144. Tripathi, S.; Mehrotra, G. K.; Dutta, P. K.; Chitosan-Silver Oxide Nanocomposite Film: Preparation and Antimicrobial Activity. *Bull. Mater. Sci.* **2011,** *34*, 29–35.
145. Maciel, V. B. V.; Yoshida, C. M. P.; Franco, T. T. Development of a Prototype of a Colourimetric Temperature Indicator for Monitoring Food Quality. *J. Food Eng.* **2012,** *111*(1), 21–27.

CHAPTER 9

NANOTECHNOLOGY IN WASTEWATER TREATMENT: A REVIEW

BAIS MADHURI*, S. P. SINGH, and R. D. BATRA

*Corresponding author. E-mail: madhuribais8@gmail.com

CONTENTS

Abstract	174
9.1 Introduction	174
9.2 Nanotechnolgy and Nanomaterials	175
9.3 Metal-Containing Nanoparticles	177
9.4 Zeolite Nanoparticles	178
9.5 Carbonaceous Nanomaterials	179
9.6 Conclusion	180
Keywords	181
References	181

ABSTRACT

Rapid pace of industrialization and its resulting by-products have affected the environment by producing hazardous wastes and poisonous gas fumes and smokes, which have been released to the environment. Nanotechnology has been extensively studied by researchers as it offers potential advantages like low cost, reuse, and high efficiency in removing and recovering the pollutants and has also proved to be one of the finest and advanced ways for wastewater treatment. Nanoparticles when used as adsorbents cause pollutant removal/separation from water, whereas nanoparticles used as catalysts for chemical or photochemical oxidation effect the destruction of contaminants. Present scientist evaluated the four classes of nanoscale materials that are functional materials for water purification: (1) dendrimers, (2) metal-containing nanoparticles, (3) zeolites, and (4) carbonaceous nanomaterials. Nanomaterials are effective for removal of metals, dyes, and pesticides from industrial wastewater and hence control the water pollution caused by textile and printing industries, and their convergence with current treatment technologies present great opportunities to revolutionize water and wastewater treatment.

9.1 INTRODUCTION

Water pollution is a major global problem that requires ongoing evaluation and revision of water resource policy at all levels. Water is typically referred to as polluted when it is impaired by anthropogenic contaminants and does not support a human use, such as drinking water. It also has impacts on economic and social costs. The rapid pace of industrialization and its resulting by-products have affected the environment by producing hazardous wastes and poisonous gas fumes and smokes, which have been released to the environment.[2] According to the 2013 Budapest Water Summit, it is expected that, by 2030, some 40% of the world's population will suffer from freshwater shortages. On the other hand, more than 50% of the total energy produced is estimated to be wasted in the form of untapped available energy and inefficient energy usage.[7] Conventional technologies have been used to treat all types of organic and toxic waste by adsorption and biological oxidation. There are various ways used commercially and noncommercially, method to fight this problem which is advancing day-by-day due to technological progress. Various physical, chemical, and biological treatment processes are used for wastewater treatment. Among these methods, currently, nanotechnology has been extensively studied by researchers as it offers potential advantages like low cost, reuse, and high efficiency in removing and recovering the pollutants and has also proved to be one of the finest and advanced ways for wastewater treatment. There are various reasons behind the success of nanotechnology and scientists are still working on further enhancement of its usage. Nanomaterials unique properties allow them to remove pollutants from the environment. Nanoparticles have very high absorbing, interacting, and reacting capabilities due to its small size with high proportion of atoms at surface. It can even be mixed with aqueous suspensions and thus can behave as colloid. Nanoparticles can achieve energy conservation due to its small size which can ultimately lead to cost savings. Nanoparticles have great advantage of treating water in depths and any location which is generally left out by other conventional technologies. They are used for softening of groundwater (reduction in water hardness), for removal of dissolved organic matter and trace pollutants from surface water, for wastewater treatment (removal of organic and inorganic pollutants and organic carbon), and for pretreatment in seawater desalination.

9.2 NANOTECHNOLGY AND NANOMATERIALS

The American Chemistry Council—Nanotechnology Panel has proposed the definition as an engineered nanomaterial is any intentionally produced material that has a size in 1, 2, or 3-dimensions of typically between 1 and 100 nm. Nanomaterials are manufactured materials with a structure between approximately 1 and 100 nm. Their unique physicochemical (e.g., size, shape) and surface (e.g., reactivity, conductivity) properties contribute to the development of materials with novel properties and technical solutions to problems that have been challenging to solve with conventional technologies.[12] The extremely small size of nanomaterial particles, typically in the range between 1 and 100 nm (billionth of a meter), creates a large surface area in relation to their volume, which makes them highly reactive, compared to non-nano forms of the same materials.

At nanoscale, material exhibits unique optical, magnetic electrical properties which have novel properties and functions because of their small and/or intermediate size, which are not seen in their conventional, bulk counterparts. Despite its lucrative applications in various fields, certain environmental and ethical concerns cloud the celebration of nanotechnology as the next technological boom.[17]

In terms of wastewater treatment, nanotechnology is applicable in detection and removal of various pollutants including heavy metal pollution that poses as a serious threat to environment because it is toxic to living organisms, including humans.

Nanoparticles when used as adsorbents cause pollutant removal/separation from water, whereas nanoparticles used as catalysts for chemical or photochemical oxidation effect the destruction of contaminants present.[20] Scientist evaluated the four classes of nanoscale materials that are functional materials for water purification: (1) dendrimers, (2) metal-containing nanoparticles, (3) zeolites, and (4) carbonaceous nanomaterials, which make them particularly attractive as separation and reactive media for water purification.[5]

Dendrimers are repetitively branched molecules. A dendrimer is typically symmetric around the core and often adopts a spherical three-dimensional morphology.[8] A dendron usually contains a single chemically addressable group called the focal point. Dendrimers are highly branched, star-shaped macromolecules with nanometer-scale dimensions.[1] Dendrimers are defined by three components: a central core, an interior dendritic structure (the branches), and an exterior surface with functional surface groups, neatly organized, hyper-branched polymer molecules that have end groups, core, and branches.[2] FeO/FeS nanocomposites that are synthesized with dendrimers as templates can be used for the construction of permeable reactive barriers for remediation of ground water. The varied combination of these components yields products of different shapes and sizes with shielded interior cores that are ideal candidates for applications in both biological and materials sciences. While the attached surface groups affect the solubility and chelation ability, the varied cores impart unique properties to the cavity size, absorption capacity, and capture-release characteristics.

Invention of dendritic polymers are providing opportunities to develop effective UF processes for purification of water contaminated by toxic metal ions, organic and inorganic solutes, and bacteria and viruses. Poly(amidoamine), or PAMAM, is a class of dendrimer which is made of repetitively branched subunits of amide and amine functionality. PAMAM belongs to the class of water-soluble polymers which is a criteria much needed for the agent in the treatment of water. They can act as flocculants for dye industry wastewater treatment. Diallo et al. (2005) tested the feasibility of PAMAM dendrimers with ethylene diamine core and terminal NH_2 groups to recover Cu(II) ions from aqueous solutions. On a mass basis, the Cu(II) binding capacities of the PAMAM dendrimers are much larger and more sensitive to solution pH than those of linear polymers with amine groups.[1]

Silica-based PAMAM dendrimer is found to have high surface functionality, which is very helpful in the adsorption of metal ions. Amine-terminated PAMAM dendrimers exhibit high affinity for adsorption of metal ions to their surface via coordination to the amine or the acid functionality. It is pH-independent in its action. Al of the ester and amino-terminated PAMAM dendrimer presented regularities in adsorption of metals like chromium, zinc, and iron. The adsorption of ester and the amino-terminated products increased with the increase in the grafting percentage and the addition of the surface functional groups.

9.3 METAL-CONTAINING NANOPARTICLES

Silver, iron, gold, titanium oxides, and iron oxides are some of the commonly used nanoscale metals and metal oxides cited by the researchers that can be used in environmental remediation. Silver nanoparticles, for example, have proved to be effective antimicrobial agents and can treat wastewater-containing bacteria, viruses, and fungi. Nanoscale titanium dioxide can also kill bacteria and disinfect water when activated by light. Nanosized metal oxides show great removal efficiency of heavy metal in wastewater, owing to their higher surface areas and much more surface active sites than bulk materials. But, it is very difficult to separate them from the wastewater due to their high surface energy and nanosize. So, many researchers turn to design polymer-based nanosorbents decomposition of organic compounds can be enhanced by noble metal doping into TiO_2 due to enhanced hydroxyl radical production and so forth. Metal oxide nanoparticles also can be impregnated onto the skeleton of activated carbon or other porous materials to achieve simultaneous removal of arsenic and organic co-contaminants, which favors point-of-use applications.[14]

The size and shape of NMs and NMOs are both important factors which affect their performance. Oxide nanoparticles can exhibit unique physical and chemical properties due to their limited size and a high density of corner or edge surface sites. They have high reactivity and photolytic properties.[4] They are considered good adsorbent for water purification because they have large surface area and their affinity can be increased by using various functionalized groups. Effectiveness of MgO nanoparticles and magnesium (Mg) nanoparticles as biocides against Gram-positive and Gram-negative bacteria (*Escherichia coli* and *Bacillus megaterium*) and bacterial spores (*Bacillus subtillus*) was demonstrated by Stoimenov and his colloquies in 2002.[22] Silver loaded nano-SiO_2 composite coated with cross-linked chitosan has high biocidal activity against *E. coli* and *Staphylococcus aureus*. Zinc oxide nanoparticles have been used to remove arsenic from water. Metal oxide nanocrystals can be compressed into porous pellets without significantly compromising their surface area when moderate pressure is applied.[14] Some adsorption processes for wastewater treatment have utilized ferrites and a variety of iron-containing minerals, such as hematite, lepidocrocite, and magnetite.

9.4 ZEOLITE NANOPARTICLES

Zeolite is a crystalline hydrated aluminosilicate of alkaline and earth metals. It is to point out that zeolites act as strong adsorbents and ion exchangers. Zeolites have a porous structure that can accommodate a wide variety of cations such as Na^+, K^+, Ca^{2+}, Mg^{2+}, and others. These positive ions are rather loosely held and can be exchanged for others in a contact solution. They can be acquired from natural sources or fabricated in laboratories. Synthetic zeolites are usually made from silicon–aluminum solutions or coal fly ash and are used as sorbents or ion-exchange media in cartridge or column filters. Zeolite nanoparticles can be prepared by laser-induced fragmentation of zeolite. Nanoparticles of zeolites are effective sorbents and ion-exchange media for metal ions which is evaluated as an ion-exchange media for the removal of heavy metals from acid-mine wastewaters. Nanoparticles embedded zeolite is prepared for better treatment of dioxin-contaminated water that can make potable one for regular use. Zeolites have been reportedly used in the removal of heavy metals such as Cr(III), Ni(II), Zn(II), Cu(II), and Cd(II) from metal electroplating and acid-mine wastewaters.

9.5 CARBONACEOUS NANOMATERIALS

As one of the inorganic materials, carbon-based nanomaterials[22] are used widely in the field for removal of heavy metals in recent decades, due to its nontoxicity and high sorption capacities. Activated carbon is used first as sorbents, but it is difficult to remove heavy metals at ppb levels. Then, with the development of nanotechnology, carbon nanotubes (CNTs), fullerene, and graphene are synthesized and used as nanosorbents. Carbonaceous nanomaterials can serve as high capacity and selective sorbents for organic solutes in aqueous solutions carbon materials are a class of significant and widely used engineering adsorbent. Carbon-based nanoparticles act as sorbents because they have high capacity and selectivity for organic solutes in aqueous solutions. CNTs are major building blocks of this new technology. CNTs have great potential as a novel type of adsorbent due to their unique properties such as chemical stability, mechanical and thermal stability, and the high surface area, which leads to various applications including hydrogen storage, protein purification, and water treatment. Removal of heavy metals from industrial wastewater leads to the biggest challenge nowadays. To reduce environmental problems, the CNTs are promising candidates for the adsorption of heavy metals.[11] The large specific surface areas, as well as the high chemical and thermal stabilities, make CNTs an attractive adsorbent in wastewater treatment. The adsorption properties of the CNTs to a series of toxic agents, such as lead, cadmium, and 1,2-dichlorobenzene have been studied and the results show that CNTs are excellent and effective adsorbent for eliminating these harmful media in water.[19] Multiwalled CNTs have been tested for the adsorption of coexisting contaminants, namely, 2,4,6-trichlorophenol and Cu(II). In particular, the large specific surface areas, as well as the high chemical and thermal stabilities, make CNTs an attractive adsorbent in wastewater treatment.

Activated carbon, CNTs, has studied for the sorption capacity, based on Co^{2+} and Cu^{2+}. The results show that carbon nanomaterials have significantly higher sorption efficiency comparing with activated carbons. Meanwhile, Stafiej and Pyrzynska[24] find solution conditions, including pH and metal ions concentrations, could affect the adsorption characteristics of CNTs, and the Freundlich adsorption model agree well with their experimental data. CNTs are rolled up graphene sheets with a quasi-one-dimensional structure of nanometer-scale diameter. In these last 20 years, CNTs have attracted much attention from physicists, chemists, material scientists, and electronic device engineers because of their excellent structural, electronic, optical, chemical, and mechanical properties.[16]

Graphene is another type carbon material as nanosorbent, which is a kind of one or several atomic layered graphites, possesses special two-dimensional structure, and good mechanical and thermal properties. Wang et al.[1] synthesized the few-layered graphene oxide nanosheets. These graphene nanosheets are used as sorbents for the removal of Cd^{2+} and Co^{2+} ions from aqueous solution; results indicate that heavy-metal ion sorption on nanosheets is dependent on pH and ionic strength, and the abundant oxygen-containing functional groups on the surfaces of grapheme oxide. Nanosheets played an important role on sorption. Kim et al.[31] reported magnetite–graphene adsorbents with a particle size of ~10 nm give a high binding capacity for As^{3+} and As^{5+}, and the results indicate that the high binding capacity is due to the increased adsorption sites in the graphene composite.

Just as graphene triggered a new gold rush, three-dimensional graphene-based macrostructures (3D GBM) have been recognized as one of the most promising strategies for bottom-up nanotechnology and become one of the most active research fields during the last 4 years. In general, the basic structural features of 3D GBM, including its large surface area, which enhances the opportunity to contact pollutants, and its well-defined porous structure, which facilitates the diffusion of pollutant molecules into the 3D structure, enables 3D GBM to be an ideal material for pollutant management due to its excellent capabilities and easy recyclability.[21]

9.6 CONCLUSION

Industrialization and population are the main reasons for increase in amount wastewater. Several methods are employed to ensure a sustained supply of water for the requisite purpose. Nanotechnology for water and wastewater treatment is gaining momentum globally. Nanotechnology is also being looked upon to provide an economical, convenient, and ecofriendly means of wastewater remediation. Nanomaterials are the backbone of the nanotechnology revolution. The efficiency of applications of nanotechnology for water purification should be the availability of the market that can provide large quantities of nanomaterials at economically reasonable price. Nanomaterials are effective for removal of metals, dyes, and pesticides from industrial wastewater and can also control water pollution caused by textile and printing industries. Convergence of nanotechnology with current treatment technologies presents great opportunities to revolutionize water and wastewater treatment.

KEYWORDS

- **nanoparticle**
- **zeolites**
- **nanomaterial**
- **dendrimers**
- **wastewater treatment**

REFERENCES

1. Tiwari, D. K.; Behari, J.; Sen, P. Application of Nanoparticles in Waste Water Treatment. *World Appl. Sci. J.* **2008,** 418.
2. Khin, M. M.; Sreekumaran, N.; Jagadeesh Babu, V.; Rajendiran, M. A Review on Nanomaterials for Environmental Remediation. *Energy Environ. Sci.* **2012,** 8094.
3. Pranjali, G.; Madathil, D.; Brijesh Nair, A. N. Nanotechnology in Waste Water Treatment—A Review. *Int. J. ChemTech Res.* **2013,** 2304.
4. Wang, X.; Yifei, G.; Li, Y.; Meihua, H.; Jing, Z.; Xiaoliang, C. Nanomaterials as Sorbents to Remove Heavy Metal Ions in Waste Water. *Treatment Environmental & Analytical Toxicology*, p. 2.
5. Tyagi, P. K.; Singh, R.; Vats, S.; Kumar, D.; Tyagi, S. Nanomaterials Use in Wastewater Treatment International Conference on Nanotechnology and Chemical Engineering (ICNCS' 2012) 21–22 December 2012, Bangkok, 66.
6. Pooi, L.; Xiaodong, C. Nanomaterials for Energy and Water Management.
7. Will, S. Absorbent Nanomaterials for Environmental Remediation.
8. Nora, S.; Diallo Mamadou, S. Nanomaterials and Water Purification: Opportunities and Challenges.
9. Qu, X.; Alvarez Pedro, J. J.; Li, Q. Applications of Nanotechnology in Water and Wastewater Treatment. *Water Res.* **2013,** *47*, 3932.
10. Kumar, M.; Yoshinori, A. Chemical Vapor Deposition of Carbon Nanotubes—A Review on Growth Mechanism and Mass Production. *J. Nanosci. Nanotechnol.* **2010,** *10*, 3739.

11. Mubarak, N. M.; Sahu, J. N.; Abdullah, E. C.; Jayakumar, N. S. Removal of Heavy Metals from Wastewater Using Carbon Nanotubes.
12. Kavaiya, A.; Kumar, R. Water pollution Prevention and Treatment using Nanotechnology.
13. Shrivastava, V. S. Metallic and Organic Nanomaterials and their Use in Pollution Control: A Review.
14. Synthesis and Applications of Carbon Nanotubes and Their Composites.
15. Physical and Chemical Properties of Carbon Nanotubes—*Satoru Suzuki*.
16. Wang, X.; Guo, Y.; Yang, L.; Han, M.; Zhao, J.; Cheng, X. Nanomaterials as Sorbents to Remove Heavy Metal Ions in Wastewater Treatment.
17. Rajesh, P. P.; Gomathi, S. P. E-waste Water Treatment Through Dendrimer Conjugated Magnetic Nanoparticles.
18. Li, H.; Zhao, Y. M.; Ahmad, W. B.; Zhu, Y. Q.; Peng, X. J.; Luan, Z. K. Carbon Nanotubes—The Promising Adsorbent in Wastewater Treatment.
19. Fernández, M.; Garcia, J. Metal Oxide Nanoparticles.
20. Yu, E. K.; Itkis, M.; Fedorov, A. V.; et al. Oxygen Reduction by Lithiated Graphene and Graphene-Based Materials.
21. Stoimenov, P. K.; Klinger, R. L.; Marchin, G. L.; Klabunde, K. J. Metal Oxide Nanoparticles as Bactericidal Agents. *Langmuir* **2002,** *18*, 6679–6686.
22. Lenz, A.; Selegård, L.; Söderlind, F.; Larsson, A.; Holtz, P.-O.; Uvdal, K.; Ojamäe, L.; Käll, P.-O. ZnO Nanoparticles Functionalized with Organic Acids: An Experimental and Quantum-Chemical Study

CHAPTER 10

A REVIEW ON PREPARATION OF CONDUCTIVE PAINTS WITH CNTs AS FILLERS

SAHITHI RAVULURI, MANSI KHANDELWAL, HARSHIT BAJPAI, G. S. BAJAD, and R. P. VIJAYAKUMAR[*]

[1]Visvesvaraya National Institute of Technology, Nagpur 440010, Maharashtra, India

[*]Corresponding author.

CONTENTS

Abstract		184
10.1	Introduction and Literature Review	184
10.2	Synthesis of Carbon Nanotubes	185
10.3	Chemical Vapor Deposition Technique	185
10.4	Functionalization	187
10.5	Dispersion of CNTs	190
10.6	Paint Preparation	195
10.7	Conducting Paints with CNTs	197
10.8	Test Methods for Electrical Resistivity of Liquid Paint	198
10.9	Spraying of Electrically Conductive Paints	200
10.10	Conclusion	201
Keywords		201
References		201

ABSTRACT

Solar paints or conducting paints have replaced the solar panels as it requires high installation and maintenance costs. Uses of cadmium and other nanoparticles as fillers to efficiently conduct solar energy have been extensively reported in the literature. However, these nanoparticles have adverse effect on the environment. In the present chapter, literature on carbon nanotube (CNT) synthesis and its use as fillers to induce conductive properties on paints is discussed in detail.

10.1 INTRODUCTION AND LITERATURE REVIEW

Solar paints or conducting paints have replaced the solar panels as it requires high installation and maintenance costs. Uses of cadmium and other nanoparticles as fillers to efficiently conduct solar energy have been extensively reported in the literature. However, these nanoparticles have adverse effect on the environment. In the present chapter, literature on carbon nanotube (CNT) synthesis and its use as fillers to induce conductive properties on paints is discussed in detail.

Conductive paints are prepared by dispersing the material which can induce conductive property in the paint matrix. The conductive paint should be economical, environmental friendly, and efficiently absorb the sunlight. The economic feasibility of solar energy utilization depends upon efficient collection, conversion, and storage. The efficient utilization of solar energy for various process applications requires the use of the collector systems which first capture as much as possible of incoming radiation and deliver a high fraction of the captured energy for propagation. The conversion efficiency of a collector system is limited by the thermal losses from the heated absorber due to conduction, convection, and radiation. The losses become increasingly significant at higher temperatures. Thus, the effective utilization of solar radiation can be achieved by an efficient and low cost "solar selective coating."[1]

The solar paints prepared using graphene and other materials which can absorb sunlight were reported by Matthew et al. Nanoparticles like cadmium were used as fillers for the preparation of conducting solar paints. However, such solar paint cannot be applied on all solid state devices.[2] Zhu et al. reported that the copper nanoparticle-filled CNTs may be used as fillers. CNTs are long cylinders of covalently bonded carbon atoms which possess extraordinary electronic and mechanical properties. CNTs are available in the form of single and multiwalled structures: single-wall carbon nanotubes (SWCNTs) which are the fundamental cylindrical structure and multiwall carbon nanotubes (MWCNTs) which are made of coaxial cylinders, having interlayer spacing close to that of the interlayer distance in graphite (0.34 nm). These cylindrical structures are only few nanometers in diameter, but length wise it can be tens of microns long with most end capped with half of a fullerene molecule.[3] The main objective of the present study is to understand the techniques of CNT synthesis and its dispersion techniques in the paint matrix to obtain conductive paints.

10.2 SYNTHESIS OF CARBON NANOTUBES

CNTs are prepared by three different methods such as arc discharge (C_{60} fullerenes are produced by this method which is perhaps the easiest way to produce CNTs) and laser ablation (an intense laser pulse is targeted to vaporize a carbon rod in presence of an inert gas). This method mainly produces single-walled CNTs and chemical vapor deposition (CVD) method. CVD technique is widely used as it is carried out at a much lower reaction temperature, and also good control over length and structure of the nanotubes is attainable. It also offers the advantage to adjust the reaction conditions to produce the desired CNTs (single-walled or multi-walled). CVD can be carried out with or without the presence of catalyst.[3] The techniques of CVD are discussed below in detail.

10.3 CHEMICAL VAPOR DEPOSITION TECHNIQUE

10.3.1 *CATALYTIC CHEMICAL VAPOR DEPOSITION*

In this technique, a mixture of hydrocarbons, metal catalyst along with inert gas, is introduced into the reaction chamber. At around 700–900°C and atmospheric pressure, nanotubes form on the catalyst substrate by the decomposition of the hydrocarbon.[4] These CNTs can be collected upon cooling the system to room temperature. In the case of a liquid hydrocarbon, liquid is heated in a flask and an inert gas (usually argon) is purged through it which in turn carries the hydrocarbon vapor into the reaction zone. In case of solid hydrocarbons, it can be kept in the low-temperature zone of the reaction tube. The volatile materials directly turn from solid to vapor and perform CVD, while passing over the catalyst kept in the high-temperature zone. Like CNT precursors, catalyst used in CVD can also be in any form (solid, liquid, or gas), which may be suitably placed in the reactor or externally fed. Alternatively, catalyst-coated substrates can be used to catalyze the CNT growth. The size of the catalyst metal particles used in this technique has influence on the diameter of the nanotubes. Two growth mechanisms can be expected from this technique (tip growth and base growth) depending upon the adhesion between the catalyst particles and the substrate (if it is weak, tip growth and if it is strong, base growth). It is well known that hydrocarbons can be easily broken at high temperatures (pyrolysis), but in the presence of suitable metal catalyst, hydrocarbons can be decomposed at lower temperatures (catalytic pyrolysis).

The key of CNT growth by CVD is to achieve hydrocarbon decomposition on the metal surface alone and prohibit spontaneous aerial pyrolysis and this is done by the proper selection of catalyst and hydrocarbon materials, vapor pressure of hydrocarbon, concentration of the catalyst, and the CVD reaction temperature.[5] For the use of CNTs to induce conductive properties in paints, CNTs are to be made conductive. This can be done by incorporating metallic conductive nanoparticles into the CNTs in a separate functionalization step (endohedral functionalization) or by directly using the metallic particles as a catalyst during the preparation of CNTs using CVD.[6] Zhu et al. reported the formation of bamboo-like CNTs using a copper foil by CVD from ethanol. They found that the yield and size of CNTs increases with increasing the temperature. The CNTs which were prepared at around 700°C had copper droplet tip and those at 800–900°C had a copper nanoparticle inside. They also observed that the enhancement of CNT growth with increase in the duration up to 30 min and prolonged duration up to 60 min did not shown any increase in the yield of CNTs to an appreciable level. It was observed and proposed that a carbon film first deposits on the top surface of the copper foil while the top surface of the copper foil partially melted and migrated across the carbon film where CNTs are formed.[4]

10.3.2 NON-CATALYTIC CHEMICAL VAPOR DEPOSITION

This method is a modification of the above technique, wherein uniform, well-aligned (MWCNTs grow in the channels of alumina porous oxide films. These alumina templates are commercially available or can be prepared in the laboratory by anodic oxidation of aluminum. By changing the process parameters, required thickness and pore diameters of the alumina films can be obtained which in turn affects the growth of CNTs in the pores. At high temperature (generally 700–900°C), on passing hydrocarbon gas into the tubular reactor which has an alumina template inside it, carbon deposits on the inner walls of the alumina template. After this process, the CNT-filled template is taken out of the tubular reactor and washed with hydrofluoric acid to separate the CNTs from the template. The resulting nanotubes are straight and have uniform diameter and thickness (thickness depends on the length of the deposition process).[7]

10.4 FUNCTIONALIZATION

The performance of CNT depends on the dispersion of CNTs in the matrix and interfacial interaction between the CNT and the matrix. The carbon atoms on the CNT walls are chemically stable because of the aromatic nature of the bond and as a result, the reinforcing CNTs are inert and can interact with the surrounding matrix mainly through van der Waals forces of attraction. Hence, the efficient load transfer is not possible across the CNT–matrix interface. It can be attained by the modification in the surface properties of the CNTs by either chemical or physical functionalization.[8]

10.4.1 CHEMICAL FUNCTIONALIZATION

It is based on the covalent linkage of functional entities and the carbon of CNTs. Direct covalent sidewall functionalization is associated with a change of hybridization from sp^2 to sp^3 and also the loss of π conjugation on the graphene layer. This can be obtained by reaction with some molecules of high chemical reactivity such as fluorine. It was reported that the fluorination of purified SWCNTs occurs at the temperatures up to 325°C and the process is reversible using anhydrous hydrazine which can remove the fluorine.

The fluorinated CNTs have C–F bonds that are weaker than those in alkyl fluorides and thus provide substitution sites for additional functionalization. Successful replacement of the fluorine atoms by amino, alkyl, and hydroxyl groups has been reported.[8]

Defect functionalization is another method for covalent functionalization of CNTs. An oxidative process using strong acids (such as nitric acid) or by using strong oxidants (such as $KMnO_4$) can induce defects on the side walls as well as open ends of the CNTs. These defects are stabilized by bonding with carboxylic acid or hydroxyl groups. These functional groups have rich chemistry and the CNTs can be used as precursors for further chemical reactions. The chemically functionalized CNTs have strong interfacial bonds with induced high mechanical and functional properties. However, this method has few drawbacks; first, during the functionalization reaction (especially in the ultrasonication process), a large number of defects are created on the sidewalls of CNTs and in some extreme cases, CNTs are fragmented into smaller pieces. This leads to decrease in the mechanical properties of CNTs as well as disruption of π electron system in the nanotubes. Second, the concentrated acids or oxidants used in covalent functionalization are not environment friendly.[9]

10.4.2 PHYSICAL FUNCTIONALIZATION

Physical functionalization has been put forward as an alternative method for tuning the interfacial properties of nanotubes. The physical functionalization techniques reported by Korneva are discussed below.

10.4.2.1 POLYMER WRAPPING

The suspension of CNTs in the presence of polymers leads to the wrapping of the polymer around the CNTs to form super molecular complexes of CNTs. This process is achieved through van der Waals interactions and π–π stacking between CNTs and polymer chains containing aromatic rings.

10.4.2.2 SURFACTANT ADSORPTION

Polymer surfactants have also been employed to functionalize CNTs. The effects of different types of surfactants (nonionic, anionic, and cationic) were studied. The physical adsorption of the surfactant on CNT surface lowered the surface tension of the CNT, effectively preventing the formation of aggregates and hence improving the dispersion. Moreover, the surfactant-treated CNTs overcome the van der Waals attraction by electrostatic or stearic repulsive forces. The efficiency of this method depends on the properties of surfactants, medium chemistry, and the base matrix. It was reported that cationic surfactants are favorable for water soluble polymers and for the water insoluble polymers, CNT dispersion was promoted by nonionic surfactant. The treatment of nonionic surfactants is based on strong hydrophobic attraction between the solid surface and the tail group of surfactant. Once the surfactant is adsorbed onto the filler surface, the surfactant molecules are self-assembled into micelles above a critical micelle concentration (CMC).[8]

10.4.2.3 ENDOHEDRAL METHOD

In this method, guest atoms or molecules are stored in the inner cavity of the CNTs by the capillary effect. The incorporation of guest atoms takes place at defect sites localized at the ends or on the sidewalls. The combination of these two materials (CNTs and guest molecules) is particularly useful to integrate the properties of the two components in hybrid materials for use in applications such as catalysis, energy storage, nanotechnology, and molecular scale devices. The filling of CNTs can be done in two ways.[7]

10.4.2.3.1 Filling by Exploiting the Phenomenon of Spontaneous Penetration

Capillary absorption is caused by extra pressure given by the Laplace equation of capillarity.

$$P_a - P_m = \frac{2\gamma\cos\theta}{R}$$

where P_a is the atmospheric pressure, P_m is the pressure under the meniscus, γ is the surface tension at the liquid air interface, θ is the liquid solid contact angle, and R is the inner radius of the nanotube.

From Laplace equation, we see that the difference $(P_a - P_m)$ will be positive whenever θ is less than 90° and negative when vice versa.

Since the meniscus forms spontaneously, this extra pressure will pull the liquid into the nanotube if $(P_a - P_m)$ is greater than zero, provided the pressure in the reservoir is atmospheric. In short, to fill the nanotubes with some liquid, the contact angle must be less than 90°. If the contact angle is greater than 90°, extra pressure must be applied to the liquid in order to impregnate the nanotube. The contact angle is related to the surface energies of the constituent materials as

$$\cos\theta = \frac{\gamma_{sv} - \gamma_{sl}}{\gamma_{lv}}$$

where γ represents surface energy and subscripts sv, sl, and lv represent solid vapor, solid liquid, and liquid vapor, respectively.

For liquids, the surface energy is exactly the surface tension, that is, the spontaneous force acting at the imaginary cut to resist the surface extinction. It was reported that the metal carbides and metal oxides fill the nanotubes and not the pure metals. The reason behind it is that the carbides and oxides of metals have a surface tension lower than the surface tension of pure metals. So, it was assumed that the nanotubes were filled with metal carbides. Benjamin et al. reported that the liquids with surface tension less than 180 m N/m can moisten the inner cavity of a nanotube at atmospheric pressure.

Alternatively, in the wet chemistry method, the nanotubes were first filled with metals salts and heated in furnace. The temperature of the furnace depends on the type of metal salt used. The metal salts decompose into their respective elements at suitable temperatures. It was found that 70% of the resultant nanotubes were filled with a very small amount of nanoparticles.[7]

10.5 DISPERSION OF CNTs

The commercially available CNTs are usually in the form of heavily entangled bundles, which gives difficulties in dispersion. Dispersion of CNTs is not only a geometrical problem (dealing with the length and size of the CNTs) but also relates to a method on how to separate individual CNTs from CNT agglomerates and stabilize them in polymer matrix to avoid secondary agglomeration.[8] Incorporation of CNTs into a matrix requires exceptionally large quantity of particles. The high aspect ratio of fillers gives difficulties in uniform dispersion of these particles.[10] Thus, the proper dispersion method is to be adopted which depends upon the matrix, in which, the CNTs are to be dispersed. The extent of dispersion may be obtained by using Raman spectroscopy, X-ray diffraction studies, etc.

10.5.1 CRITERIA FOR DISPERSION

During mixing, the filler aggregates are subjected to shear stresses from the medium with which they are mixed (e.g., paint matrix, solvent) and this creates local shear stresses that are responsible for dispersion. A mixing process can be interpreted as the delivery of a mechanical energy into the solution to separate the aggregates. The opposing factor to separation is ultimately the binding energy which holds the aggregates together and so the supplied energy (or the local energy density) from the chosen mixing technique should be greater than the binding energy of the CNT aggregates (the energy per local volume of the contact) and should be lower than the energy which can fracture a single nanotube.[10]

Two types of dispersion are mentioned – mechanical dispersion and methods which are designed to alter the surface energy of solids either physically (non-covalent treatment) or chemically (covalent treatment).[11] Chemical methods use surface functionalization of CNT to improve their chemical compatibility with the target medium (solvent or polymer solution/melt), enhancing wetting or adhesion characteristics and reducing their tendency to agglomerate. However, aggressive chemical functionalization, such as the use of neat acids at high temperatures, might introduce structural defects resulting in inferior properties for the tubes.[12] Non-covalent treatment is particularly attractive because of the possibility of adsorbing various groups on CNT surface without disturbing the π system of the graphene sheets. In the last few years, the non-covalent surface treatment by surfactants or polymers has been widely used in the preparation of both aqueous and organic solutions to obtain high-weight fraction of individually dispersed CNTs.[11]

10.5.2 MECHANICAL DISPERSION OF CNTs

The different mechanical dispersion techniques suggested by Peng et al. are ultrasonication, high shear mixer, calendaring, ball milling, stirring, and extrusion techniques.

Ultrasonication is the act of using ultrasound energy to agitate particles in a solution. It is usually achieved by using an ultrasonic bath or an ultrasonic probe. The principle behind ultrasonication is that when ultrasound waves propagate through series of compression, attenuated waves are induced in the medium through which the waves are passed. The production of such shock waves promotes the peeling of individual nanoparticles (mostly peel off the nanoparticles located at the outer part of the bundle) and thus results in the separation of individual nanoparticles from bundles. After the removal of external shear stress, the dispersed CNTs in solution would reconfigure themselves to a new equilibrium state of low energy through reaggregation (driving force for reaggregation is provided by the van der Waals forces of attraction). This process will take place unless surfactants are added to provide steric hindrance or static charge repulsion in order to stabilize the particles.[13] Literature studies show that the application of weak shear subsequent to a well-mixed state can significantly accelerate the reaggregation.[14] Sonication at higher rate to be avoided as it may lead to the damage of CNTs. This method is quite suitable to disperse CNTs in liquids having low viscosity.

A high shear mixer disperses a phase into a continuous phase with which it would generally be immiscible. It uses a rotating impeller or high-speed rotor or series of impellers to create shear and flow. Calendaring technique is generally used when the dispersion matrix is viscous. In this method, high shear stresses are applied to disentangle CNT bundles and to distribute them into the matrix, while a short residence time will limit the breakage of individual nanotubes. However, there are several concerns in using calendaring technique for CNT dispersion. For example, generally the minimum gap between the rollers is maintained about 1–5 μm, which is comparable to the lengths of CNTs, but is much larger than the diameter of individual CNTs. Such dimensional disparities between the roller gap and the dimensions of CNT may suggest that calendaring can better disperse the large agglomerated CNTs into smaller agglomerates at submicron level (although some individual CNTs may be disentangled out from the agglomerates). Ball milling is a grinding technique which is used to grind materials into extremely fine powder for use in paints, ceramics, etc. In this technique, a high pressure is generated locally because of the collisions between tiny and rigid balls in a concealed container. In stirring technique, the size, shape, and speed of the propeller control dispersion results. In this technique, CNTs tend to re-agglomerate as they are only separated but not stabilized during separation. The extrusion process consists of twin screws which rotate at high speed and creates a high shear forces which help in dispersion. This technique is particularly used for producing CNT nano-composites with high filler content.[8]

10.5.3 DISPERSING CNTs USING SURFACTANTS

Strano et al. proposed a mechanism of nanotube isolation from a bundle with combined assistance of ultrasonication and surfactant adsorption. The role of ultrasonic treatment is likely to provide high local shear forces, particularly to the nanotube bundle end. Once the spaces or gaps at the bundle ends are formed, they are propagated by surfactant adsorption, ultimately separating the individual nanotubes from the agglomerates. Generally, ionic surfactants are preferable for CNT/water soluble solutions and nonionic surfactants are proposed when organic solvents are used.

10.5.4 WATER-SOLUBLE DISPERSIONS OF CNTs

CNTs can be dispersed in water with the help of surfactants having relatively high HLB (hydrophyle–lyphophyle balance). This is a straightforward non-covalent method and is classically employed to disperse both organic and inorganic particles in aqueous solutions. The nature of surfactant, its concentration, and type of interaction are known to play a key role in the phase behavior of CNTs.

Among the ionic surfactants, SDS[15] and dodecyl-benzene sodium sulfonate (NaDDBS)[16] were commonly used to decrease CNT aggregation in water. The benzene ring along the surfactant is one of the main reasons for high dispersive efficiency of NaDDBS and even better efficiency of Dowfax surfactant (anionic alkyldiphenyl-oxide disulfonate), the latter having twice the charge of NaDDBS and a di-benzene group.[17] π-Stacking interactions of the benzene rings onto the CNT surface are believed to increase the adsorption ratio of surfactants as well as of other highly aromatic molecules and rigid conjugated polymers.[18]

According to the so-called unzipping mechanism,[19] a surfactant has to get into the small gaps between the bundle and the isolated tube, so as to prevent them from re-agglomerating. Surfactants with too bulky hydrophobic groups will be hindered to penetrate into the inter-tube region and show less de-bundling efficiency. However, bulky hydrophilic groups were reported to have an advantage in the case of nanotubes suspended with nonionic surfactants probably due to the improved steric stabilization provided by longer polymeric groups.[20]

10.5.5 DISPERSIONS OF CNTs IN ORGANIC SOLVENTS

Compared to water-soluble systems, limited research work has been carried out with surfactant-assisted dispersions in organic solvents. As opposed to aqueous solutions, hydrophobic CNTs are expected to be wetted by organic solvents, and therefore to less self-assemble in bundles. However, CNTs showed solubility only in a limited number of solvents, namely, DMF, dimethyl acetamide, and dimethyl pyrrolidone.[21] Unfortunately, immersion of SWCNT in DMF was found to damage the structure of CNT.[22]

10.5.6 CHARACTERIZATION OF CNT DISPERSION

Besides the difficulty in obtaining stable and homogeneous dispersions of CNTs, another complication is finding a valid method to evaluate their state of dispersion. Agglomerates of CNTs could be visualized directly or with the help of mechanical or electric response of CNT-based materials which indirectly indicate the state of filler dispersion.[12] Thus, if the dispersion is poor, the mechanical properties may decrease relatively to pure polymer. Visualization of CNT-based samples by optical microscopy enables to access mainly micrometer-sized bundles (agglomerates), while atomic force microscopy (AFM) is used to monitor suspended CNT at nanoscale level.[23] However, by AFM, one can probe only a few nanotubes at a time, which might not be the representative of whole sample. Imaging of CNT-based polymeric composites by scanning electron microscopy or TEM (transmittance electron microscopy) often demands pretreatment using gold or carbon sputtering or microtome slicing of the sample, which might cause a defect in the original pattern.

Solutions of CNTs are best viewed with cryo-TEM, which is ideally suited for imaging of samples which are wet.[20,24,25] Characterization of MWCNT suspensions by particle size analyzer based on dynamic light scattering technique was reported to be effective in indicating the bundle (agglomerate) size reduction with surfactant adsorption.[11] However, in that work, the measurement was given assuming the particles to be spherical, resulting in an average value for the length of the CNT particles. As a result, the reported dimensions of nanotubes exhibited average sizes in the range of micrometer, though the dynamic light scattering technique in principle can be applied to rod-type and disk-type particles.[26,27]

The discovery of nanotube fluorescence[15] has led to a precise method of detecting individual CNT dispersion.[19] A nanotube in an aligned bundle does not emit because of energy transfer to neighboring nanotubes in the dispersion, particularly to the metallic ones. Thus, the dispersion process can be monitored by examining transient fluorescent emission as a function of various parameters, like the type of surfactant used, sonication time, and surfactant concentration and functionalization. On various occasions, small angle neutron scattering, Raman spectroscopy, and size exclusion chromatography were applied to evaluate CNT dispersions. UV–vis scanning of the tube suspensions enabled to plot CNT concentration with increasing sediment time and to conclude on surfactant stabilization efficiency.

10.6 PAINT PREPARATION

Paint is pigmented liquid composition containing drying oils in combination with resins which after application to a substrate in a thin layer converts to a solid film. Paints are formulated according to their proposed use. General paint preparation requires binder, solvent, additives, pigments, and an extender.

Binder, commonly called the vehicle, is the film-forming component of paint. It holds the pigment particles distributed throughout the paint but it should not dissolve or affect the pigment in the paint. It is dispersed in a carrier (water or organic solvent either in molecular form or as colloidal dispersions). Binders provide good adhesion of the coating to the substrate and also strongly influence properties such as gloss, durability, flexibility, and toughness. These can be categorized based on the mechanism for drying or curing. Although drying may refer to evaporation of solvent or thinner, it usually refers to oxidative cross-linking of binders and equivalent to curing. Some paints are formed only by solvent evaporation but most rely on cross-linking process.[28] Examples of binders include synthetic or natural resins such as acrylics, epoxies, melamine resins, polyesters, etc. Polyurethanes would fall into the category where paints cure by polymerization by way of a chemical reaction and cure into a cross-linked film.[29]

Solvent serves to dissolve the polymer and adjust the viscosity of paint and it is also used to control flow and application properties and in some cases even affect the stability of paint when in liquid state. It does not become a part of paint film as it is volatile and evaporates during drying of paint when applied on substrate.

Pigments are granular solids which impart color, opacity, film cohesion, and sometimes corrosion inhibition to paint. Additives assist in improving the quality of paint and extenders are used in conjugation with pigments to extend the properties of the binders. Fillers impart toughness and texture, reduce cost of paint, and even help in incorporating special properties to paint.

Conventionally protective paint systems consist of primer, undercoat, and top coat. Depending on the properties to be induced in the paint matrix, the paint layers are modified accordingly. Primer wets the surface and provides good adhesion for subsequently applied coats. Undercoats are applied to build the total film thickness, the thicker the coating, the longer the life of the paint. These are specially designed to enhance the overall protection and when highly pigmented decrease permeability to oxygen and water. Finish coat provides appearance and surface resistance to the system. It must also provide the first line of defense against various conditions of exposure like sunlight, weather, etc.[30] Depending on the filler to be used in paint to obtain the necessary properties, the binder is chosen. To obtain uniformity of filler in paint, a dispersing agent is used whose selection depends on the properties of filler.

When CNTs are used as fillers, they need to be dispersed properly since they agglomerate more and hence, a suitable dispersing agent should be used which properly disperses CNTs and stabilizes them in the dispersed state. Selection of dispersing agent also depends on the type of binder (hydrophobic or hydrophilic). HLB number is used to predict the properties of dispersing agent and hence can be used as a tool to select it. This can be achieved by matching the dispersing agent's HLB number to that of the material being dispersed. (CNTs behave like hydrophobic particles and tend to clump together in liquids. The appropriate dispersant is chosen for the CNT and the paint medium [aqueous or organic] and the dispersant is dissolved into the paint medium to form solution.)[31]

In order to prepare a conductive paint with CNTs as fillers, the paint composition should be modified such that it carries the conductive properties of CNTs. Coupling agents and complexing agents can be added to paint matrix to serve the purpose. Complexing agent results in lower electrical resistivity of conductive paint composition. 2,4-Pentadione is the generally preferred complexing agent while preparing conductive paint using CNTs. Coupling agents provide uniform distribution of the conductive filler particles and enhance the flow characteristics, that is, reduce the viscosity of the paint. Titanate and silane-coupling agents are some of the preferred coupling agents.[32] Probe sonicator, high shear mixer, etc. are used for dispersing the surfactant (which disperses CNTs) with the paint matrix.

10.7 CONDUCTING PAINTS WITH CNTs

To prepare a conductive paint (with CNTs as fillers), first, CNTs are to be synthesized by chemical vapor deposition technique with the incorporation of copper nanoparticles. The addition of nanoparticles in CNTs is obtained by using a copper foil during the process of CNT synthesis.[4] Powdered form of CNTs are not preferred as it leads to health issues.[33] The as prepared CNTs are to be dispersed properly in a suitable solvent (using ultrasonicator). The solvent should be compatible with paint constituents and also should be environment friendly. The rate of ultrasonication to be controlled in such a way that the aspect ratio of CNTs (which give unique properties to it) should not be reduced due to the shear forces applied on CNTs. The prepared nano-fluid having CNT particles is to be properly mixed with paint (using ultrasonicator). With variation in volume fraction of CNTs in the paint matrix, its conductivity varies. Conductivity for the different volume fraction of CNTs is to be checked and the volume fractions to be optimized.

In another method, CNTs are to be prepared using suitable catalyst by CVD method and then obtained CNTs are to be functionalized using metal oxides (copper oxide) using endohedral method of functionalization. During this functionalization step, the metal oxide particles get filled in CNTs by capillary action (initially metal oxide particles are to be dispersed in suitable solvent then CNTs are to be added with suitable pressure so that the particles get incorporated in the CNTs through capillary action). Then, these functionalized CNTs are to be dispersed in suitable solvent using ultrasonicator. The extent of dispersion may be analyzed using Raman spectroscopy and the efficiency of prepared paint needs to be studied.

The methods of preparation of copper nanoparticle-filled CNTs are reported in literature. Copper nanoparticle-filled multiwalled CNTs can be synthesized by chemical reduction method using chemical vapor deposition technique and copper chloride. Cu/MWCNT-based nano-fluids are synthesized by dispersing nano-composites of Cu/MWCNTs in deionized water and ethylene glycol. A maximum thermal conductivity at a very low-volume fraction of Cu/MWCNTs can be obtained by homogeneous dispersion of Cu on the MWCNTs.[34]

The process of preparation of vertically aligned CNTs that are completely encapsulated by a dense network of copper nanoparticles involves the conformal deposition of pyrolytic carbon. These stabilized arrays can be functionalized using oxygen plasma treatment (to improve wettability) and then are infiltrated with an aqueous, supersaturated Cu salt solution. After drying, the salt forms a stabilizing crystal network throughout the array. The calcination and H_2 reduction leads to deposition of Cu nanoparticles on the CNT surfaces.[35]

10.8 TEST METHODS FOR ELECTRICAL RESISTIVITY OF LIQUID PAINT

Some standard methods are prescribed according to ASTMD 5682-95 to determine the specific resistance of liquid paints and other related materials in the range of 0.6–2640 MΩ cm. These methods, methods A and B, measure direct current through concentric cylinder electrodes immersed in a specimen (here, the liquid paint). Test method A describes a procedure for making resistance test with a commonly used paint application test assembly and test method B describes a procedure for making resistance test with a conductivity meter. Conductivity meter permits evaluation of liquid paints in the resistance range of 0.05–20 MΩ. Methods A and B are suitable for testing paints' compatibility with various electrostatic spray-coating applications. Contamination of specimen, high humidity, temperature conditions (resistivity varies with temperature), and electrification time (ions migration which causes current flow decreases with time) are likely to affect the test results. Very small amounts of water, acids, or polar solvents will lower the resistance of high resistivity solvents and paints.

Various standards are mentioned for testing paint properties like adhesion (tape test, X-cut) (ASTM D3359), flexibility (ASTM D522), impact resistance (ASTM D2794 and ASTM D4226), thickness (NIST standards), abrasion resistance (ASTM D4060), thermal shock, viscosity (ASTM D445, ASTM D5293 and ASTM D4684), and density (ISO 10012).

10.8.1 METHOD A

A paint application test assembly is designed to measure electrical resistance of all types of conductive paint formulations. The meter is provided with a dual range selection to provide greater accuracy in measuring low-resistance paints. The reagents and materials used for method A are low-resistivity cell constant standards, potassium chloride (1000 MΩ/cm^4), and cleaning solvents and solutions which are chosen on the basis of the paint tested. The probe must be dry and free from contaminants before and after tests. The probe must be standardized to measure the exact cell constant for maximum accuracy (a cell constant (K) of 132 may be used for routine measurements).

It is made sure that the probe is thoroughly cleaned before starting the test. The paint test probe is then inserted into the jack in the lower right side of the meter case of the test assembly and the scale set switch is set to scale B. The mode-select switch is adjusted to zero adjust position and the zero adjust knob is rotated until the dial indicator needle is centered on the adjust position. Now, the mode-select switch is moved to the paint test position. The probe is immersed vertically into a well-mixed specimen of the testing material until the holes at the bottom of the probe are submerged. The paint resistance is read from scale B. If it is less than 0.5 on scale B, the scale-select switch is moved to the scale A position and the mode-select switch is again adjusted to zero adjust position and the zero adjust knob is rotated until the dial indicator needle is centered on the adjust position and the paint resistance is read from scale A. The scale value at 10 s after immersion is read. Any slow drift that occurs after this time is ignored. The mega ohms reading on the tester is converted to resistivity in MΩ cm by multiplying by the constant K. The probe is cleaned thoroughly and the apparatus and specimen are allowed to stand for 1 h (the lids on the specimen should be tightly closed to prevent any loss of volatiles). The measurement is repeated making sure to remix the specimen. The result is reported as a mean value of the two measurements (ASTM D 5682-95).

10.8.2 METHOD B

The measurement of electrical resistivity of solvents and paint formulations is done by a conductivity meter in the resistance range of 0.05–20 MΩ. The probe must be standardized to measure the exact cell constant for maximum accuracy [a cell constant (K) of 132 may be used for routine measurements].

The probe is thoroughly cleaned before the tests and the measuring cable of the probe is connected to the socket in the back of the instrument. The measuring cell is immersed into a well-mixed specimen which should reach the two holes in the probe. Ten seconds after the measuring button is pressed, the measured value is displayed in mega ohms which is converted to specific resistivity in MΩ cm by multiplying by 132.5 cm. The probe is cleaned thoroughly and the apparatus and specimen are allowed to stand for 1 h. The measurement is repeated making sure to remix the specimen. The result is reported as a mean value of the two measurements (ASTM D 5682-95).

10.9 SPRAYING OF ELECTRICALLY CONDUCTIVE PAINTS

Conductive paints cannot be handled or applied in the same manner as conventional paints as these carry heavy metal fillers which tend to settle quickly for which continuous agitation is required. An air-driven mixture for the paint pot or a recirculation loop or both will help in keeping the paint constantly moving and prevent particles from settling down. If a recirculation loop is not provided, the paint tries to settle in the spray lines and these lines must be fully purged before spraying can resume. In high-volume spray applications, it is sometimes less expensive to use the equipment with a recirculation loop than to purge and dispose of conductive paint.

Another issue in spraying conductive paint may be "dry spray." Dry spray occurs when the paint does not level correctly, causing particles to not lie down. The potential cause of dry spray is that the percentage of solids is too high, and other is that too much solvent evaporates between atomization and contact with a substrate. This can be fixed by decreasing the distance between the spray nozzle and the substrate or by addition of more solvent to the paint.

Coating thickness is also an important factor to be considered when spraying conductive paints. The more conductive the filler, the thinner a coating required to achieve the paints' full shielding potential. Uniform thickness is significantly more important when spraying a conductive coating.[36]

10.10 CONCLUSION

CNTs with metallic particles show conductive properties and when they are used as fillers in paints, they tend to acquire those conductive properties and such paints can be used as solar paints. Various techniques of CNT preparation with metallic particles induced in them, functionalization techniques of CNTs, dispersion techniques of CNTs, paint preparation methods, testing methods for electrical resistivity of paint, and techniques of applying conductive paint to substrate have been discussed in this report. Thus, the suitable technique for the paint preparation with addition of CNTs is to be chosen for attaining the required properties.

KEYWORDS

- **Solar paint**
- **CNTs**
- **endohedral functionalization**
- **testing methods**

REFERENCES

1. Agnihotri, O. P.; Gupta, B. K. *Solar Selective Surfaces*; John Wiley & Sons, 1981; pp 159–168.
2. Genovese, M. P.; Lightcap, I. V.; Kamat, P. V. Sun-believable Solar Paint: A Transformative One-step Approach for Designing Nanocrystalline Solar Cells. *Am. Chem. Soc.* **2012,** *6,* 865–872.
3. Choudhary, V.; Gupta, A. *Carbon Nanotubes—Polymer Nanocomposites*; InTech Publishers, 2012; pp 65–67.

4. Zhu, J.; Jia, J.; Kwong, F.-L.; Ng, D. H. L. Synthesis of Bamboo Like Carbon Nanotubes on a Copper Foil by Catalytic Chemical Vapor Deposition from Ethanol. *Carbon* **2012**, *50*, 2504–2512.
5. Kumar, M.; Ando, Y. Chemical Vapor Deposition of Carbon Nanotubes: A Review on Growth Mechanism and Mass Production. *J. Nanosci. Nanotechnol.* **2010**, *10*, 3739–3758.
6. Bajad, G. S.; Tiwari, S.; Vijayakumar, R. P. Synthesis and Characterization of CNTs Using Polypropylene Waste as Precursor. *Mater. Sci. Eng. B* **2015**, *194*, 68–77.
7. Korneva, G. Functionalization of Carbon Nanotubes, A Thesis Submitted to the Faculty of Drexel University, 2008.
8. Ma, P.-C.; Siddiqui, N. A.; Marom, G.; Kim, J.-K. Dispersion and Functionalization of Carbon Nanotubes for Polymer-based Nanocomposites. *Composites A: Appl. Sci. Manuf.* **2010**, *41A*(10), 1345–1367.
9. Gebhardt, B. *Type Selective Functionalization of Single Walled Carbon Nanotubes*, 2012.
10. Huang, Y. Y.; Terentjev, E. M. A Review on Dispersion of Carbon Nanotubes: Mixing, Sonication, Stabilization and Composite Properties. *Polymers* **2012**, *4*(1), 275–295.
11. Vaisman, L.; Daniel Wagner, H.; Marom, G. The Role of Surfactants in Dispersion of Carbon Nanotubes. *Adv. Colloid Interface Sci.* **2006**, *128–130*, 37–46.
12. Hilding, J.; Grulke, E. A.; Zhang, Z. G.; Lockwood, F. J. *Dispers. Sci. Technol.* **2003**, *24*, 1.
13. Jogi, B. F.; Sawant, M.; Kulkarni, M.; Brahmankar, P. K. Dispersion and Performance Properties of Carbon Nanotubes (CNTs) Based Polymer Composites: A Review. *J. Encapsul. Adsorpt. Sci.* **2012**, *2*, 69–78.
14. Lin-Gibson, S.; Pathak, J. A.; Grulke, E. A.; Wang, H.; Hobbie, E. K. Elastic Flow Instability in Polymer-dispersed Nanotubes. *Phys. Rev. Lett.* **2004**, *92*, 048302.
15. O'Connell, M. J.; Bachilo, S. M.; Huffman, C. B.; Moore, V. C.; Strano, M. S.; Haroz, E. H. *Science* **2002**, *297*, 593.
16. Camponeschi, E.; Florkowski, B.; Vance, R.; Garett, G.; Garmestani, H.; Tannenbaum, R. *Langmuir* **2006**, *22*, 1858.
17. Tan, Y.; Resasco, D. E. *J. Phys. Chem. B* **2005**, *109*, 14454.
18. Chen, J.; Liu, H.; Weimer, W. A.; Halls, M. D.; Waldeck, D. H.; Walker, G. C. *J. Am. Chem. Soc.* **2002**, *124*, 9034.
19. Strano, M. S.; Moore, V. C.; Miller, M. K.; Allen, M. J.; Haroz, E. H.; Kittrell, C. *J. Nanosci. Nanotechnol.* **2003**, *3*, 81.
20. Moore, V. C.; Strano, M. S.; Haroz, E. H.; Hauge, R. H.; Smalley, R. E. *Nano Lett.* **2003**, *3*, 1379.
21. Kim, B.; Lee, Y.-H.; Ryu, J.-H.; Suh, K.-D. *Colloid Surf. A: Physicochem. Eng. Asp.* **2006**, *273*, 16.
22. Monthioux, M.; Smith, B. W.; Burteaux, B.; Claye, A.; Fisher, J. E.; Luzzi, D. E. *Carbon* **2001**, *39*, 1251.
23. Islam, M. F.; Rojas, E.; Bergey, D. M.; Johnson, A. T.; Yodh, A. G. *Nano Lett.* **2003**, *3*, 269.
24. Kedem, S.; Schmidt, J.; Paz, Y.; Cohen, Y. *Langmuir* **2005**, *21*, 5600.
25. Dror, Y.; Pyckhout-Hintzen, W.; Cohen, Y. *Macromolecules* **2005**, *38*, 7828.
26. Ivakhnenko, V.; Eremin, Y. J. *Q. Spectrosc. Radiat. Transf.* **2006**, *100*, 165.

27. Pecora, R. *Pure Appl. Chem.* **1984,** *56*, 1391.
28. Berendsen, A. M.; Berendsen, A. M. *Marine Painting Manual*. Graham & Trotman: London, 1989.
29. Kopeliovich, D. Composition of Paints.
30. Bayliss, D. A.; Chandler, K. A. *Steelwork Corrosion Control*. Elsevier Applied Science, 1991.
31. Griffin William, C. Classification of Surface-Active Agents by HLB. *J. Soc. Cosmet. Chem.* **1949,** *1*(5), 311–326.
32. Patent on Conductive Paint Composition, US 4490282 A, 1984.
33. Berkei, M. Conductive Coatings Using Carbon Nanotubes. *CHEManag. Eur.* **2011,** 7–8.
34. Jha, N.; Ramaprabhu, S. Synthesis and Thermal Conductivity of Copper Nanoparticle Decorated Multiwalled Carbon Nanotubes Based Nanofluids. *J. Phys. Chem. C* **2008,** *112*(25), 9315–9319.
35. Stano, K. L.; Chapla, R.; Carroll, M.; Nowak, J.; McCord, M.; Bradford, P. Copper-encapsulated Vertically Aligned Carbon Nanotube Arrays. *ACS Appl. Mater. Interfaces* **2013,** *5*(21), 10774–10781.
36. Hagar, J. *The Art of Spraying of Electrically Conductive Paint*, 2014.

PART III
Biopolymers and Their Applications

CHAPTER 11

REMOVING HEAVY METALS FROM INDUSTRIAL WASTEWATER USING ECONOMICALLY MODIFIED BIOPOLYMERS AND HYDROGEL ADSORBENTS

VEDANT MANOJKUMAR DANAK and
YASHAWANT P. BHALERAO

Department of Chemical Engineering, Shroff S. R. Rotary Institute of Technology (SRICT), Ankleshwar 393002, Gujarat, India

CONTENTS

Abstract		208
11.1	Introduction to Heavy Metals	208
11.2	Sources of Heavy Metals in the Industrial Wastewater	209
11.3	Mcl Standards for the Heavy Metals	210
11.4	Adsorption on Modified Biopolymers	211
11.5	Natural Biopolymers and Modified Natural Biopolymer Adsorbents	212
11.6	Modified Bio-Adsorbents Prepared by Chemical Reaction of Crosslinking Agent with the Polysaccharides	213
11.7	Advantages and Limitations of Chitosan-Based and Starch-Based Modified Biopolymer Adsorbents	215
11.8	Adsorption of Heavy Metals on Hydrogels	216
11.9	Future Scope for Research/Business	219
11.10	Conclusion	220
Keywords		221
References		221

ABSTRACT

In the present scenario, due to the discharge of large amount of metal-contaminated wastewater, heavy metals such as Cd, Cr, Cu, Ni, As, Pb, and Zn are the most hazardous among the chemical intensive industries. But here's the problem arises. Because of their high solubility in the aquatic environments, there is a chance that that it can be absorbed by the living organisms and they enter in the food chain, which may cause severe very serious health diseases. So, one can understand why there is a need of removing these heavy metals from the industrial wastewater before sending it to the environment. There are certain conventional treatment processes available such as chemical precipitation, ion exchange, etc., but they have the disadvantages like incomplete removal, high energy requirements, and the main disadvantage that limits their use is nothing but the production of sludge of higher toxicity.

11.1 INTRODUCTION TO HEAVY METALS

"Which heavy metals are found in industrial wastewater?"

"Why there is a need to remove those heavy metals?"

"Why the current technologies are facing problems?"

"Why the latest technologies/methods are most promising and economic?"

"What can I do to adopt these latest methods with my existing plant?"

"Will this latest development lead the nation ahead?"

These types of questions come to the minds of the Chemical Engineers and Environmental Engineers while discussing about the topic. This introductory chapter aims to address these and other fundamental questions about the need and importance of the research.

In the present scenario, due to the discharge of large amount of metal-contaminated wastewater, heavy metals such as Cd, Cr, Cu, Ni, As, Pb, and Zn are the most hazardous among the chemical intensive industries. But here's the problem arises. Because of their high solubility in the aquatic environments, there is a chance that that it can be absorbed by the living organisms and they enter in the food chain, which may cause severe very serious health diseases.[1] So, one can understand why there is a need of removing these heavy metals from the industrial wastewater before sending it to the environment. There are certain conventional treatment processes available such as chemical precipitation, ion exchange, etc., but they have the disadvantages like incomplete removal, high energy requirements, and the main disadvantage that limits their use is nothing but the production of sludge of higher toxicity.[2]

"Now what to do?"

"How to deal with this problem?"

"Don't you think there's a need of developing new methods for removing heavy metals from industrial wastewater?"

Well, to overcome the disadvantages of the conventional processes, many researchers worked on developing the methods/processes for removing heavy metals from industrial wastewater all around the world. Many processes are being invented for the same. But, from all the processes invented in the recent time, adsorption has become one of the major alternatives.[3] Now again, the need arises in selecting proper adsorbent for adsorbing heavy metals on it. One will have the questions like in the next section:

11.2 SOURCES OF HEAVY METALS IN THE INDUSTRIAL WASTEWATER

"Are the existing adsorbents are having low cost and eco-friendly?"

"Do all they have high metal-binding capacities?"

"Are all the adsorbents having organic or biological origin?"

Usually, these type of questions will come in to the mind of the readers while discussing about this process. But,

"Why we are more emphasized on the 'biopolymers'?"

"Can we use it directly?"

"Why the research is going on making modified biopolymers?"

This chapter presents an overview of the various modified biopolymers and hydrogel adsorbents which are invented in recent time. The maximum contaminant level (MCL) standards for the heavy metals in industrial wastewater are also mentioned. The advantages along with disadvantages of the various adsorbents in application are also discussed. The future scope of the use of certain modified biopolymers and hydrogel adsorbents is being discussed along with the case study on how to use these adsorbents with the existing plant in the industry and how it can be adopted with other wastewater treatment plants.

"What are the heavy metals?"

"From where do all the heavy metals come into the environment?"

The metals whose density exceeds 5 g/cm^3 are generally termed as heavy metals.

The industrial wastewater streams containing heavy metals may come into the environment[7] from different industries, such as

- electroplating industry;
- metal surface treatment process industry;
- Printed Circuit Board manufacturing industries;
- wood processing industry—where a chromated copper-arsenate wood treatment produces arsenic-containing wastes;
- inorganic pigment manufacturing industry that produces pigments containing chromium compounds and cadmium sulfide;
- petroleum-refining industries which generates conversion catalysts contaminated with nickel, vanadium, chromium; and
- photographic operations producing film with high concentrations of silver and ferro cyanides.

11.3 MCL STANDARDS FOR THE HEAVY METALS

United States Environmental Protection Agency (US EPA) has established certain MCL standards, which are summarized in Table 11.1.

TABLE 11.1 MCL Standards for the Heavy Metals by US EPA.

Heavy metals	MCL (mg/L)
Arsenic	0.05
Cadmium	0.01
Chromium	0.05
Copper	0.25
Nickel	0.2
Zinc	0.8
Lead	0.006
Mercury	0.00003

The Central Pollution Control Board of India (CPCB), under Environmental Protection Act (1986), in their notification based on 30 March 2012 defined limiting concentrations of heavy metals in wastewater, especially for the electroplating industries as shown in Table 11.2.[4]

TABLE 11.2 Limiting Concentration of Heavy Metals for Electroplating Industry Notified by CPCB of India.

Heavy metals	Limiting concentration (mg/L)
Nickel	3
Hexavalent chromium	0.1
Total chromium	2
Copper	3
Zinc	5
Lead	0.1
Cadmium	2
Total metal	10

11.4 ADSORPTION ON MODIFIED BIOPOLYMERS

"What is adsorption process?"

"How is it helpful in removing heavy metals from industrial wastewater?"

"Why biopolymers are used as adsorbent with some modification?"

"Why there is a need to modify the natural biopolymers?"

Well, you may find the answer of above questions in Section 11.4.

Adsorption is a mass transfer process by which a substance is transferred from liquid phase to the surface of the solid or adsorbent and becomes bound by chemical/physical interactions.[5]

Biopolymers are industrially attractive because

- Widely available and safe environmental point of view.
- Possess a number of different functional groups such as hydroxyl and amines, which helps in increasing the efficiency of metal ion uptake and also helps to maximize chemical loading possibility.

Then why it is necessary to modify the biopolymers?

It is because of the reason that the active binding sites of these biopolymers are not readily available for adsorption. Transport of metal contaminants to the binding sites plays a very important role in the process design. Therefore, it is necessary to provide the physical support and increase the accessibility of the metal-binding sites for process applications.[8]

11.5 NATURAL BIOPOLYMERS AND MODIFIED NATURAL BIOPOLYMER ADSORBENTS

Chitosan has the highest adsorption capacity for several metal ions.[9]

Chitin (2-acetamido-2-deoxy-b-D-glucose-(*N*-acetyl glucan)), which is a main structural components of insects, fungi, algae, and marine invertebrates like crabs and shrimps.[10–12] Worldwide, the solid waste from processing of shellfish, crabs, shrimps, and krill constitutes large amount of chitinaceous waste.

"Why chitosan is used as base of the modified adsorbent?"

"Why chitin is not being used as base of the modified adsorbent?"

Well, it is because of the fact that the chitosan chelates are in 5–6 times greater amount of metals than chitin.

11.5.1 ADSORPTION ON MODIFIED BIOPOLYMERS

11.5.1.1 CHITOSAN-BASED MODIFIED BIO-ADSORBENTS

1. Chitosan-coated palm oil shell charcoal: The combination of the useful properties of palm oil shell charcoal and that of natural chitosan could introduce a composite matrix with many applications and a very good adsorption capability. To overcome the mass transfer limitations, researchers have developed a chitosan-based modified bio-adsorbent by making a synthesized bio-adsorbent by coating chitosan on palm oil shell charcoal. Using synthetic wastewater, the successive removal of Cr from it was experimented (Fig. 11.1).

FIGURE 11.1 The figure demonstrates the pictorial view of palm oil shell, palm oil shell charcoal, and chitosan which is after all coated on the palm oil shell charcoal and can be used as adsorbent.

2. Nonporous glass beads immobilized with chitosan can also be used as adsorbent for the removal of heavy metals from the industrial wastewater.[13]

These materials have high adsorption capacity for Pb(II), Cr(III), Cd(II), and Hg(II).

11.5.2 STARCH-BASED MODIFIED BIO-ADSORBENT

The chemical modification of starch allows preparation of cyclodextrin that is also being used as an adsorbent for the removal of heavy metals from the industrial wastewater.

Modified bio-adsorbents prepared by chemical reactions. Cyclodextrin molecules are natural macrocyclic polymers, formed by the action of an enzyme on the starch.

The most characteristic features of cyclodextrin as a modified bio-adsorbent are as follows:

- The interior cavity of the molecule provides a relatively hydrophobic environment into which polar pollutant can be trapped.
- It has the ability to form inclusion compounds with various molecules, especially aromatics.

But, there are certain limitations of use of cyclodextrin as modified bio-adsorbent such as

It exhibits poor mechanical properties which limit its use in columns as it creates column fouling in the large columns; however, it can be used in the small columns.

11.6 MODIFIED BIO-ADSORBENTS PREPARED BY CHEMICAL REACTION OF CROSSLINKING AGENT WITH THE POLYSACCHARIDES

Through chemical reactions, particular with crosslinking and grafting reactions, polysaccharides can give interesting macromolecular superstructures, for example, gels and hydrogel network, polymeric resin, beads, membranes, fibers, and composite materials. These polysaccharide-based materials can then be used as adsorbents.

The various materials thus obtained are as tabulated in Table 11.3.

TABLE 11.3 Some selected examples of polysaccharide-based materials which are widely used in the wastewater treatment for removal of heavy metals.[15–19]

Polysaccharide	Crosslinking agent	Materials obtained	Pollutant
Chitosan	*GLA	Beads	Ni^{2+}, Cu^{2+}, Zn^{2+}
	*GLA	Membranes	Ni^{2+}, Cu^{2+}
	Benzoquinone	Beads	Cu^{2+}, Zn^{2+}
	EPI, GLA, **EDGE	Beads	Cu^{2+}
Starch	$POCl_3$	Beads	Pb^{2+}, Cu^{2+}, Cd^{2+}

*GLA, Glutaraldehyde; **EDGE, ethylene glycol diglycidyl ether.

Well, till now, what we have seen is about removing the heavy metals by adsorption process on modified biopolymers, but we haven't discussed about removing the selected heavy metals. Say you have an industry and you have the effluent coming out from the industry that contains only some selected heavy metals, provided that you want to only remove some selected heavy metals from your industrial wastewater then what you would do? Right?

"What if I want to remove particular metal ions from the industrial wastewater?"

"Is it possible to do so?"

So, in order to resolve that problem, throughout the world, many researchers have worked on that and are still working for getting the better adsorption capacity with having minimum cost too. The following research studies suggest the modified bio-adsorbents for removing particular heavy metals from the industrial wastewater:

- A cross-linked starch graft copolymers containing amine groups can be used as the adsorbents for the particular removal of Pb(II) and Cu(II) ions. Zhang and Chen[20] proposed in their research study that 2 h of adsorption tends to be sufficient for reaching the adsorption equilibrium.
- For removing several hundred ppm of divalent metal ions (Cu, Pb, Cd, Hg), Kim and Lim[19] proposed that an adsorbent prepared by crosslinking starch with $POCl_3$ can be used. In their studies, it was found that these divalent metal ions can be effectively removed from water by dispersing only 1% of the modified starch within a few minutes.

So, don't you think that this research study is helpful? Because it leads us toward the minimum amount of adsorbent required, less process time, and the main advantage is that this process is eco-friendly leading toward giving the safe environment to our loved ones!

But again, the problem arises as after the discussion, one will have a question in mind:

"Is the modified bio-adsorbent what we are using is cytotoxic?"

"If yes then, what is the solution?"

"If it is not cytotoxic, can I use it for the other application too?"

Well, Shyu[21,22] reported that the beads which are prepared by cross-linking reaction with GLA are cytotoxic. But he also suggested that instead of GLA, if we use tripolyphosphate, for preparing the chitosan beads then, the beads prepared are less cytotoxic than those cross-linked by GLA, and they also have the quick gelling ability, which helps in removing heavy metals from the industrial wastewater.

11.7 ADVANTAGES AND LIMITATIONS OF CHITOSAN-BASED AND STARCH-BASED MODIFIED BIOPOLYMER ADSORBENTS

11.7.1 CHITOSAN-BASED MODIFIED BIO-ADSORBENTS

Advantages

- Low-cost natural polymers;
- extremely cost-effective;
- environment friendly;
- outstanding metal-binding capacities;
- a high-quality-treated effluent can be obtained; and
- easy regeneration if required.

In the recent studies, it was found that a maximum of 88% of Cr(IV) was desorbed from the metal loaded chitosan with 0.1 M H_2SO_4 without any physical damage to the adsorbent.[27]

Limitations

- Nonporous adsorbents;
- pH-dependence; and
- requires chemical modifications to improve its performance.

11.7.2 STARCH-BASED MODIFIED BIO-ADSORBENTS

Advantages

- Very abundant natural biopolymer and widely available in many countries;
- economically attractive and feasible;
- remarkable high swelling capacity in water;
- amphiphilic cross-linked adsorbent; and
- easy to prepare with relatively inexpensive reagents.

Limitations

- Low surface area;
- its use in adsorption column is limited since the characteristics of particles introduce hydrodynamic limitations and column fouling.

Now, let us discuss about another new adsorbent, that is, hydrogels in the next upcoming section.

11.8 ADSORPTION OF HEAVY METALS ON HYDROGELS

"What is this hydrogels?"

"How they can be made?"

"Why they are helpful in removing heavy metals from industrial wastewater?"

"What are the advantages and limitations of using it?"

While discussing about the title of this topic, one will have the questions similar like as mentioned above. Right? Then, let us see in detail about the various hydrogels as an adsorbent. Hydrogel is a cross-linked hydrophilic polymer which is capable of expanding their volume due to high swelling in water. Various hydrogels are widely used in the wastewater treatment.

Figure 11.2 shows the schematic representation of polymerization/cross-linking reaction that results in three-dimensional network formation of cationic hydrogel.

Hydrogels have been defined as two or multicomponent system consisting of three-dimensional network of polymer chains and water, which fills the space between macromolecules.

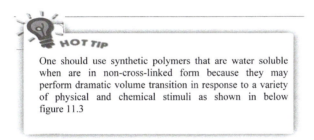

FIGURE 11.2 Three-dimensional network formation of cationic hydrogel.[23] (Reprinted with permission from Barakat, M. A.; Sahiner, N. Cationic Hydrogels for Toxic Arsenate Removal from Aqueous Environment. *J. Environ. Manage.* **2008**, *88*, 955–961. © 2008 Elsevier.)

In swollen state, mass fraction of water in a hydrogel is much higher than the mass fraction of polymer resulting in the increase of the efficiency of treating industrial wastewater-containing heavy metals.

> **HOT TIP**
> One should use synthetic polymers that are water soluble when are in non-cross-linked form because they may perform dramatic volume transition in response to a variety of physical and chemical stimuli as shown in below figure 11.3

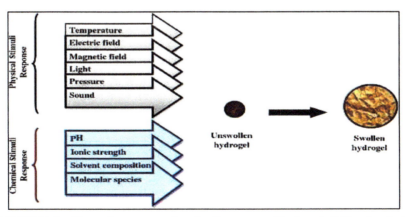

FIGURE 11.3 Figure demonstrates the stimuli response of hydrogels.[26] (Reprinted with permission from Ahmed, E. M. Hydrogel: Preparation, Characterization and Applications: A Review. *J. Adv. Res.* **2013**, *6*, 105–121. © 2013 Elsevier.)

Now, let us see the swelling and deswelling transition of pH-responsive hydrogel and temperature-responsive hydrogel, because the temperature and pressure are among the physical stimuli that we have earlier discussed and have effect on volume transition that can be understood by seeing Figure 11.4.

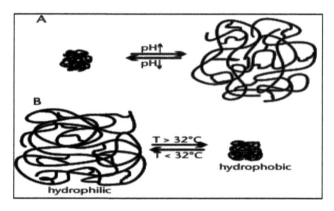

FIGURE 11.4 Schematic sketch of the swelling and deswelling transition of pH-responsive hydrogel (A) and temperature-responsive hydrogel (B).[28]

Poly(3-acrylamidopropyl) tri-methyl ammonium chloride hydrogel is prepared by Barakat and Sahiner,[23] in their studies, it was found that maximum binding capacity increases with pH increase to above 6.

Poly(ethylene glycol di-methacrylate-*co*-acrylamide) hydrogel beads are also in the use for treating industrial wastewater.[24]

Poly(vinyl pyrrolidone-*co*-methylacrylate) hydrogel is another example of hydrogel being used as an adsorbent.[25]

Now, let us see one case example on how to implement this modified bio-adsorbent/hydrogel in the existing equipment which are being used for treating industrial wastewater.

CASE EXAMPLE 11.1

Problem statement:

Assume that you are the owner of the effluent treatment plant (ETP), and your plant was established say 15 years ago. Now, as the new industries are developing rapidly day by day and to your effluent treatment plant, so as more amount of effluent will come to your ETP. On the effluent analysis you found that the common effluent is having the heavy metals. Now, as per the norms given by the government, the effluent should meet the standard for the MCL level for heavy metals, that is, your effluent should contain heavy metals within limit. So how would you proceed? How you can use these modified bio-adsorbent/hydrogel with your existing equipment in order to satisfy dual needs within one unit which should be financially feasible too?

Answer:

Let us say, in the ETP, many preliminary, secondary, and tertiary treatment process are being carried out. Now for our convenience, let us assume that we have only Fenton process, which is used to reduce the COD of the effluent, but as it is common effluent, the effluent would contain heavy metals and that needs to be removed as it is harmful to the environment as we have seen at the starting of this chapter. So, one can setup or modify the unit as shown in Figure 11.5. Isn't it so simple? Do you Agree or not?

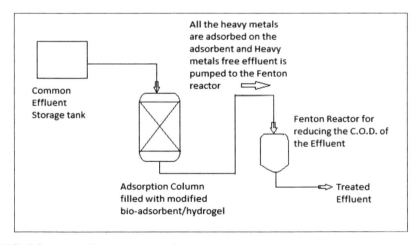

FIGURE 11.5   Effluent treatment plant.

11.9   FUTURE SCOPE FOR RESEARCH/BUSINESS

As we have seen in this chapter, you will find that less research work is done specially in the case of hydrogels, so one can do further research in this area. As far as modified bio-adsorbents are concerned, relatively more studies have been done, but still researchers are working on preparing modified bio-adsorbent which are less cytotoxic and have the required strength too and are having highest adsorption capacity but with a necessary condition cost-effectiveness and with an easy regeneration.

However, the price of chitosan in the global market is from around 2 to 300 USD depending upon which quality of chitosan you want because chitosan is not only used in the wastewater treatment but is also widely used in the drug delivery system. If an entrepreneur want to start his/her own business on making chitosan-based modified bio-adsorbent and has the proper location of the plant near seashores, one can earn so many profits with an advantage as many pharmaceutical industries are using chitosan, so one can also supply it to them.

Now, talking about technical aspects of this study, much work is necessary for better understanding of adsorption phenomenon and to demonstrate the possible technology at the industrial scale. As we imagine, the future trend use of these adsorbents, technical applicability, plant simplicity will play a major role in selection of this adsorption process for the removal of the heavy metals from the industrial wastewater.

11.10 CONCLUSION

As you have observed, over the time period of past two decades, the environmental regulations have become more stringent as improved quality of treated effluent is required for saving the environment and to make environment safe and maintain its flora and fauna and wholesomeness. So, after reading this chapter, you will definitely come to a conclusion that the modified bio-adsorbent and hydrogels are recently studied and the adsorption process is sustainable in removing heavy metals from heavy metal-contaminated industrial wastewater. On the basis of the advantages, we have seen the use of these modified bio-adsorbent will reduce the waste generation which is now all the industries are bearing as in the form of sludge produced after the treatment process. But this problem is now solved after studying this chapter. Right?

KEYWORDS

- heavy metals
- MCL standards
- adsorption
- modified biopolymer adsorbent
- hydrogel adsorbent
- application at the industrial scale

REFERENCES

1. Babel, S.; Kurniawan, T. A. Low-cost Adsorbents for Heavy Metals Uptake from Contaminated Water: A Review. *J. Hazard. Mater.* **2003,** *B97,* 219–243.
2. Eccles, H. Treatment of Metal-contaminated Wastes: Why Select a Biological Process? *Trends Biotechnol.,* **1999,** *17,* 462–465.
3. Leung, W. C.; Wong, M. F.; Chua, H.; Lo, W.; Leung, C. K. Removal and Recovery of Heavy Metals by Bacteria Isolated from Activated Sludge Treating Industrial Effluents and Municipal Wastewater. *Water Sci. Technol.* **2000,** *41*(12), 233–240.
4. Central Pollution Control Board, Govt. of India. Schedule-I: Standards for Emission or Discharge of Environmental Pollutants from Various Industries, Electroplating Industry. Notification dated on 30 March 2012.
5. Babel, S.; Kurniawan, T. A. Low-cost Adsorbents for Heavy Metals Uptake from Contaminated Water: A Review. *J. Hazard. Mater.* **2003,** *B97,* 219–243.
6. Babel, S.; Kurniawan, T. A. Various Treatment Technologies to Remove Arsenic and Mercury from Contaminated Groundwater: An Overview. In: Proceedings of the First International Symposium on Southeast Asian Water Environment, Bangkok, Thailand, 24–25 October, 2003; pp 433–440.
7. Sorme, L.; Lagerkvist, R. Sources of Heavy Metals in Urban Wastewater in Stockholm. *Sci. Total. Environ.* **2002,** *298,* 131–145.
8. Saifuddin; Nomanbhay, M., Palanisamy, K. Removal of Heavy Metal from Industrial Wastewater Using Chitosan Coated Oil Palm Shell Charcoal. *Electron. J. Biotechnol.* **2005,** *8*(1).
9. Deshpande, M. V. Enzymatic Degradation of Chitin and its Biological Applications. *J. Sci. Ind. Res.* **1986,** *45,* 277–281.
10. Olin, T. J.; Rosado, J. M.; Bailey, S. E.; Bricka, R. M. Low Cost Sorbents Screening and Engineering Analysis of Zeolite for Treatment of Metals Contaminated Water and Soil Extracts—Final Report. Report SERDP, 1996; 96-387, Prepared for USEPA and SERDP.
11. Bailey, S. E.; Olin, T. J.; Bricka, R. M.; Adrian, D. D. A Review of Potentially Low Costs Sorbents for Heavy Metals. *Water Res.* **1999,** *33*(11), 2469–2479.

12. Bailey, S. E.; Olin, T. J.; Bricka, R. M. Low-Cost Sorbents: Literature Summary. *Technical Report SERDP-97-1*, U.S. Army Engineer Waterways Experiment Station: Vicksburg, MS, 1997.
13. Liu, X. D.; Tokura, S.; Nishi, N.; Sakairi, N. A Novel Method for Immobilization of Chitosan onto Non-porous Glass Beads through a1,3-Thiazolidine Linker. *Polymer* **2003**, *44*, 1021–1026.
14. Yi, Y.; Wang, Y.; Liu, H. Preparation of New Crosslinked with Crown Ether and their Adsorption for Silver Ion for Antibacterial Activities. *Carbohydr. Polym.* **2003**, 53, 425–30.
15. Juang, R. S.; Shiau, R. C. Metal Removal from Aqueous Solutions Using Chitosan-enhanced Membrane Filtration. *J. Membr. Sci.* **2000**, *165*, 159–167.
16. Krajewska, B. Diffusion of Metal Ions through Gel Chitosan Membranes. *React. Funct. Polym.* **2001**, *47*, 37–47.
17. McAfee, B. J.; Gould, W. D.; Nadeau, J. C.; Da Costa, A. C. A. Biosorption of Metal Ions Using Chitosan, Chitin, and Biomass of *Rhizopus oryzae*. *Sep. Sci. Technol.* **2001**, *36*, 3207–3222.
18. Wan Ngah, W. S.; Endud, C. S.; Mayanar, R. Removal of Copper(II) Ions from Aqueous Solution onto Chitosan and Cross-linked Chitosan Beads. *React. Funct. Polym.* **2002**, *50*, 181–190.
19. Kim, B. S.; Lim, S. T. Removal of Heavy Metal Ions from Water by Cross-linked Carboxymethyl Corn Starch. *Carbohydr. Polym.* **1999**, *39*, 217–223.
20. Zhang, L. M.; Chen, D. Q. An Investigation of Adsorption of Lead(II) and Copper(II) Ions by Water-insoluble Starch Graft Copolymers. *Colloids Surf. A: Phys. Eng. Aspects* **2002**, *205*, 231–236.
21. Mi, F. L.; Sung, H. W.; Shyu, S. S.; Su, C. C.; Peng, C. K. Synthesis and Characterization of Biodegradable TPP/genipin Co-crosslinked Chitosan Gel Beads. *Polymer* **2003**, *44*, 6521–6530.
22. Lee, S. T.; Mi, F. L.; Shen, Y. J.; Shyu, S. S. Equilibrium and Kinetic Studies of Copper(II) Ion Uptake by Chitosan–Tripoly Phosphate Chelating Resin. *Polymer* **2001**, *42*, 1879–1892.
23. Barakat, M. A.; Sahiner, N. Cationic Hydrogels for Toxic Arsenate Removal from Aqueous Environment. *J. Environ. Manage.* **2008**, *88*, 955–961.
24. Kesenci, K.; Say, R.; Denizli, A. Removal of Heavy Metal Ions from Water by Using Poly(Ethylene Glycoldimethacrylate-*co*-acrylamide) Beads. *Eur. Polym. J.* **2002**, *38*, 1443–1448.
25. Essawy, H. A.; Ibrahim, H. S. Synthesis and Characterization of Poly(Vinylpyrrolidone-*co*-methylacrylate) Hydrogel for Removal and Recovery of Heavy Metal Ions from Wastewater. *React. Funct. Polym.* **2004**, *61*, 421–432.
26. Ahmed, E. M. Hydrogel: Preparation, Characterization and Applications: A Review. *J. Adv. Res.* **2013**, *6*, 105–121.
27. Bhuvaneshwari, S.; Sruthi, D.; Sivasubramanian, V.; Kanthimathy, K. Regeneration of Chitosan after Heavy Metal Sorption. *NISCAIR-CSIR, India* **2012**, *71*(04), 266–269.
28. Chen, Q.; Zhu, L.; Zhao, C.; Zheng, J. Hydrogels for Removal of Heavy Metals from Aqueous Solution. *J. Environ. Anal. Toxicol.* **2012**. DOI:10.4172/2161-0525.S2-001.
29. Crini, G. Recent Development in Polysaccharides Based Materials used as Adsorbents in Wastewater Treatment. *Prog. Polym. Sci.* **2005**, *30*, 38–70.

CHAPTER 12

DEVELOPMENT OF BIOPOLYMERS FOR DETERGENT

DHAKITE PRAVIN[1*], BURANDE BHARATI[2], and GOGTE BHALCHANDRA[3]

[1]*Department of Chemistry, S.N. Mor Arts, Commerce and Smt. G.D. Saraf Science College, Tumsar, Bhandara, India*

[2]*Department of Applied Chemistry, Priyadarshini Indira Gandhi College of Engineering, Nagpur, India*

[3]*Department of Oil Technology, Laxminarayan Institute of Technology, R.T.M. Nagpur University, Nagpur, India*

*Corresponding author. E-mail: pravinchemkb@gmail.com

CONTENTS

Abstract		224
12.1	Introduction	224
12.2	Experimental	225
12.3	Preparation of Biopolymer	226
12.4	Methods of Physicochemical Analysis	229
12.5	Preparation of Liquid Detergent	231
12.6	Result and Discussion	234
12.7	Conclusion	235
Keywords		236
References		236

ABSTRACT

In an age of increasing oil prices, global warming, and other environmental problems, the change from fossil feedstock to renewable resources can considerably contribute to a sustainable development in the future. Especially, plant-derived fats and oils bear a large potential for the substitution of currently used petrochemicals, fine chemicals can be derived from these resources in a straightforward fashion. The price and availability of the petroleum products is souring every day and alternatives resources have to be searched. At present, detergents industries are totally using petroleum based active ingredients such as linear alkyl benzene sulphonate and alpha olefin sulphonate in 20–30% in detergent formulations. Efforts are necessary to substitute these petroleum-based products by ecofriendly products such as coconut oil and rosin which is of vegetable origin so they can be labeled as biopolymer for green environment. Biopolymer is used as polymeric surfactants in liquid detergent formulations as partial as well as total substitute for linear alkyl benzene sulphonate and sodium lauryl sulphate. The special feature of these liquid detergents is freedom from conventional linear alkyl benzene sulphonate and sodium tripoly phosphate. Some formulations, which are technically excellent and yet cost-effective have been identified for commercial production.

12.1 INTRODUCTION

The petrochemical-based surfactant has somewhat tarnished image because of their past association with environmental pollution, foaming river, and eutrophication were often linked with surfactant. The new class of surfactant was introduced that is polymeric surfactant. Most of polymeric surfactants are biodegradable in nature and prepared from renewable sources.[1,2] The present work is attempt to synthesis biopolymer. The new concept of polymeric surfactant got popularity in last 25 years. The six important features that a polymer can deliver are calcium and magnesium sequestration, clay soil dispersancy, clay soil removal, calcium carbonate inhibition, and prevention of soil redeposition, fabric anti-incrustation.[3–5]

The molecular weight is important in case of biopolymer and has wide impact on detergency properties such as hardness of water, clay removal, percent detergency, surface tension foam volume, and pH. The M_w = 3500–4500 is ideal for getting excellent detergent properties.

In the present paper, the preparation of biopolymer and its characterization and usefulness in detergent formulation is discussed. The main aim of this paper is to focus on preparation of biopolymer based on vegetables oil.

Biopolymers are produce in closed kettle than the open kettles because in open kettle, there is a considerable loss of phthalic anhydride from sublimation. However, this loss is not serious in a properly closed kettle and does not occur at all in the solvent method.

12.2 EXPERIMENTAL

12.2.1 RAW MATERIAL ANALYSIS[6,7]

12.2.1.1 ANALYSIS OF COCONUT OIL

The analysis of coconut oil was done as per the ASTM Standard methods. The oil was analyzed for its acid value, hydroxyl value, saponification value, viscosity, iodine value, and specific gravity.

12.2.1.2 ANALYSIS OF ROSIN

Rosin was analyzed for its acid value by ASTM Standard method.

12.2.1.3 ANALYSIS OF SORBITOL

The sorbitol was analyzed for moisture content, percentage content of sulfated ash as per ASTM Standard methods.

12.2.1.4 ANALYSIS OF MALEIC ANHYDRIDE AND PHTHALIC ANHYDRIDE

These anhydrides were analyzed for their acid values using ASTM Standard method.

12.3 PREPARATION OF BIOPOLYMER[8]

Biopolymers were investigated in this research work by selecting different compositions based on coconut oil, sorbitol, phthalic anhydride, maleic anhydride, rosin, and benzoic acid.

12.3.1 COMPOSITION OF BIOPOLYMER

Compositions selected for the preparation of biopolymer are given in Table 12.1. The percentage of coconut oil was taken 16. The chain stoppered compound like benzoic acid and rosin was used. The content of maleic anhydride was taken 2.5% while phthalic anhydride was taken 5.0%.

TABLE 12.1 Composition of Novel Biopolymer (% by Weight).

Ingredient	Composition (wt%)
Coconut oil	16.00
Sorbitol	36.00
Phthalic anhydride	05.00
Maleic anhydride	02.50
Benzoic acid	02.50
Rosin	38.00

Catalyst used: 1.5% sodium bisulphate and 0.5% sodium bisulphite on weight of total mass.

12.3.2 STIOCHOMETRICAL CALCULATION FOR INDIVIDUAL INGREDIENTS IN

12.3.2.1 BIOPOLYMER

The functional groups (–OH and –COOH) present in the reactants of the batch of 100 g can be calculated as follows

12.3.3 CALCULATION FOR –OH GROUPS

a) Sorbitol

Mol. Wt. of sorbitol = 182.2
One mole sorbitol contains six –OH groups
No. of –OH groups in 360.0 g of sorbitol =06 × 360.0/182.2= 11.86.

12.3.4 CALCULATION FOR –COOH GROUP

a) Rosin

Mol. Wt. of rosin= 56,100/A.V.
Acid value of rosin is 166.5
Mol. Wt. of rosin=56,100/166.5 = 336.9
One mole rosin contains one –COOH group
No. of –COOH group in 380 g of rosin=1×380/336.9 = 1.128.

b) Maleic anhydride

Mol. Wt. of maleic anhydride=98
One mole of maleic anhydride contains two –COOH groups.
No. of –COOH group in 25.0 g of maleic anhydride =2 ×25.0/98 = 0.510.

c) Phthalic anhydride

Mol. Wt. of phthalic anhydride=148
One mole of phthalic anhydride contains two –COOH groups.
No. of –COOH group in 50.0 g of phthalic anhydride = 2 ×50.0/148 = 0.6756.

d) In benzoic acid

Mol. Wt. of benzoic acid=122.12
One mole of benzoic acid contains one –COOH group.
No. of –COOH group in 25 g of benzoic acid =1×25/122.12= 0.2047

Total No. of –COOH group = 2.518
Total No. of –OH group= 11.86.

12.3.5 THE REACTOR

The preparation of biopolymer was carried out in a glass reactor. The reactor consists of two parts. Lower part of the rector is a round bottom vessel with very wide mouth. The capacity of the flask is about 2 L. The upper part of the reactor is its lid, having four necks with standard joints. A motor-driven stirrer was inserted in the reactor through the central neck, while another neck was used for thermometer. A condenser was fitted with the reactor through the third neck. And the fourth neck was used for dropping the chemicals in to the reactor. The reactor was heated by electric heating mantle having special arrangement for smooth control of the temperature of the reactor. The speed of the stirrer was controlled by a regulator. The reaction vessel and its lid were tied together with the help of clamps.

12.3.6 REACTOR PROGRAMING

The flow chart representation has been given in Table 12.2.

TABLE 12.2 Heating Schedule of Novel Biopolymer.

Order of addition of ingredient	Temp. (°C)	Time of heating (h:min)
coconut oil, sorbitol, rosin		
Malelic anhydride, benzoic	200	0.2:00
Acid, sod. bisulphate, and sod.		
Bisulphite		
↓		
Lowering the temp.	120	
↓		
Phthalic anhydride	230–240	03:00
↓		
Cool down temp.	200–210	02:00

First Step:

Initially coconut oil, sorbitol, rosin, maleic anhydride, benzoic acid, and catalyst were taken in glass reactor. The mass was heated slowly and steadily to 200°C in about half an hour.

Second Step:

The reaction was carried steadily for 2 h. Now the temperature was brought down to 120°C. Now, phthalic anhydride was added steadily and (3:1) xylene:butanol solvent by weight was added at this lower temperature. Now temperature was raised slowly to 230–240°C. The reaction was continued for 3 h.

Third Step:

The reaction was continued further for 2 h at a lower temperature of 200–210°C. The batch was taken out after lowering the temperature to 120°C (benzoic acid was added as a chain stopper because it is difficult to control the reaction at such a lower oil length of 16%).

Fourth Step:

Acid value and viscosity is observed periodically and reaction is terminated when desired acid value and viscosity is attained. Total water removed is measured. Batch is withdrawn and weighted to find out % yield.

12.4 METHODS OF PHYSICOCHEMICAL ANALYSIS

12.4.1 ACID VALUE

The no. of milligrams of KOH required for neutralization of 1 g of material under consideration of test.

Acid value of the samples was determined by ASTM Standard method, using following formula.

$$\text{Acid value} = \frac{56.1 \times V \times N}{W}$$

where, V = volume of alcoholic KOH solution

N = normality of alcoholic KOH solution

W = weight in g of biopolymer.

12.4.2 SAPONIFICATION VALUE

The saponification value is the amount of alkali necessary to saponify a definite quality of the sample. It is expressed as the number of milligrams of KOH required to saponify 1 g of the sample.

$$\text{Saponification value} = \frac{56.1 \times (B-S) \times N}{\text{Weight of the sample}}$$

where B = blank titration reading
S = sample titration reading
N = Normality of 0.5 N HCl.

12.4.3 VOLATILE CONTENT

It was determined by weighing out 2–3 g of biopolymer sample into a petri dish and heating for about 3 h. At 105°C, it was cooled, weighed, and reheated for further half an hour to check that the weight was constant. The volatile content is expressed as a percentage of the original polymer.

12.4.4 SOLID CONTENT

The percentage of solid in the biopolymer was calculated as follows:

% Solid = (100 − Volatile content).

12.4.5 DETERMINATION OF MOLECULAR WEIGHT VISCOSITY-AVERAGE METHOD[9–11]

This method depends upon the principle that the limiting viscosity number that is proportional to molecular weight. The molecular weight is related with limiting viscosity number (n), specific viscosity (nsp), and ratio of amount of nonvolatile to the amount of volatile (C).

12.4.6 DETERMINATION OF HYDROPHILIC LIPOPHILIC BALANCE[12,13]

It is defined as the ratio of hydrophilic group to hydrophobic group. HLB of the sorbitol-based polymer is calculated by the saponification method. This method includes finding out the saponification number of a polymer and acid number of acid present in the reaction mass. The value in substitutes in the formula given by Griffin (1954) is as follows:

The HLB is calculated by using saponification method describe by Griffin (1949) by the formula.

$$\text{HLB} = 20 \times \left(1 - \frac{\text{SV}}{\text{AV}}\right)$$

where SV is the saponification value of the polymer and AV is the acid value of the raw material.

12.5 PREPARATION OF LIQUID DETERGENT

The compositions of selected liquid detergent are given in Table 12.3. Required amount of novel biopolymer and other ingredient like sodium lauryl sulphate, sorbitol foam booster sodium sulphate, and urea were taken in a 500-ml beaker and after being homogenized by running the stirrer for about half an hour, a clear solution of liquid detergent was obtained after 1 h. This clear liquid solution was filtered and packed in superior grade air-tight container.

TABLE 12.3 Compositions of Liquid Laundry Detergents Based on Novel Biopolymer.

Sr. No.	Ingredients	LD1	LD2	LD3	LD4	LD5
1	Neutralized acid slurry (75% solid)	4.8	3.6	2.4	1.2	0.0
2	Sodium lauryl sulphate	5.9	4.5	3.1	1.55	0.0
3	Alpha olefin sulphonate (76% solid)	7.1	7.1	7.1	7.1	7.1
4	Sodium sulphate	5	5.0	5.0	5.0	5.0
5	Urea	2.5	2.5	2.5	2.5	2.5
6	Sodium lauryl ether sulphate (71% solid)	10	10.0	10.0	10.0	10.0
7	Sorbitol (70% solid)	7.0	7.0	7.0	7.0	7.0
8	Neutralized novel biopolymer 80% solid	2.4	4.8	7.2	9.6	12.00
9	Distilled water	55.3	55.5	55.7	56.05	56.4

TABLE 12.4 Physico-chemical Properties of Novel Biopolymer.

Sr. No.	Physical property	Observation
1	Acid value of polymer	69.62
2	pH value	1.21
3	Saponification (Sap) value	198.4
4	Solid (%)	71.84
5	Color	Dark brown
6	Consistency	Thick
7	Solubility of biopolymer:	
	(i) In hot water	Partially soluble
	(ii) In alcohol	Soluble
	(iii) In xylene:butanol	Soluble
	(iv) In NaOH	Soluble
8	HLB ratio of biopolymer	18.31
9	Molecular weight of biopolymer	3241
10	Viscosity (by Ford Cup Method) (70:30)	160 s

TABLE 12.5 Analysis of Laundry Liquid Detergents.[14]

Sr. No.	Sample	%Solid	pH
1	**LD1**	44.7	9.36
2	**LD2**	44.5	8.14
3	**LD3**	44.3	9.46
4	**LD4**	44.1	7.87
5	**LD5**	43.9	8.94
6	**CD1**	46.80	7.54
7	**CD2**	45.14	7.98

70% KOH solution was prepared for neutralization of polymer and for adjusting pH.

TABLE 12.6 Physiochemical Properties[14] of Liquid Laundry Detergents Based on Novel Biopolymer.

Property	Conc. (%)	LD1	LD2	LD3	LD4	LD5	CD1	CD2
Foam volume (cm³) (cylinder method)	5	700	600	550	450	500	500	400
	1	800	700	650	550	600	600	500
Surface tension(dyne/cm) Stalagnometer test[10]	5	39.56	40.21	40.25	41.65	42.36	42.85	43.89
	1	36.52	37.89	38.52	38.96	39.12	39.65	41.48

LD, Liquid detergent; CD, commercial detergent.

TABLE 12.7 Stain Removing Properties[14] on Cotton, Polyester, and Terricot Sample.

Sample	Polyester cloths				Terricot cloths				Cotton cloths			
	Soil	Tea	Coffee	Spinach	Soil	Tea	Coffee	Spinach	Soil	Tea	Coffee	Spinach
LD1	85.96	76.05	87.50	88.23	86.68	76.81	91.42	88.13	86.56	79.68	86.95	95.58
LD2	73.68	74.84	78.12	80.58	81.96	71.01	82.85	79.66	79.10	71.87	79.71	89.70
LD3	82.45	67.60	82.81	82.35	85.24	69.56	85.71	76.27	83.58	75.00	81.15	89.70
LD4	75.43	71.83	81.25	79.41	81.96	71.01	85.71	74.57	79.10	70.31	76.81	85.29
LD5	82.45	66.78	82.81	82.35	81.96	66.67	82.65	74.57	83.58	73.43	79.71	88.23
CD1	89.47	80.28	87.50	86.76	93.44	79.71	90.00	83.05	89.55	64.06	85.60	93.54
CD2	77.19	76.05	85.93	79.41	81.96	76.81	88.57	74.57	80.59	60.93	82.60	83.87

Note: Stains of soil, tea, coffee, and spinach were tested at 1% concentration in distilled water applied as per standards methods. R_0 = reflectance measured on clean cotton, polyester, terricot cloths are 100; R_s = reflectance measured on soil-stained cotton, polyester, and terricot cloths are 33, 43, and 39, respectively; R_s = reflectance measured on tea-stained cotton, polyester, and terricot cloths are 36, 29, and 31, respectively; R_s = reflectance measured on coffee-stained cotton, polyester, and terricot cloths are 31, 36, and 30, respectively. R_s = reflectance measured on spinach-stained cotton, polyester, and terricot cotton cloths are 38, 32, and 41, respectively.

12.6 RESULT AND DISCUSSION

1. A 2-kg batch of novel biopolymer has been prepared based on coconut oil, sorbitol, maleic anhydride, phthalic anhydride, rosin, and benzoic acid. The proportion of benzoic acid (chain stopper) and catalyst sodium bisulphate and sodium bisulphite has been maintained constant.
2. The synthesis was carried out in glass reactor by batch process. The time of heating and order of addition of various ingredients has been planned as given in Table 12.2. After 1.5 h, a small quantity of water was added to maintain paste-like consistency of the mass. The reaction conditions have been standardized to get molecular weight of 3000–3500. The total time of heating was 5 h and 30 min. The heating temperature range was 120–150°C.
3. The composition of polymer and physicochemical analysis is given in Table12.1 and 12.4, respectively.
4. An ingredient like rosin and coconut oil which are of natural origin fulfills the demands of the biodegradable polymers. In our country, vegetable-based polyol like sorbitol is abundantly available and cheap. Our detergent industry is totally dependent on petroleum-based products like linear alkyl benzene sulphonate, so in this research work, we have tried to develop polymeric surfactant based on sorbitol and natural ingredients.
5. In first liquid detergent compositions LD-1, combination of acid slurry, sodium lauryl sulphate, alpha olefin sulphonate, and sodium lauryl ether sulphate has been used. In successive progressive compositions LD-2 to LD-7, acid slurry and sodium lauryl sulphate has been progressively replaced by selected novel biopolymer so that the final compositions are free from soft acid slurry and sodium lauryl sulphate and contain 12% novel biopolymer.
6. The proportion of alpha olefin sulphonate and sodium lauryl ether sulphate has been maintained at a constant level of 7.1% and 10%, respectively. A small proportion of urea and sodium sulphate has been incorporated.
7. The desired pH of 7.5–9 has been active by neutralized of sample with 30% KOH solution.

Development of Biopolymers for Detergent   235

8. The sample is free from sodium tripolyphosphate and harsh alkali sodium carbonate and builders and fillers which unnecessarily increase the bulk weight of detergent and create pollution.
9. A small proportion of sorbitol (7%) has been incorporated for clearly, homogeneity, and smooth feel of liquid detergent. In all the samples about 55–56%, water has been incorporate.

12.7 CONCLUSION

1. The novel biopolymer based on coconut oil, sorbitol, maleic anhydride, phthalic anhydride, and benzoic acid has been synthesized. The mole ratio, temperature, stirring, the proportion of catalyst, time of heating, and order of addition of ingredients has been standardized to get novel renewable polymer in molecular weight ranges from 3000 to 3500 (Table 12.2).
2. The pH of the sample is acidic. The HLB ratio, molecular weight, and viscosity of sample suggest that they can be used with advantage for making liquid detergent.
3. In formulation of liquid detergents, constant properties of sodium lauryl ether sulphate (10%) and alpha olefin sulphonate (7%) have been used. In the first composition, acid slurry (6.0%) and sodium lauryl sulphate (7.3%) have been used. Both these ingredients are successively replaced by 2.8–14.2% novel biopolymer. In the final compositions, acid slurry and sodium lauryl sulphate have been totally replaced by novel biopolymer.
4. In India, the common people are more prone to use powder and cake detergent. The use of liquid detergent should be promoted to avoid pollution. Many types of filler, silicates, and sodium tripolyphosphates are used in these properties. The liquid detergent is practically free from these ingredients.
5. The main barrier in using liquid detergents is their high cost and psychologies of the user. Our products based on novel raw materials are quite cheap. To give cost estimations, our liquid detergent is just cost of Rs. 30–40 per liter. The masses need to be educated for more use of liquid laundry detergent. This will make sure water resources are pollution free.
6. In some instances, the stain-removing property in cotton cloths is slightly better than polyesters and terricot cloths.

KEYWORDS

- biopolymer
- polymeric surfactants
- ecofriendly products
- liquid detergents
- commercial production

REFERENCES

1. Carbohydrite Polymer. http//www.elasewere.
2. Carneiro, M. J.; Fernande, A.; Figneiredo, C. M.; Fortes, A.G.; Frietas, A. M. *Carbohydr. Polym.* **2001,** *45*, 135–138.
3. Dontulwar, J. R.; Gogte, B. B. *Asian J. Chem.* **2004,** *16*(3–4), 1385–1390.
4. Annual Book of ASTM Standard. *ASTM Standard Method for Acid Value of Organic Coating Materials*; American Society for Testing and Material, 1981, Vol 6.01d; pp 1639–1670.
5. Melhlen Bacher, V. C. *The Analysis of Facts and Oils*; Garrard Publication: Champaign, IL, 1960; pp 299–308.
6. Annual Book of ASTM Standard. *Annual Book of ASTM Standard*; American Society for Testing and Material, Vol6.01D, 1981;pp 1639–1661.
7. ASTM Standard Method. *For Non-volatile Content of Biopolymer*; 1980, 6.01D; pp 1259–1261.
8. Gogte, B. B.; Agrawal, R. S. *J. Soap, Detergent, Toiletr. Rev*, **2003,** *34*, 25–28.
9. Gogte, B. B.; Agrawal, R. S. Starch Sorbitol Based Co-polymer. *J. Soaps Detergent Toilet Rev.* **2003,** *25*, 28.
10. Gogte, B. B.; Dontulwar, J. R.; Borikar, D. K. *Carbohydr. Polym.* **2006,** *65*, 207–210.
11. Gogte, B. B.; Dontulwar, J. R.; Borikar, D. K. *Carbohydr. Polym.* **2006,** *63*, 375–378.
12. Encyclopedia of Polymer Science and Technology. *Encyclopedia of Polymer Science and Technology*; John Wiley and Sons, 1982; pp 182–191.
13. Griffin, W. C. *J. Soc., Cos Chem.* **1954,** 249.
14. Gogte, B. B.; Kharkate, S. K. *J. Chem. Eng. World* **2006,** 50–53.

CHAPTER 13

THE THERAPEUTIC ROLE OF THE COMPONENTS OF *Aloe vera* IN ACTIVATING THE FACTORS THAT INDUCE OSTEOARTHRITIC JOINT REMODELING

ABHIPRIYA CHATTERJEE and PATIT PABAN KUNDU[*]

Advanced Polymer Laboratory, Department of Polymer Science and Technology, University of Calcutta, 92 APC Road, Kolkata 700009, India

[*]*Corresponding author. E-mail: ppk923@yahoo.com.*

CONTENTS

Abstract		238
13.1	Introduction	238
13.2	Components of *Aloe vera*	239
13.3	Few Components of *Aloe vera* That Can Be Beneficial in Treating Osteoarthritis	241
13.4	Osteoarthritis	242
13.5	Progression of the Disease	245
13.6	Factors Contributing to Cartilage Repair	250
13.7	Therapeutic Role of Components of *Aloe vera* in Treating Osteoarthritis	255
13.8	Conclusion and Future Perspective	258
Keywords		258
Reference		259

ABSTRACT

Aloe vera is a perennial succulent plant, each and every component of which has its own individual and established beneficial properties. It is used as laxative to immunostimulant, anti-inflammatory agent, antiseptic, moisturizer, anti-ageing, wound healing and as antioxidant. In recent years, many studies have indicated the role of *A. vera* in treating arthritis, especially osteoarthritis, which is the most common disease among elderly people. Nine out of 10 aged person have complains about joint pain, swelling, stiffness of joint, and lack of movement which are the common symptoms of osteoarthritis. *A. vera* can be considered as a potential therapeutic agent in treatment of osteoarthritis. *A. vera* can activate a series of factors essential in cartilage tissue repair, collagen synthesis, and anti-inflammation. *A. vera* not only contributes in treating the disease but is also easily available and cost-effective.

13.1 INTRODUCTION

Aloe vera has proved its medical efficacy since early civilization. With every passing year, new spheres of beneficial properties of *A. vera* have come to the lime light. Since early times, *A. vera* has been used as laxative, for healing wounds, in cosmetics, etc.[1] But with time and further experiments, numerous other therapeutic properties of *A. vera* have been discovered. Nowadays, when the demand of natural medicine has highly increased, patients want medicine with maximum benefit but minimum side effects. *A. vera* has surely gained popularity in this field. *A. vera* gel is said to contain most of the therapeutic properties, with few very important beneficial components. Though the mucilaginous gel is composed of 99–99.5% of water, the remaining 1–0.5% of dry constituent is rich with ingredients like glucomannans, amino acids, vitamins, sterols, carbohydrate, enzymes, etc.[2-4] The major polysaccharide in *A. vera* gel, acemannan,[5-7] has gained particular recognition for its beneficial role in immunostimulation, detoxifying cell debris, cell regeneration, and many more, though all the other components are also equally efficient.

Considering all the beneficial ingredients in *A. vera*, it can be presumed that *A. vera* can be used as a treatment in osteoarthritis. Osteoarthritis is a chronic degenerative disease that causes pain in the joints, swelling and crippled movement of the joint. Age, sudden injury, and obesity cause erosion in the cartilage tissue that covers the joint bones. Loss of water content in the extracellular matrix restricts smooth movement of the joints, resulting in degraded cartilage tissue, synovitis, impaired collagen, and proteoglycan synthesis. Though not particularly an inflammatory disease, it shows all the symptoms related to inflammation. Recent studies have revealed that inflammation is a major aspect in the pathophysiology of osteoarthritis. Moreover, the cytokines released in course of the inflammatory response also activate a series of factors like matrix metalloproteinases (MMPs), ADAMTs that contribute to chondrocyte apoptosis, collagen degradation, proteolysis of aggrecan, and extracellular matrix degeneration.[8] Many over the counter medicines available for osteoarthritis are suggested by physicians, but none of them can be considered as a treatment for this particular disease as most of them are pain killers and steroids and come with huge side effects.

A. vera contains many biologically potent ingredients that have the potential to treat osteoarthritis. Components like bradykinase, sterols, have anti-inflammatory effect, acemannan which is the most abundant polysaccharide in *A. vera* gel, stimulates immune response by releasing macrophage and cytokines, these components can detoxify the damaged area and serve as pain killers. Acemannan also stimulates monocytes to cleave macrophage in M1 and M2, where M1 carry out phagocytes to remove cell debris and induce inflammation, M2 activates anti-inflammatory cytokines, growth factors that promote cartilage remodeling and extracellular matrix repair.[9]

So, the aim of this review is to revisit the pathophysiology of osteoarthritis and the therapeutic role of *A. vera* in treating osteoarthritis successfully by either reversing or eliminating the causes.

13.2 COMPONENTS OF *ALOE VERA*

A. vera with its uncountable beneficial properties has been famously called "the magical plant." Back in early civilization, *A. vera* has been used as laxative,[10,11] in beauty treatments and in wound healing.[12,13] *A. vera* is a perennial xerophyte growing mainly in regions of India, Africa, Europe, and America where there is erratic water availability. But *A. vera* has very high adaptive quality, so if provided with proper care, it can be grown in any region of the world. This high adaptive property of *A. vera* enhances its specialty more as a highly beneficial plant.[14] The plant belongs to

Liliaceae family and is specialized in storing large amount of water in their tissues. The leaves are fleshy, triangular in shape and the sides are serrated. *Aloe vera* leaf is composed of three parts; the outer rind, the middle latex, and the inner mucilaginous gel.[15] The inner gel of *A. vera* is composed of many components, all with its individual usefulness. Extraction of *A. vera* gel is a delicate process where temperature, centrifugation speed has to be specific,[16] as the components of *A. vera* gel tend to loose their biological properties to high temperature, improper centrifugation, and time. Centrifugation was done at 5°C temperature, 10,000 rpm speed for 30 min (http://www.aloecorp.com)[17,18] To preserve all the properties of the gel extract for longer period of time, usually preservatives are added. But preservatives can have adverse negative effects to our health and can even contaminate the actual structural and biochemical properties of the gel extract. Freeze-dried gel extract is a dry, light powder which is devoid of any preservative and additive. Freeze drying is done in temperature varying from −45 to 30°C, the frozen gel is kept in high vacuum where water sublimes from frozen gel as temperature increases.[19] *A. vera* gel has manifold benefits in medical and pharmaceutical world. Properties like immunostimulant, anti-inflammation, anti-diabetic, moisturizer, anti-arthritis, antioxidant, wound healing, etc. have come to the lime light recently.[4,20–23] A schematic diagram in Figure 13.1 is showing the properties briefly.

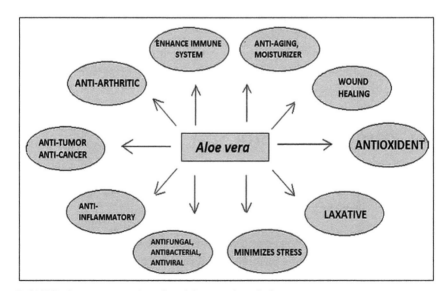

FIGURE 13.1 Presentation of useful properties of *Aloe vera*.

13.3 FEW COMPONENTS OF *ALOE VERA* THAT CAN BE BENEFICIAL IN TREATING OSTEOARTHRITIS[2,5,23–29]

Active components	Compounds	Function
Polysaccharides	Pure mannan Acetylated mannan (acemannan) Acetylated glucomannan Glucogalactomannan Galactan Pectic substance Xylan Cellulose Mannose Glucose	Acemannan is the most abundant polysaccharide in *Aloe vera* gel. It is said to have immunostimulatory effect, acts as antiarthritic and anti-inflammatory agent Mannan-6-phosphate can help in collagen deposition and can stimulate release of growth factors Stimulate fibroblast that promotes tissue regrowth
Hormones	Auxins Gibberellins	Wound healing and anti-inflammatory Fortifies body's natural immunoreaction to pathogen
Organic compounds and lipids	Arachidonic acid Steroids (campesterol, cholesterol, β-sitosterol) Triglycerides Triterpenoid Saponins Lignin Salicylic acid Uric acid	Salisylic acid acts as anti-inflammatory and antibacterial agent Saponin can be used as cleanser and moisturizer Lignin increases deep penetrating properties of other components Have anti-inflammatory function
Vitamins	B1, B2, B6, C β-Carotene Choline Folic acid Ascorbic acid	Antioxidant Neutralizes free radicals
Enzymes	Bradykinase Alkaline phosphatase Amylase Carboxypeptidase Catalase Cyclooxidase Cyclooxygenase Lipase Oxidase Phosphoenolpyruvate Carboxylase Superoxide dismutase	Help in metabolism and breakdown of sugar and fat Bradykinase has proven anti-inflammatory effects Possess antifungal and antiviral activity but toxic at high concentrations

Table *(Continued)*

Chromones	C-glucosyl	Acts as anti-inflammatory
	Isoaloeresin D	Antiallergic
	Isorabaichromone	
	Neoaloesin A	

13.4 OSTEOARTHRITIS

Osteoarthritis is a chronic degenerative disease that causes the breakdown of cartilage in the joints. Symptoms include pain, stiffness, and swelling. Osteoarthritis causes progressive breakdown and wearing away of the cartilage, leaving the bone ends unprotected. As this occurs, the joint can become painful, stiff, and difficult to move and eventually swollen.[30]

This is the most common type of rheumatism, mostly seen in people of age group above 40 years. A study conducted in USA reported that about 90% of the populations between the age group of 40–90 years are suffering from this chronic disease.[31]

To understand the pathophysiology of the disease properly, it is necessary to have a sound knowledge about the morphology of the articular cartilage so that the gradual step by step changes induced by the progression of the disease becomes clear.

13.4.1 STRUCTURE OF THE CARTILAGE

The articular cartilage is divisible into three structurally and morphologically different zones, depending on the arrangement of type chondrocyte cells, II collagen fibers, and proteoglycan. Zone I which is known as superficial or tangential zone has flattened chondrocytes with collagen arranged parallel to the surface and this zone is in direct contact with the synovial fluid. Superficial zone comprises about 10–20% of the cartilage thickness and it protects the other zones from stress and force during movement. The density of chondrocyte cells in this zone is very high compared to the other. The superficial zone is immediately followed by the transitional zone which occupies about 40–60% of the total volume of the cartilage. In this zone, the density of chondrocytes is comparatively low and is spherical in shape; the collagens are obliquely arranged to the surface intervened by proteoglycan in between as shown in Figure 13.2. This zone provides resistance to huge force falling on the joints during movement. Next is the basal zone which has much less density of chondrocyte cells and higher density of collagen and proteoglycan. Collagen is arranged perpendicular to the surface and this arrangement works as a cushion that can withstand maximum of the compressive force.[32,33]

Thus, the cartilage comprises the chondrocyte cells and extracellular matrix which is composed of water, collagen, proteoglycan, and few non-collagenous proteins and glycoproteins (Fig. 13.2).

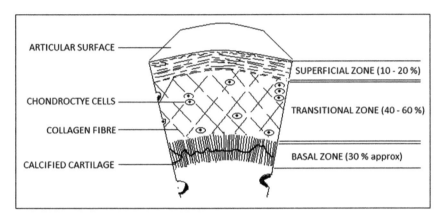

FIGURE 13.2 Structure of the articular cartilage.

13.4.2 CHONDROCYTE CELLS

The chondrocyte is the specialized cell type in the articular cartilage. Chondrocytes are highly modified, metabolically active cells that have the most important part to play in the development, maintenance, and repair of the extracellular matrix. Origin of chondrocytes is from mesenchymal stem cells and constitutes about 2% of the total volume of articular cartilage.[34] Chondrocytes have different shape, number, and size that vary depending on the different anatomical regions of the articular cartilage. The chondrocytes in the zone I are flatter and smaller and generally have a greater density than that of the cells of other zones, deeper in the matrix. Each chondrocyte establishes a specialized microenvironment and is responsible for the turnover of the ECM in its immediate surrounding. Chondrocytes maintain a balance between matrix synthesis and breakdown that helps in normal tissue metabolism. Though chondrocytes do not form cell-to-cell contacts for direct signal transduction and communication between cells, but they definitely respond to a variety of stimuli like growth factors, mechanical loads, and hydrostatic pressures. Chondrocytes have limited replication capability which prevents rapid healing caused by any injury, but if provided with proper chemical and mechanical environment, the cells can survive.[35]

13.4.3 COLLAGEN

Collagen constitutes about 10–20% of wet weight of the articular cartilage. The major component of fibrillar collagen in articular cartilage is type II collagen; it constitutes 90–95% of total collagen present in cartilage. Type II collagen forms a highly cross-linked and interconnected network of collagen fibrils. Other than type II collagen, types IX and XI collagen are the most abundant among the minor types of collagen and both are present in equal amounts. Type IX collagen is short fibrillar collagen with a proteoglycan moiety and forms cross-links with type II collagen along the surface of collagen fibrils and integrates with proteoglycan in the extracellular matrix. Type XI collagen mainly regulates the diameter of the fibril of type II collagen, to form copolymers.[36] Though the rate of metabolism of collagen is slow, but in condition of injury or disease, the rate of metabolism increases excessively and can exceed the ability of chondrocytes to produce a well-organized replacement matrix. In this type of conditions, the matrix undergoes more rapid mechanical failure and deterioration, which results in degeneration and the development of osteoarthritis. High levels of collagenases, which degrade collagen, have been reported responsible in the pathogenesis of osteoarthritis.[37]

13.4.4 PROTEOGLYCANS

They are protein polysaccharide molecules that constitute 10–20% of the wet weight and provide a compressive strength to the articular cartilage. The fluid and electrolyte balance in the articular cartilage is maintained by proteoglycan.[38] Chondrocytes produce these proteoglycans and secrete in the matrix. The subunits of proteoglycan, which are known as glycosaminoglycans, are of mainly two types, chondroitin sulfate and keratin sulfate. There are two major classes of proteoglycans found in articular cartilage, large aggregating proteoglycan monomers, or aggrecans and small proteoglycans, including decorin, biglycan, and fibromodulin.[38] Aggrecan is the most abundant proteoglycan in the matrix and is composed of a central core protein bound to glycosaminoglycans by sugar bonds. The core protein stabilizes the structure with a central hyaluronic acid chain to form an intricate structure of the glycosaminoglycans molecule. There are two types of the chondroitin sulfate, type 4 and type 6. Type 6 remains constant throughout life, whereas type 4 decreases with the age. Depletion in proteoglycan level has been observed in experimental arthritis.[38]

13.5 PROGRESSION OF THE DISEASE

The two most observed causative agents leading to osteoarthritis are age and obesity. Sudden injury or past trauma and genetics are the rare causes that could lead to the disease. Use and over use of the joints over years causes loss of plasticity and fluidity of the joints. Decrease in water content results in reduction of proteoglycan level which in turn makes the collagen fibers susceptible to degeneration and finally cartilage degradation.

The main function of the cartilage is to distribute the weight on the joints uniformly. During movement when pressure falls on the joints, the viscous synovial fluid between the cartilages is pushed in and out, thus lubricating the joints and helping in smooth movement. Due to obesity as the pressure on the joints increases, extra effort is needed to keep lubricating the joints and the cartilages constantly rub against each other, eventually causing erosion in the cartilage. This wear and tear in the cartilage and surrounding synovial fluid triggers inflammatory response for prevention of infection and tissue regeneration.[39]

It has been observed that the ratio of women suffering from osteoarthritis is much higher than male. Moreover, women who has reached menopause are the main sufferers. The Framingham knee osteoarthritis study suggests that the lack of estrogen production after menopause contributes to the pathogenesis of the disease. Estrogen is known for its anti-inflammatory property, so it can be presumed that during normal times when the level of estrogen is high, it prevents the effects of inflammation, like pain and swelling. But as the level of estrogen drops, these symptoms express themselves.[40] Normal cartilage maintains a state of homeostasis where regeneration and degeneration occurs simultaneously to maintain a healthy equilibrium. In osteoarthritis joints, this equilibrium is disturbed, imbalance between repair and degradation is seen which leads to structural damage, inflammation in the synovium, and pain. Figure 13.3 gives a general idea of the difference seen before and after the effect of the disease osteoarthritis.

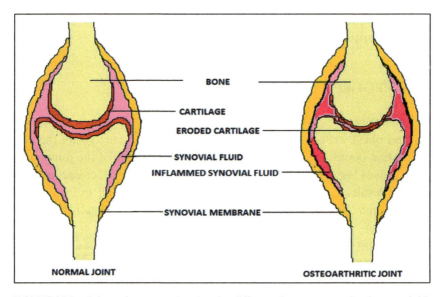

FIGURE 13.3 Schematic presentation showing difference between normal and osteoarthritic joint.

Osteoarthritis occurs as a cascade of structural, biochemical, and metabolic changes as follows:

13.5.1 STRUCTURAL CHANGES

The hyaline cartilage, coating the surface of the joints, is mainly composed of chondrocyte cells embedded in extracellular matrix which is constituted by water, type II collagen, and proteoglycan (Fig. 13.2). During movement, the cartilage distributes the weight on the joint uniformly, thus reducing stress. A viscous synovial fluid present in between the articular cartilage acts as a lubricant and reduces friction between cartilages during movement. This synovial fluid has high concentration of hyaluronic acid. In general, a healthy cartilage maintains a state of homeostasis where there is a constant process of resorption and regeneration of cartilage matrix. But in case of osteoarthritis, the balance between this resorption and regeneration is disturbed. The cell tries to repair itself but is unable to regenerate a functional matrix and the shock-absorbing function is gradually lost.[39]

In the early stage of osteoarthritis, degeneration of chondrocytes and type II collagen leads to fibrillation. As the homeostasis between the anabolism and catabolism in the cartilage is lost, constant apoptosis of the chondrocytes occurs mainly in the surface of the articular cartilage. In an attempt to self-repair, chondrocytes rapidly proliferates, forming irregular sub-articular cysts. As the disease advances, deep clefts are formed in cartilage, nearby matrix gets depleted of metachromatic material, indicating loss of proteoglycans. A rapid loss of aggrecan, the most predominant proteoglycan in cartilage, results in excessive load on the collagen networks which slowly starts losing its tensile strength and degrade. All these ultimately lead to the destruction of extracellular matrix and loss of cartilage. The degraded fragments of the cartilage get deposited in the synovium which results in thick and hypertrophied synovium and it also triggers the inflammatory response leading to synovitis as shown in Figure 13.3.[20,41]

13.5.2 COMPLEMENT SYSTEM

Researchers at Stanford University of Medicine observed that osteoarthritic patients were diagnosed with high level of proteins that were usually secreted by the body when under bacterial or viral attack. These are the proteins of the complement system that attack damaged osteoarthritic joints, as they would have attacked any other antigen harmful to body. This active complement system carries off a cascade of events as shown in Figure 13.4 leading to inflammation and severe pain experienced in osteoarthritis.

Complement system activates a cluster of proteins called membrane attack protein (MAC). In osteoarthritic joints, MAC activates inflammatory response that causes cartilage degradation and leads to impaired movement.[42]

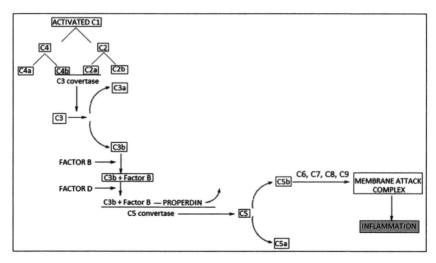

FIGURE 13.4 Schematic representation of the pathway of complement cascade. C(1–9)—complement components.

13.5.3 INFLAMMATORY RESPONSE

Synovial inflammation or synovitis can be a leading factor in irregular functioning of the chondrocytes that causes imbalance between the catabolic and anabolic activity of chondrocytes, in extracellular matrix remodeling in the cartilage. Though osteoarthritis is not defined as an inflammatory disease because the synovial fluid lacks neutrophils, the symptoms of inflammation, like pain, swelling, stiffness, and impaired movement, are prevalent in the patients.[43] Even active lymphocytes like B cells and T cells and pro-inflammatory mediators like macrophage and cytokines have been regularly detected in osteoarthritic joints.[44] From recent studies, it is evident that inflammatory process involves the Toll-like receptors and activation of complement cascade which lead to the synthesis and release of cytokines and chemokines.[45] As shown in Figure 13.5, in response to the pro-inflammatory macrophages, cytokines like interleukin-1β (IL-1β) and tumor necrosis factor-α (TNF-α) are secreted. This enhances the production of prostaglandins E2 by influencing the activities of cyclooxygenase 2, microsomal PGE synthetase-1, and soluble phospholipase A2, which stimulate the production of nitric oxide (NO) in chondrocytes by activating NO synthetase. NO blocks the synthesis of collagen and proteoglycan by activating MMP and also promotes apoptosis of chondrocytes by producing free radicals. IL-β and TNF-α, synthesized by activated synoviocytes, mononuclear cells, or articular cartilage, upregulate the gene expression of metalloproteinase.[46,47] These cytokines block the chondrocyte compensatory synthesis pathway that can restore the integrity of degraded extracellular matrix. In patients with osteoarthritis, the receptor for IL-β (IL-1R1) has shown to be increased in the chondrocytes and fibroblast-like synoviocytes.[47] IL-1β blocks the chondrocytes to synthesize main building blocks of extracellular matrix, that is, type II collagen and aggrecan. IL-1β and TNF-α stimulate ADAMTS-4 and 5 (A Disintegrin And Metalloproteinase with a Thrombospondin 4 and 5) productions which are mainly responsible for proteolysis of aggrecan molecules.[48] The schematic flowchart given in Figure 13.5 provides a stepwise representation of their way of function. Even these cytokines induced rapid chondrocytes apoptosis. All these factors synergize together and result in inflammation, chondrocyte apoptosis, and finally extracellular matrix degradation.

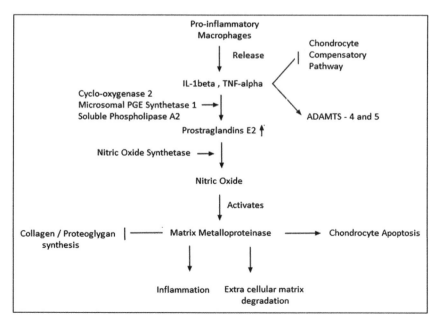

FIGURE 13.5 Schematic representation of pathway of action in inflammatory response. ADAMTS—A Disintegrin And Metalloproteinase with a Thrombospondin; IL-1β—Interleukin-1 β; TNF-α—tumor necrosis factor α.

13.5.4 EFFECT OF MMPS

It is evident from many studies that MMPs play an important role on matrix degradation, degeneration of collagen, and proteoglycan.[49] Although different types of MMPs are synthesized by different genes, some of the matrix metalloproteinases are synthesized in chondrocytes and synovial cells itself in osteoarthritic joints.[50]

The major MMPs that are responsible for extracellular matrix degradation are MMP-1, MMP-8, and MMP-13, but among these, MMP-13 is most potential in degradation of type II collagen, the main collagen component in cartilage. Other than this, MMP-2, 3, and 9 cleave substrates like gelatin, fibronectin, elastin, proteoglycan, and type IX and XI collagen.[51] MMP-3 even plays an important role in activation of other metalloproteinases.[52] Moreover, the different types of membrane type metalloproteinases (MT-MMPs) contribute to the activation of Pro-MMP-2 and Pro-MMP-13 which are considered as important players in breakdown of type II collagen in the cartilage[53,54]

13.5.5 BRADYKININ

Bradykinin belongs to the family of kinins, which are significantly associated with pro-inflammatory mechanisms and pain. Bradykinin binds to its activated receptor, B2, and induces activation of phospholipase C and its signaling pathways, like protein kinase C and phospholipase A phosphorylation,[55] NO synthase activation, and ultimately release of inflammatory mediators. B2 receptors in chondrocytes and synovial fluid bind to bradykinin and stimulate the release of inflammatory cytokines like IL-6, IL-8, and MMP-3 that is involved in collagen degradation. Presence of bradykinin in synovial fluid, fibroblast, and endothelial cell lining has been detected in patients with osteoarthritis. Moreover, it can also interact with IL-1β signaling and increase cyclooxygenase 2 expression which results in increased prostaglandin production.[56]

13.6 FACTORS CONTRIBUTING TO CARTILAGE REPAIR

It is previously mentioned that a cascade of immune reactions leads to the pathogenesis of osteoarthritis. Just like that, several factors are also there, which when stimulated properly can play major roles in cartilage repair and remodeling.

13.6.1 MACROPHAGE

Macrophages are a type of white blood cell that play a critical role in nonspecific defense and also help to initiate specific defense mechanisms by inducing inflammatory response in injured tissue. Macrophage has multifacet functions in tissue remodeling too. There are two phenotypes of macrophage, M1 and M2;[57] both have their individual properties but interestingly have antagonistic functions. M1 macrophages activated by classical pathway consist of immune-effector cells which have an acute inflammatory response. These are highly aggressive against bacteria and carry out phagocytosis by producing large amount of lymphokines. The anti-inflammatory M2 macrophages, activated by alternative pathway, reportedly have three subtypes – M2a, M2b, and M2c, which have various different functions, including regulation of immunity, angiogenesis, and tissue repair/wound healing. In osteoarthritic joint repair, M2 plays a major role.[9] M2 activates cytokines IL-4, IL-10, and IL-13, all of which have anti-inflammatory functions. These cytokines downregulate the synthesis of NO, thus lowering the level of inflammatory cytokines and matrix degrading protein, MMP. Along with anti-inflammatory cytokines, M2 can also stimulate the activation of growth factors like transforming growth factor β (TGF-β), bone morphogenetic protein (BMP), etc. which are the main components in matrix degeneration, collagen synthesis, and chondrocyte differentiation. Thus, macrophage has an overall function in both pathogenesis and repair of osteoarthritic joints.[58,59]

13.6.2 GROWTH FACTORS

Numerous extracellular stimuli affect the chondrocytes by influencing the biosynthesis and catabolic activity such as mechanical stress. Ageing, sudden injury, and any disease can cause imbalance in regulatory factors which can hinder proper maintenance and repair of tissue. All these result in deleterious changes in gene expression altering extracellular matrix constitution and tissue degeneration. As a consequence, it causes an accelerated erosion of the articular surface, leading to end stage of osteoarthritis.

In order to maintain growth and development and restore homeostasis in the cartilage, numerous growth factors work together in a healthy articular cartilage. Growth factors are biologically active polypeptides, synthesized in our body that can promote cell division, differentiation, and growth.

Thus, in osteoarthritic joint, growth factors can stimulate cartilage regeneration. The multifaceted functions of growth factors in articular cartilage include synthesis of type II collagen, proteoglycan, promote differentiation of chondrocyte cell in mesenchymal stem cells, and down-regulate the effects of cytokine IL-1 and MMPs.[60] Considering all these functions of growth factors, if proper stimulation is provided, they can have the potential in treating degraded cartilage in osteoarthritis. Different growth factors with different functions work in concert to regenerate eroded cartilage and regain its previous healthy state.

13.6.2.1 PLATELET-DERIVED GROWTH FACTOR

Platelet-derived growth factor (PDGF) is the first growth factor present in a wound and initiates connective tissue healing through the promotion of collagen and protein synthesis. The most important specific activities of PDGF include angiogenesis, chemotaxis for fibroblasts, and collagen synthesis. The primary effect of PDGF can be its mitogenic activity on mesoderm-derived cells like fibroblasts, vascular muscle cells, and chondrocytes. Studies supported the role of PDGF in wound healing or stimulation of matrix synthesis in growth plate chondrocytes.[61]

13.6.2.2 TRANSFORMING GROWTH FACTOR β

TGF-β is an important member of the growth factor family which has major role in cartilage repair and extracellular matrix remodeling. TGF-β usually remains in inactivated form and needs stimulation in the form of any injury or disease, which can trigger immune response. Activated TGF-β usually forms hetero- or homodimers with single disulphide bond. Activated TGF-β regulates cell proliferation, migration, differentiation, and apoptosis. In osteoarthritic cartilage, the isoforms of TGF-β, TGF-β1, TGF-β2, and TGF-β3 can bind to different promoter sequences, inhibit catabolism of extracellular matrix, and promote chondrocyte differentiation.[62,63]

TGF-β1 and 3 has a role in maintaining the balance of anabolism in cartilage and synthesis of extracellular matrix. In vitro studies suggested that in normal cartilage, the level of TGF-β was high, which was drastically decreased in osteoarthritic cartilage. Even if TGF-β pathway was blocked in normal cartilage, it became more prone to damage and susceptible to osteoarthritis.[64] This indicates that TGF-β is a major factor in maintaining cartilage homeostasis. TGF-β can suppress the expression of cytokines like IL-1 and 6 and induce extracellular matrix synthesis and chondrogenesis. TGF-β signaling is mediated by Smad (mothers against decapentaplegic homolog) via activin-like kinase 1 and 5 receptors.[65–67] In spite of all these beneficial activities of TGF-β, it can also be harmful as it highly increases chondrocyte proliferation which can result in formation of osteophyte.[62]

13.6.2.3 BONE MORPHOGENETIC PROTEIN-2

Many experimental data indicated that BMP-2 could stimulate matrix synthesis and was capable of reversing chondrocyte dedifferentiation to some extent, by increasing synthesis of cartilage-specific collagen type II in dedifferentiated and OA chondrocytes.[68] Studies conducted in defective cartilage of animal models has shown BMP-2 to enhance cartilage repair and also increase chondrocyte localization.[69,70] There is a similarity between the function of BMP-2 and TGF-β1 in their effect on mesenchymal stem cells with increased extracellular production and decreased expression of collagen type 1.[71] In a mouse model of IL-1-induced cartilage degeneration, BMP-2 enhanced cartilage matrix turnover by increased aggrecan degradation and increased collagen type II and aggrecan expression[70,71]

19.6.2.4 OSTEOGENIC PROTEIN-1/BONE MORPHOGENETIC PROTEIN-7

Several in vitro studies conducted on normal as well as osteoarthritic cartilage highlighted the immense functional attributes of osteogenic protein-1 (OP-1)/BMP-7. Apart from maintaining homeostasis in the cartilage, by controlling the rate of catabolism and anabolism, it can also promote anabolism in osteoarthritic cartilage, where high rate of catabolism causes chondrocyte destruction. OP-1 stimulates the release of extracellular matrix-specific proteins, collagen, and proteoglycan. The most notable property of OP-1 is that it upregulates the rate of metabolism in cartilage chondrocytes but does not promote excessive chondrocyte proliferation leading to osteophyte formation.[72,73] Moreover, OP-1 activates the gene expression of TIMP (tissue inhibitor of metalloproteinase) that inhibits the expression of MMP, which is the most important component responsible for extracellular matrix degradation. OP-1 can inhibit the inflammatory cytokines IL-1β, IL-6, and TNF-α which can stimulate the release of matrix metalloproteinase 1 and 13.

An important aspect about OP-1 that makes it valuable in cartilage repair is that its function is not affected by age like other growth factors. Age is an important causative factor in osteoarthritis and with age, the functional properties of many growth factors diminish. OP-1 is an exception.[60,74]

OP-1 if coupled with insulin-like growth factor-1 (IGF-1) enhances the repair mechanism of damaged cartilage further. Chondrocyte cell survival, extracellular protein synthesis, and chondrocyte proliferation increase double fold resulting in rapid cartilage remodeling.[60,72,74]

13.6.2.5 INSULIN-LIKE GROWTH FACTOR-1

The function of IGF-1 in articular cartilage metabolism has been an interesting subject of investigation in cartilage-related diseases.[75] IGF-1 induced a series of anabolic effects and decreased catabolic responses when applied in monolayer or explant culture of normal cartilage from variety of species.[76,77] An in vivo study in rats showed that chronic deficiency of IGF-1 led to articular cartilage lesions which proved the fact that IGF-1 is necessary to maintain cartilage integrity.[74] Chondrogenic differentiation of mesenchyme stem cells is stimulated by IGF-1, but it is reported to be enhanced further when a combination of IGF-1 and TGF-β1 is used. In animal models, IGF-1 has led to enhanced repair of extensive cartilage defects and protection of the synovial membrane from chronic inflammation. However, with age and in osteoarthritis, chondrocytes fail to respond to IGF-1.[67,77,78] In spite of the fact that the ability of IGF-1 to decrease catabolism in cartilage goes down with age and in osteoarthritic cartilage, a combination with BMP-7 enhances the function of IGF-1. Activity of BMP-7 does not go down with age or injury, so its combination with IGF-1 resolves the age-related lack of property in IGF-1.[74,79,80]

13.6.3 WNT-SIGNALING PATHWAY

Wnt-signaling pathway has been identified as a key regulator in bone, cartilage, and joint development and maintaining homeostasis. It has been studied that at late embryonic and post natal development, the level of β-catenin synthesized by chondrocytes stimulated bone formation.[81,82]

Recently, Wnt/β-catenin-signaling pathway has been reported to play a major role in relating cartilage and subchondral bone in its remodeling. During osteoarthritis, when the chondrocyte cells are hypertrophic, expression of β-catenin regulates the synthesis of RANKL (receptor activator of nuclear factor kappa-B ligand),[83] which can control osteoclast activity in subchondral growth plates.[84] Wnt signaling can occur through a canonical pathway, involving stabilization of β-catenin and its translocation to the nucleus to regulate gene expression. Another way of regulation can be through a β-catenin independent noncanonical pathway, involving calcium and calcium-calmodulin kinase II.[85–89]

During the progression of osteoarthritis, the observed altered activity of Wnt signaling in chondrocytes affected the osteoplastic activity in subchondral bone growth plates, leading to sclerosis or osteophyte formation at the edges of joints. Alteration in Wnt-signaling pathway appeared to modulate key regulatory factors for remodeling of subchondral bone, resulting in its aberrant behavior as observed in the case of osteoarthritis.[90,91]

13.7 THERAPEUTIC ROLE OF COMPONENTS OF *ALOE VERA* IN TREATING OSTEOARTHRITIS

It has been well established from various studies that inflammation plays a major role in pathophysiology of osteoarthritis. The degenerative cartilage and synovium stimulate the immune system which in turn activates a cascade of responses such as the complement system, inflammation, and activation of various matrix degenerative factors. Normally, physicians recommend anti-inflammatory drugs, steroids, or tropically administered cortisone injections. These over the counter medicines surely ease pain caused due to osteoarthritis but come with undesirable side-effects and are also unable to repair damaged tissue. *A. vera* not only eases pain but has also proved itself potential in repairing damaged tissue, having anti-inflammatory function and detoxify by dispersing degenerated tissues with its deep penetrating property.

Aloe vera has been known for its anti-inflammatory, immunostimulatory, and cell or tissue repairing properties. Ample studies have been done which proves this fact that there is more than one component in *A. vera* which contributes to healing injured or degenerative tissues. Studies suggested that mannose-6-phosphate can bind to fibroblast receptors and induce fibroblast proliferation which can promote collagen deposition and tissue reorganization.[92] Mannan-rich polysaccharide of *A. vera* gel serves the same function as mannose-6-phosphate.

It has been mentioned before that bradykinin has an active role in inducing inflammation in osteoarthritic cartilage. A peptidase, namely bradykinase, present in *A. vera* gel extract has shown strong anti-inflammatory effect.[93,94] In vitro study on carrageenin-induced paw oedema of rat showed that bradykinase along with few other components like C-glucosyl chromone has rapidly reduced inflammation in injured cartilage and decreased pain.[95] Bradykinase may have inhibited the binding of bradykinin to B2 receptor present in synovial fluid. This has prevented bradykinin to stimulate the release of inflammatory cytokines via cyclooxygenase pathway. Bradykinase has shown to reduce the production of prostaglandin E2 from arachidonic acid. Eventually, a drastic decrease in cyclooxygenase and NO synthesis has been observed, which led to downregulation of MMP, the collagen degrading component in inflamed cartilage.[96] Other than bradykinase, few different plant sterols in *A. vera* gel like compesterol, β-sitosterol, cholesterol, and lupeol have notable anti-inflammatory and analgesic effect and aspirin-like compounds have anti-microbial property,[95,96] as shown in adjuvant-induced arthritic inflammatory rat model.[95,97]

Antioxidants like vitamin C and ascorbic acid present in *A. vera* gel can play an important role in eliminating free radicals and increase collagen synthesis. Free radicals are unstable oxygen molecules that cause cell damage, cell death, and break collagen/proteoglycan cross-link. Pro-inflammatory cytokine-induced inflammation that releases free radicals damages near by cells in the form of chain reaction. Antioxidants trap and neutralize these free radicals, thus prevent further cell damage.[98]

Acemannan has well-established immunostimulatory property, it has been found to stimulate the release of white blood cell activity. These white blood cells are necessary for detoxifying the degenerated tissue particles in the synovial fluid and activating cytokines to induce inflammation. On the other hand, it activates different other factors which promote cartilage repair. Acemannan can activate macrophages which are very essential in tissue repair; they increase cell and tissue growth, fibroblast proliferation, and activity. They are highly specialized in removal of dead cell/tissue or tissue debris by the process of phagocytosis. This debris, if not removed, results in chronic inflammation which is the main causative agent of prolonged pain in osteoarthritis. There is also an interrelation between acemannan-activated macrophages that induce inflammation and macrophages that has function in anti-inflammation, detoxification, and most importantly tissue repair. These are the most important aspects in treatment of osteoarthritis.[99]

Monocytes activate macrophage cells in the blood; it can be presumed from various studies that acemannan stimulates the differentiation of monocytes, which activate macrophages. Macrophage presents two different phenotypes, macrophage 1 (M1) and macrophage 2 (M2). Interestingly, M1 and M2 have antagonistic functions (Fig. 13.6). M1 exhibits strong tumor suppressor and adjuvant properties. Activated M1 stimulates the release of various cytokines like IL-1, IL-6, and TNF-α that can enhance T-cell-mediated cytotoxicity and proliferation resulting in inflammatory response, characterized by chronic pain, swelling, etc. On the other hand, M2 stimulates the release of cytokines like IL-4, IL-10, and IL-13 which have anti-inflammatory effect. Moreover, as shown in Figure 13.6, M2 promotes tissue repair or remodeling by activating growth factors like TGF-β, BMP, IGF-1, and PDGF and stimulates Wnt-signaling pathway which promotes fibroblast growth, chondrocyte differentiation, and collagen regeneration.

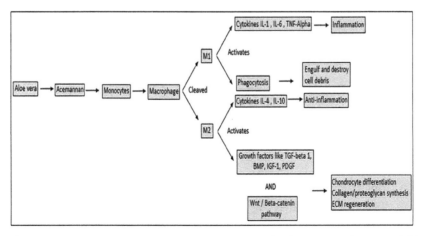

FIGURE 13.6 Schematic representation of the pathway of action of *A. vera* in osteoarthritic cartilage repair.

13.8 CONCLUSION AND FUTURE PERSPECTIVE

Though huge percentage of population in every nation is suffering from osteoarthritis, still no established treatment or cure has been found. In most of the cases, physicians recommend regular exercise, pain killers to ease the pain, cortisol injections, and in acute cases, joint replacement. All these treatments come with its own set of side effects and are really costly. *A. vera* is a succulent xerophyte growing almost in every corner of the world, apart from being numerously beneficial for human health; its easy availability is also an advantage. *A. vera* plays a major role in treatment of osteoarthtitis. *A. vera* has well-established anti-inflammatory, antioxidant, immunostimulatory, pain relieving properties, but recently, it has attracted much attention by its role in tissue repair and regeneration. Components of *A. vera* stimulate chondrocyte differentiation, increase collagen, and proteoglycan synthesis and restore extracellular matrix. The main polysaccharide, acemannan in *A. vera*, stimulates the growth factors via macrophage which acts on osteoarthritic cartilage to promote repair and remodeling. Thus, *A. vera* can be considered as a potential therapy for osteoarthritis. *A. vera* may be applied tropically on area of injured cartilage to ease the pain and inflammation but oral administration of *A. vera* may be much more beneficial as it will be specific in action and may have quicker results. But many aspects like pH sensitivity and colon digestive profile may affect the efficacy of the components of *A. vera*.

If *A. vera* can be delivered by coating, it with any biodegradable, nonionic polymer, its colon release profile, and pH sensitivity can be improved, resulting in high beneficial effect. Considering the huge capabilities of *A. vera* in medical field, further research can bring new spheres in curing many chronic diseases which have no established treatment till date, just like osteoarthritis.

KEYWORDS

- *Aloe vera*
- side effects
- mucilaginous gel
- **extracellular matrix degeneration**
- **inflammation**

REFERENCE

1. Davis, R. H. *Aloe vera: A Scientific Approach*. Vantage Press: New York, 1997. ISBN: 0-533-12137-X.
2. Hamman, J. H. Composition and Applications of Aloe vera Leaf Gel. *Molecules* **2008**, *13*(8), 1599–1616. DOI:10.3390/molecules13081599.
3. Atherton, P. *The Essential Aloe vera: The Actions and the Evidence*, 2nd ed., 1997.
4. Ro, J. Y.; Lee, B.; Kim, J. Y.; Chung, Y.; Chung, M. H.; Lee, S. K.; et al. Inhibitory Mechanism of Aloe Single Component (Alprogen) on Mediator Release in Guinea Pig Lung Mast Cells Activated with Specific Antigen–Antibody Reactions. *J. Pharmacol. Exp. Ther.* **2000**, *292*, 114–121.
5. Femenia, E. S.; Sanchez, S. S.; Rossello, C. Compositional Features of Polysaccharides from *Aloe vera* (*Aloe barbadensis* Miller) Plant Tissues. *Carbohydr. Polym.* **1999**, *39*(2), 109–117.
6. Djeraba, A.; Quere, P. *In Vivo* Macrophage Activation in Chickens with Acemannan, a Complex Carbohydrate Extracted from *Aloe vera*. *Int. J. Immunopharmacol.* **2000**, *22*(5), 365–372.
7. Lee, J. K.; Lee, M. K.; Yun, Y. P.; Kim, Y.; Kim, J. S.; Kim, Y. S.; Kim, K.; Han, S. S.; Lee, C. K. Acemannan Purified from *Aloe vera* Induces Phenotypic and Functional Maturation of Immature Dendritic Cells. *Int. J. Immunopharmacol.* **2001**, *1*(7), 1275–1284.
8. Cawston, T. E.; Wilson, A. J. Understanding the Role of Tissue Degrading Enzymes and their Inhibitors in Development and Disease. *Best Pract. Res. Clin. Rheumatol.* **2006**, *20*, 983–1002.
9. Mantovani, A.; Biswas, S. K.; Galdiero, M. R.; Sica, A.; Locati, M. Macrophage Plasticity and Polarization in Tissue Repair and Remodeling. *J. Pathol.* **2012**, *229*, 176–185 (wileyonlinelibrary.com). DOI:10.1002/path.4133.
10. Park, Y. I.; Jo, T. H. Perspectives of Industrial Application of *Aloe vera*. In: *New Perspectives on Aloe*; Park, Y. I., Lee, S. K., Eds.; Springer Science + Business Media: New York, NY, 2006; pp 191–200.
11. Grindlay, D.; Reynolds, T. The *Aloe vera* Phenomenon: A Review of the Properties and Modern Uses of the Leaf Parenchyma Gel. *J. Ethnopharmacol.* **1986**, *16*(2–3), 117–151. DOI:10.1016/0378-8741(86)90085-1. PMID:3528673.
12. Subhash, A. V.; Suneela, S.; Anuradha, Ch.; Bhavani, S. N.; Srinivas Minor Babu, M. The Role of *Aloe vera* in Various Fields of Medicine and Dentistry. *J. Orofac. Sci.* **2014**, *6*(1), 5–9. DOI:10.4103/0975-8844.132564.
13. Balasubramanian, J.; Narayanan, N. *Aloe vera*: Nature's Gift. *Species* **2013**, *2*(6), 3–4.
14. Steenkamp, V.; Stewart, M. J. Medicinal Applications and Toxicological Activities of *Aloe* Products. *Pharm. Biol.* **2007**, *45*(5), 411–420. DOI:10.1080/13880200701215307.
15. Brown, J. P. "A Review of the Genetic Effects of Naturally Occurring Flavonoids, Anthraquinones and Related Compounds. *Mutat. Res.* **1980**, *75*(3), 243–277.
16. Chandegara, V. K.; Varshney, A. K. Effect of Centrifuge Speed on Gel Extraction from *Aloe vera* Leaves. *J. Food Proc. Technol.* **2014**. http://dx.doi.org/10.4172/2157-7110.1000295.
17. Ramachandra, C. T.; Srinivasa Rao, P. Processing of *Aloe vera* Gel: A Review. *Am. J. Agric. Biol. Sci.* **2008**, *3*(2), 502–510. ISSN 1557-4989.

18. Ni, Y.; Turner, D.; Yates, K. M.; Tizard, I. Isolation and Characterisation of Structural Components of *Aloe vera* L. Leaf Pulp. *Int. Immunopharmacol.* **2004**, *4*, 1745–1755.
19. Waller, T. A.; Pelley, R. P.; Strickland, F. M. Industrial Processing and Quality Control of *Aloe barbadensis* (*Aloe vera*) Gel. In: *Aloes: The genus Aloe*; Reynolds, T., Ed.; CRC Press: London, 2004, pp 139–205.
20. Dicesare, P. E.; Abramson, S. B. Pathogenesis of Osteoarthritis. In: *Kelley's Textbook of Rheumatology*, vol. II, 7th ed.; Harris, E. D., Budd, R. C., Genovese, M. C.; et al. Ed.; Elsevier Saunders, 2005; pp 1493–1513.
21. Stolk, L. M.; Hoogtanders, K. Detection of Laxative Abuse by Urine Analysis with HPLC and Diode Array Detection. *Int. J. Clin. Pharmacy* **1999**, *21*(1), 40–43. DOI:10.1023/A:1008647424809. PMID:10214668.
22. West, D. P.; Zhu, Y. F. Evaluation of *Aloe vera* Gel Gloves in the Treatment of Dry Skin Associated with Occupational Exposure. *Am. J. Infect. Control* **2003**, *31*, 40–42.
23. Shelton, M. *Aloe vera*—Its Chemical and Therapeutic Properties. *Int. J. Dermatol.* **1991**, *30*, 678–683. http://dx.doi.org/10.1111/j.1365-4362.1991.tb02607.x. PMID:1823544.
24. Ni, Y.; Tizard, I. R. Analytical Methodology: The Gel-analysis of Aloe Pulp and Its Derivatives. In: *Aloes The Genus Aloe*; Reynolds, T., Ed.; CRC Press: Boca Raton, 2004; pp. 111–126.
25. Ni, Y.; Turner, D.; Yates, K. M.; Tizard, I. Isolation and Characterization of Structural Components of *Aloe vera* L. Leaf Pulp. *Int. J. Immunopharmacol.* **2004**, *4*(14), 1745–1755.
26. Dagne, E.; Bisrat, D.; Viljoen, A.; Van Wyk, B.-E. Chemistry of *Aloe* Species. *Curr. Organ. Chem.* **2000**, *4*, 1055–1078.
27. Choi, S.; Chung, M.-H. A Review on the Relationship between *Aloe vera* Components and their Biologic Effects. *Semin. Integrat. Med.* **2003**, *1*, 53–62.
28. de Rodríguez, D. J.; Hernández-Castillo, D.; Rodríguez- García, R.; Angulo-Sanchez, J. L. Antifungal Activity *In Vitro* of *Aloe vera* Pulp and Liquid Fraction against Plant Pathogenic Fungi. *Ind. Crops Prod.* **2005**, *21*(1), 81–87.
29. Surjushe, A.; Vasani, R.; Saple, D. G. *Aloe vera*: A Short Review. *Indian J. Dermatol.* **2008**, *13*, 1599–1616.
30. Felson, D. T.; Chaisson, C. E.; Hill, C. L.; Totterman, S. M. S.; Elon Gale, M.; Skinner, K. M.; Kazis, L.; Gale, D. R. The Association of Bone Marrow Lesions with Pain in Knee Osteoarthritis. *Ann. Int. Med.* **2001**, *134*, 541–549.
31. Sinusas, K. Osteoarthritis: Diagnosis and Treatment. *Am. Fam. Phys.* **2012**, *85*(1), 49–56.
32. Ulrich-Vinther, M.; Maloney, M. D.; Schwarz, E. M.; Rosier, R.; O'Keefe, R. J. Articular Cartilage Biology. *J. Am. Acad. Orthopaed. Surg.* **2003**, *11*, 421–430.
33. Sophia Fox, A. J.; Bedi, A.; Rodeo, S. A. The Basic Science of Articular Cartilage: Structure, Composition, and Function. *Sports Health* **2009**, *1*(6), 461–468. DOI:10.1177/1941738109350438.
34. Martin, J. A.; Buckwalter, J. A. Articular Cartilage Ageing and Degeneration. *Ports Med. Arthrosc. Rev.* **1996**, *4*, 263–275.
35. Martin, J. A.; Buchwalter, J. A. Human Chondrocytes Senescence and Osteoarthritis. *Biotechnology* **2002**, 39, 145–152.

36. Buckner, P.; Mendler, M.; Steinmann, B.; et al. The Structure of Human Collagen Type IX an its Organization in Fetal and Infant Cartilage Fibrils. *J. Biol. Chem.* **1988**, *263*, 16911–16917.
37. Maroudas, A. Physiochemical Properties of Articular Cartilage. In: *Adult Articular Cartilage*; Freeman, M. A. R., Ed.; Cambridge University Press: Kent, UK, 1979; pp 215–290.
38. Buckwalter, J. A.; Mankin, H. J. Articular Cartilage: Part I: Tissue Design and Chondrocytes Matrix Interactions. *J. Bone Joint Surg.* **1997**, 79-A, 600–611.
39. Juhl, C.; Christensen, R.; Roos, E., Lund, H.. *Dose-response of Exercise Therapy for Patients with Osteoarthritis of the Knee. a Meta-regression Analysis (MEREX)*. Institute of Sports Science and Biomechanics, University of Southern Denmark, Odense, Denmark, 2014.
40. Felson, D. T. The Epidemiology of Knee Osteoarthritis: Results from the Framingham Osteoarthritis. *Study Semin. Arthritis Rheum.* 1990.
41. Doherty, M.; Jones, A.; Cawston, T. Osteoarthritis. In: *Oxford Textbook of Rheumatology*, 3rd ed.; Isenberg, D. A.; et al., Eds. Oxford University Press: Oxford, 2004, pp. 1091–1118.
42. Wang, Q.; Rozelle, A. L.; Lepus, C. M.; Scanzello, C. R.; Song, J. J.; Larsen, D. M.; Crish, J. F. Identification of a Central Role for Complement in Osteoarthritis. *Nature Med.* **2011**, *17*, 1674–1679. DOI:10.1038/nm.2543.
43. Felson, D. T.; Chaisson, C. E.; Hill, C. L.; Totterman, S. M. S.; Elon Gale, M.; Skinner, K. M.; Kazis, L.; Gale, D. R. The Association of Bone Marrow Lesions with Pain in Knee Osteoarthritis. *Ann. Intern. Med.* **2001**, *134*, 541–549.
44. Benito, M. J.; Veale, D. J.; FitzGerald, O.; van den Berg, W. B.; Bresnihan, B. Synovial Tissue Inflammation in Early and Late Osteoarthritis. *Ann. Rheum. Dis.* **2005**, *64*, 1263–1267.
45. Goldring, M. B. Update on the Chondrocyte Lineage and Implications for Cell Therapy in Osteoarthritis. In: *Osteoarthritis: A Companion to Rheumatology*, 1st ed.; Sharma, L., Berenbaum, F., Eds.; Mosby, Inc., an affiliate of Elsevier, Inc.: Philadelphia, PA, 2007; pp 53–76.
46. Goldring, M. B. Are Bone Morphogenetic Proteins Effective Inducers of Cartilage Repair? Ex Vivo Transduction of Muscle-derived Stem Cells. *Arthritis Rheum.* **2006**, *54*, 387–389.
47. Goldring, M. B.; Berenbaum, F. The Regulation of Chondrocyte Function by Proinflammatory Mediators: Prostaglandins and Nitric Oxide. *Clin. Orthoped. Relat. Res.* **2004**, *427*, S37–S46.
48. Goldring, S. R.; Goldring, M. B. The Role of Cytokines in Cartilage Matrix Degeneration in Osteoarthritis. *Clin. Orthoped. Relat. Res.* **2004**, *423*, S27–S36.
49. Reboul, P.; Pelletier, J. P.; Tardif, G.; Cloutier, J. M.; Martel-Pelletier, J. The New Collagenase, Collagenase-3, Is Expressed and Synthesized by Human Chondrocytes But Not by Synoviocytes. A Role in Osteoarthritis. *J. Clin. Investig.* **1996**, *97*, 2011–2019.
50. Okada, Y.; Takeuchi, N.; Tomita, K.; Nakanishi, I.; Nagase, H. Immunolocalization of Matrix Metalloproteinase 3 (Stromelysin) in Rheumatoid Synovioblasts (B Cells): Correlation with Rheumatoid Arthritis. *Ann. Rheum. Dis.* **1989**, *48*, 645–653.

51. Ishiguro, N.; Ito, T.; Obata, K.; Fujimoto, N.; Iwata, H. Determination of Stromelysin-1, 72 and 92 kDa Type IX Collagenase, Tissue Inhibitor of Metalloproteinase-1 (TIMP-1), and TIMP-2 in Synovial Fluid and Serum from Patients with Rheumatoid Arthritis. *J. Rheumatol.* **1996,** 23, 1599–1604.
52. Ito, A.; Nagase, H. Evidence that Human Rheumatoid Synovial Matrix Metalloproteinase 3 is an Endogenous Activator of Procollagenase 1. *Arch. Biochem. Biophys.* **1998,** 267, 211–216.
53. Ishiguro, N.; Kojima, T.; Robin Poole, A. Mechanism of Cartilage Destruction in Osteoarthritis. *Nagoya J. Med. Sci.* **2002,** 65, 73–84.
54. Cole, A. A.; Chubinskaya, S.; Schumacher, B.; Huch, K.; Szabo, G.; Yao, J.; Mikecz, K.; Hasty, K. A.; Kuettner, K. E.; et al. Chondrocyte Matrix Metalloproteinase-8. Human Articular Chondrocytes Express Neutrophil Collagenase. *J. Biol. Chem.* **1996,** *271,* 11023–11026.
55. De Falco, L.; Fioravanti, A.; Galeazzi, M.; Tenti, S. Bradykinin and its Role in Osteoarthritis. *Reumatismo* **2013,** 65(3), 97–104. DOI:10.4081/reumatismo.2013.97.
56. Bellucci, F.; Meini, S.; Cucchi, P.; Catalani, C.; Nizzardo, A.; Riva, A.; et al. Synovial Fluid Levels of Bradykinin Correlate with Biochemical Markers for Cartilage Degradation and Inflammation in Knee Osteoarthritis. *Osteoarthritis Cartilage* **2013**. http://dx.doi.org/10.1016/j.joca.
57. Mantovani, A.; Sozzani, S.; Locati, M.; Allavena, P.; Sica, A. Macrophage Polarization: Tumor-associated Macrophages as a Paradigm for Polarized M2 Mononuclear Phagocytes. *Trends Immunol.* **2002,** 23, 549–555.
58. Bergenfelz, C.; Medrek, C.; Ekstrom, E.; Jirström, K.; Janols, H.; Wullt, M.; Bredberg, A.; Leandersson, K. Wnt5a Induces a Tolerogenic Phenotype of Macrophages in Sepsis and Breast Cancer Patients. *J. Immunol.* **2012,** *188,* 5448–5458.
59. Mantovani, A.; Sica, A.; Sozzani, S.; Allavena, P.; Vecchi, A.; Locati, M. The Chemokine System in Diverse Forms of Macrophage Activation and Polarization. *Trends Immunol.* **2004,** 25, 677–686.
60. Fortier, L. A.; Barker, J. U.; Strauss, E. J.; McCarrel, T. M.; Cole, B. J. The Role of Growth Factors in Cartilage Repair. *Clin. Orthop. Relat. Res.* **2011,** *469*(10), 2706–2715. 2011. Published online: 15 March.
61. Fortier, L. A.; Hackett, C. H.; Cole, B. J. The Effects of Platelet-Rich Plasmaon Cartilage: Basic Science and Clinical Application. *Oper. Tech. Sports Med.* **2011,** *19*(3):154–159.
62. Blaney Davidson, E. N.; Vitters, E. L.; van Lent, P. L.; van de Loo, F. A.; van den Berg, W. B.; van der Kraan, P. M. Elevated Extracellular Matrix Production and Degradation Upon Bone Morphogenetic Protein-2 (BMP-2) Stimulation Point Toward a Role for BMP-2 in Cartilage Repair and Remodeling. *Arthritis Res. Ther.* **2007,** *9,* R102.
63. Redini, F.; Mauviel, A.; Pronost, S.; Loyau, G.; Pujol, J. P. Transforming Growth Factor Beta Exerts Opposite Effects from Interleukin-1 beta on Cultured Rabbit Articular Chondrocytes through Reduction of Interleukin-1 Receptor Expression. *Arthritis Rheumatol.* **1993,** *36*(1), 44–50.
64. Chandrasekhar, S.; Harvey, A. K. Transforming Growth Factor-beta is a Potent Inhibitor of IL-1 Induced Protease Activity and Cartilage Proteoglycan Degradation. *Biochem. Biophys. Res. Commun.* **1988,** *157,* 1352–1359.

65. Konig, H. G.; Kogel, D.; Rami, A.; Prehn, J. H. TGF-{beta} 1 Activates Two Distinct Type I Receptors in Neurons: Implications for Neuronal NF-{kappa}B Signaling. *J. Cell Biol.* **2005,** *168,* 1077–1086.
66. Finnson, K. W.; Parker, W. L.; ten Dijke, P.; Thorikay, M.; Philip, A. ALK1 Opposes ALK5/Smad3 Signaling and Expression of Extracellular Matrix Components in Human Chondrocytes. *J. Bone Miner. Res.* **2008,** *23,* 896–906.
67. Longobardi, L.; O'Rear, L.; Aakula, S.; Johnstone, B.; Shimer, K.; Chytil, A.; Horton, W. A.; Moses, H. L.; Spagnoli, A. Effect of IGF-I in the Chondrogenesis of Bone Marrow Mesenchymal Stem Cells in the Presence or Absence of TGF-beta Signaling. *J. Bone Miner. Res.* **2006,** *21,* 626–636.
68. Badlani, N.; Inoue, A.; Healey, R.; Coutts, R.; Amiel, D. The Protective Effect of OP-1 on Articular Cartilage in the Development of Osteoarthritis. *Osteoarthritis Cartilage* **2008,** *16,* 600–606.
69. Lee, C. H.; Cook, J. L.; Mendelson, A.; et al. Regeneration of the Articular Surface of the Rabbit Synovial Joint by Cell Homing: A Proof of Concept Study. *Lancet* **2010,** 376:440–448.
70. Van Beuningen, H. M.; Glansbeek, H. L.; van der Kraan, P. M.; van den Berg, W. B. Differential Effects of Local Application of BMP-2 or TGF-beta 1 on Both Articular Cartilage Composition and Osteophyte Formation. *Osteoarthritis Cartilage* **1998,** *6,* 306–317.
71. Blaney Davidson, E. N.; van der Kraan, P. M.; van den Berg, W. B. TGF-β and Osteoarthritis. *Osteoarthritis Cartilage* **2007,** *15*(6), 597–604. DOI:10.1016/j.joca.02.005.
72. Hayashi, M.; Muneta, T.; Ju, Y. J.; Mochizuki, T.; Sekiya, I. Weekly Intra-articular Injections of Bone Morphogenetic Protein-7 Inhibits Osteoarthritis Progression. *Arthritis Res. Ther.* **2008,** *10,* R118.
73. Elshaier, A. M.; Hakimiyan, A. A.; Rappoport, L.; Rueger, D. C.; Chubinskaya, S. Effect of Interleukin-1beta on Osteogenic Protein 1-Induced Signaling in Adult Human Articular Chondrocytes. *Arthrit. Rheumatol.* **2009,** *60,* 143–154.
74. Chubinskaya, S.; Hurtig, M.; Rueger, D. C. OP-1/BMP-7 in Cartilage Repair. *Int. Orthopaed.* **2007,** *31*(6), 773–781. DOI:10.1007/s00264-007-0423-9.
75. Loeser, R. F.; Carlson, C. S.; Del Carlo, M.; Cole, A. Detection of Nitrotyrosine in Aging and Osteoarthritic Cartilage: Correlation of Oxidative Damage with the Presence of Interleukin-1beta and with Chondrocyte Resistance to Insulin-like Growth Factor 1. *Arthrit. Rheumatol.* **2002,** *46,* 2349–2357.
76. Bucholz, R. W.; Einhorn, T. A.; Marsh, J. L. In: *Bone and Joint Healing*, 6th ed.; Bucholz, R. W.; Heckman, J. D.; Court-Brown, C., Eds.; Lippincott Williams & Wilkins, 2006; pp. 300–11.
77. Hoeben, A.; Landuyt, B.; Highley, M. S.; et al. Vascular Endothelial Growth Factor and Angiogenesis. *Pharmacol. Rev.* **2004,** *56*(4), 549–580.
78. Kwon, D. R.; Park, G. J. Intra-articular Injections for the Treatment of Osteoarthritis: Focus on the Clinical Use of Several Regimens. *Osteoarthritis—Diagn., Treat. Surg.* **2012,** 67–100.
79. Loeser, R. F.; Pacione, C. A.; Chubinskaya, S. The Combination of Insulin-like Growth Factor 1 and Osteogenic Protein 1 Promotes Increased Survival of and Matrix Synthesis by Normal and Osteoarthritic Human Articular Chondrocytes. *Arthrit. Rheumatol.* **2003,** *48,* 2188–2196.

80. Morales, T. I. The Quantitative and Functional Relation between Insulin-like Growth Factor-I (IGF) and IGF-binding Proteins During Human Osteoarthritis. *J. Orthopaed. Res.* **2008**, *26*, 465–474.
81. Nalesso, G.; Sherwood, J.; Bertrand, J.; Pap, T.; Ramachandran, M.; De Bari, C.; Pitzalis, C.; Dell'Accio, F. WNT-3A Modulates Articular Chondrocyte Phenotype by Activating both Canonical and Noncanonical Pathways. *J. Cell Biol.* **2011**, *193*, 551–564.
82. Monroe, D. G.; McGee-Lawrence, M. E.; Oursler, M. J.; Westendorf, J. J. Update on Wnt Signaling in Bone Cell Biology and Bone Disease. *Gene* **2012**, *492*, 1–18.
83. Logan, C. Y.; Nusse, R. The Wnt Signaling Pathway in Development and Disease. *Annu. Rev. Cell Dev. Biol.* **2004**, *20*, 781–810.
84. Golovchenko, S.; Hattori, T.; Hartmann, C.; Gebhardt, M.; Gebhard, S.; Hess, A.; Pausch, F.; Schlund, B.; von der Mark, K. Deletion of Beta Catenin in Hypertrophic Growth Plate Chondrocytes Impairs Trabecular Bone Formation. *Bone* **2013**, *55*, 102–112.
85. Miclea, R. L.; Siebelt, M.; Finos, L.; Goeman, J. J.; Löwik, C. W. G. M.; Oostdijk, W.; et al. Inhibition of Gsk3beta in Cartilage Induces Osteoarthritic Features through Activation of the Canonical Wnt Signaling Pathway. *Osteoarthritis Cartilage* **2011**, *19*, 1363–1372.
86. Loughlin, J.; Dowling, B.; Chapman, K.; Marcelline, L.; Mustafa, Z.; Southam, L.; et al. Functional Variants within the Secreted Frizzled-related Protein 3 Gene are Associated with Hip Osteoarthritis in Females. *Proc. Nat. Acad. Sci. U.S.A.* **2004**, *101*, 9757–9762.
87. Dell'Accio, F.; de Bari, C.; El Tawil, N. M.; Barone, F.; Mitsiadis, T. A.; O'Dowd, J.; Pitzalis, C. Activation of WNT and BMP Signaling in Adult Human Articular Cartilage Following Mechanical Injury. *Arthrit. Res. Ther.* **2006**, *8*, R139.
88. Lodewyckx, L.; Cailotto, F.; Thysen, S.; Luyten, F. P.; Lories, R. J. Tight Regulation of Wingless-type Signaling in the Articular Cartilage—Subchondral Bone Biomechanical Unit: Transcriptomics in Frzb-Knockout Mice. *Arthrit. Res. Ther.* **2012**, *14*, R16.
89. Nalesso, G.; Sherwood, J.; Bertrand, J.; Pap, T.; Ramachandran, M.; De Bari, C.; Pitzalis, C.; Dell'Accio, F. WNT-3A Modulates Articular Chondrocyte Phenotype by Activating both Canonical and Noncanonical Pathways. *J. Cell Biol.* **2011**, *193*, 551–564.
90. Chun, J. S.; Oh, H.; Yang, S.; Park, M. Wnt Signaling in Cartilage Development and Degeneration. *BMB Rep. Online* **2008**, *41*, 485–494.
91. Zhu, M.; Tang, D.; Wu, Q.; Hao, S.; Chen, M.; Xie, C.; Rosier, R. N.; O'Keefe, R. J.; Zuscik, M.; Chen, D. Activation of Beta-catenin Signaling in Articular Chondrocytes Leads to Osteoarthritis-like Phenotype in Adult Beta-catenin Conditional Activation Mice. *J. Bone Miner. Res.* **2009**, *24*, 12–21.
92. Cochrane Collaboration. *Aloe vera for Treating Acute and Chronic Wounds* (Review). JohnWiley & Sons, Ltd., 2012.
93. Che, Q. M.; Akao, T.; Hattori, M.; Kobashi, K.; Namba, T. Isolation of Human Intestinal Bacteria Capable of Transforming Barbaloin to Aloe-Emodin Anthrone. *Planta Med.* **1991**, *57*(1), 15–19. http://dx.doi.org/10.1055/s-2006-960007.
94. Ito, S.; Teradaira, R.; Beppu, H.; Obata, M.; Nagatsu, T.; Fujita, K. Properties and Pharmacological Activity of Carboxypeptidase in *Aloe arborescens* Mill. var.

Natalensis Berger. *Phytother. Res.* **1993,** *7*(7), S26–S29. http://dx.doi.org/10.1002/ptr.2650070710.
95. Hanley, D. C.; Solomon, W. A.; Saffran, B.; Davis, R. H. The Evaluation of Natural Substances in the Treatment of Adjuvant Arthritis. *J. Am. Pediatr. Med. Assoc.* **1982,** *72*, 275–284.
96. Haller, J. S. A Drug for All Seasons, Medical and Pharmacological History of *Aloe*. *Bull. N. Y. Acad. Med.* **1990,** *66*, 647–659.
97. Davis, R. H.; Parker, W. L.; Samson, R. T.; Murdoch, D. P. Isolation of a Stimulatory System in an Aloe Extract. *J. Am. Pediatr. Med. Assoc.* **1991,** *81*, 473–478.
98. Lewis, W. H.; Elvin-Lewis, M. *Medical Botany, Plants Affecting Mans Health*. John Wiley Interscience: New York, NY, 1977.
99. Guyton, A. C. *Textbook of Medical Physiology*, 8th ed. WB Saunders: Philadelphia, PA, 1991.

PART IV
Industrial Waste Mitigation

CHAPTER 14

ADSORPTION OF REACTIVE DYE 21 ON FLY ASH AND MnO$_2$-COATED FLY ASH ADSORBENT: BATCH AND CONTINUOUS STUDIES

DEEPIKA BRIJPURIYA[1*], MANOJ JAMDARKAR[2], PRATIBHA AGRAWAL[3], and TAPAS NANDY[1]

[1]*NEERI, Nehru Marg, Nagpur, India*

[2]*Shri M. Mohota College of Science, Nagpur, India*

[3]*Laxminarayan Institute of Technology, Nagpur, India*

*Corresponding author. E-mail: dipikagupta.nagpur@gmail.com

CONTENTS

Abstract		270
14.1	Introduction	270
14.2	Materials and Methods	271
14.3	Results and Discussion	276
14.4	Column Study	283
14.5	Conclusions	285
Keywords		287
References		287

ABSTRACT

The adsorption of reactive blue 21 (RB 21) on fly ash and MnO_2-coated fly ash was studied in batch and column modes of operation. Various experimental parameters were studied including initial dye concentration (2.5–50 ppm), contact time, pH, and temperature (25–35°C). The experimental data were analyzed using Langmuir, Freundlich isotherm model. The adsorption RB 21 for fly ash and treated fly ash was found to follow Langmuir and Freundlich isotherm, respectively. The fly ash and MnO_2-coated fly ash were characterized by scanning electron microscopy, X-ray fluorescence, and Brunauer–Emmett–Teller. The column adsorption experiments were carried out in glass columns, filled with fly ash and sand mixtures with different proportion (1:1, 1:2, and 1:3). Removal of dye decreases with the increase in solution concentration at room temperature, showing the process to be highly dependent on the concentration of the solution. It is concluded from the present study that MnO_2-coated fly ash is a potential and active low-cost adsorbent for the removal of reactive dye from its aqueous solution and industrial wastewater.

14.1 INTRODUCTION

Dyeing process is main source of pollution in textile industry. The disposal of dye-laden wastewater poses one of the industry's major problems because such effluents contain a number of contaminants including acid or caustic, dissolved and suspended solids, toxic compounds, and coloring pigments. Dye removal is one of the biggest challenges in treating dyeing wastewater because of their complex molecular structures, which make them more stable and difficult to treat.

Color is the first contaminant to be recognized in wastewater. The presence of these dyes at 0.005 ppm in water is highly visible. The occurrence of color in water reduces aquatic diversity by blocking the passage of light through water.

Reactive dyes are the largest single group of dyes used in textiles industry and their removal is the most problematic compound among other dyes in textile wastewater. It is highly water-soluble and estimated that 10–20% of reactive dyes remains in the wastewater generated during the production process.[1]

The removal of dyes from industrial waste before they are discharged into water bodies is therefore very important from health and hygiene point of view and for environmental protection in general.[2] Different treatment methods are reported in literature, including biological, chemical, and physical method.[3] Among these methods, adsorption has been proven to be more efficient, offering advantages over conventional processes, biological and chemical processes.[4]

Adsorption is an effective method for the separation of pollutants in wastewater and an efficient treatment option. Among various factors, adsorption capacity of adsorbent also depends on the characteristics of material namely specific surface area; pore size, pore volume, and its distribution; etc.

Use of granulated-activated carbon or powdered-activated carbon is most common. However, these adsorbents are expensive and their regeneration or disposal has several problems. Thus, to make the process customer friendly, the use of several low cost adsorbents can be targeted. Locally available natural material can minimize or avoid the concerns and significantly reduce the treatment cost. This has led to search for cheaper and simpler substitutes. New approaches based on the use of natural, inexpensive adsorbent materials for effluent treatment have been reported.[5–10]

The present work was undertaken to explore the feasibility of finding an effective low cost adsorbent, from the easily available materials. Use of clay minerals,[11] bottom fly ash,[12–18] waste apricot,[19] rice husk,[20–22] sawdust,[20–23] fungi,[24] waste materials from agriculture,[25,26] pomegranate peel[27] as adsorbent is reported.

14.2 MATERIALS AND METHODS

Fly ash (Fig. 14.1 (A)) was procured from Koradi Thermal Power Plant, Nagpur, Maharashtra. It was washed several times with water to remove impurities. Then, it was dried in an electric oven at 105°C for 24 h. The fly ash was passed over the standard size mesh no. 45 (354 µm) molecular sieve (Endecott). The sand was also passed over standard size mesh no. 14 (1.41 mm). Dye was acquired from Ludhiana, Punjab. The dye used in the analysis was reactive blue 21 (RB 21). RB 21 is commonly called copper phthalocyanine dye and is readily soluble in water, ethanol, and dimethyl formamide. Properties of RB 21 are given below.

A different concentration of synthetic dye water was prepared by dissolving reactive dye in distilled water. The concentration of 2.5–50 mg/l was selected after literature survey. Each dye solution was made in small batches of 2 l in a glass container. The chemicals used in the present study were of analytical grade and the details are given in Table 14.1.

TABLE 14.1 Molecular Formula, Formula Weight, Structure, and Color CAS Details of Chemical Used.

Color index dyes	CAS No.	Molecular formula	Formula weight	Structure
Reactive blue 21	12236-86-1	$C_{41}H_{23}N_{14}O_{14}S_5Cl_1Na_4Cu_1$	1126.71	

14.2.1 TREATMENT OF FLY ASH

Treatment of fly ash for increasing the adsorption efficiency for dye removal is reported.[28,29] It was reported that MnO_2 was an effective water treatment agent. Thus, treatment of fly ash was carried out in laboratory to increase the adsorption capacity of fly ash as a low cost adsorbent. An amount of 20 g raw fly ash was mixed with 50 mmol of $MnSO_4$ (Merck) in 70 ml aqueous solution and maintained at constant temperature of 80°C using water bath for 20 min. under magnetic stirring. A volume of 350 ml $KMnO_4$ aqueous solution (33 mmol) was added gradually in the above reaction mixture. Addition was done with continuous magnetic stirring, which results in the immediate formation of brown precipitation of MnO_2. After complete addition of $KMnO_4$, the reaction mixture was kept for stirring at 80°C for 15 min. Then, it was allowed to cool at room temperature. Washing was done several times by centrifugation (Cooling Centrifuge-REMI) using distilled water and then, the MnO_2-coated fly ash (Fig. 14.1(B)) was dried at 100°C overnight.

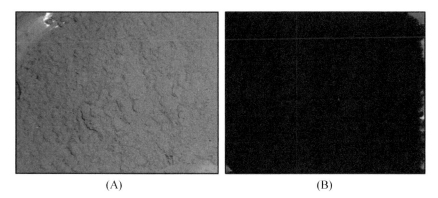

FIGURE 14.1 (A) Fly ash and (B) MnO_2-coated fly ash.

14.2.2 CHARACTERIZATION OF ADSORBENTS

Fly ash and MnO_2-coated fly ash were first characterized by different technique to know its surface area, pore size, pore volume, chemical contents, and mineralogical composition. The surface area, pore size, and total pore volume of the raw fly ash and MnO_2-coated fly ash were measured through N_2 adsorption–desorption method Brunauer–Emmett–Teller (BET) (Micrometrics). The chemical content of the raw fly ash and MnO_2-coated fly ash were analyzed using X-ray fluorescence (XRF) technique. The surface morphology and quality of raw fly ash and MnO_2-coated fly ash was characterized by a high resolution scanning electron microscopy (SEM) (JEOL, JXA-840).

14.2.3 ADSORPTION OF DYE

14.2.3.1 BATCH STUDY

Batch experiments were conducted to determine the optimum pH, dye concentration, adsorbent dose, contact time, and temperature for color removal using fly ash and MnO_2-coated fly ash for reactive dye 21 (RB 21) solution. As a general practice, all laboratory glassware used in the experiments was washed with detergent solution and then with dilute hydrochloric acid (HCl). The glassware were then thoroughly rinsed with tap water and finally with distilled water to prevent any traces of residual color.

14.2.3.2 BATCH ADSORPTION STUDIES AT VARIOUS OPERATING CONDITIONS

14.2.3.2.1 Effect of Dye Solution pH

The pH is a key factor for determining the optimal conditions for dye adsorption onto fly ash. pH studies were conducted by shaking 100 ml of RB 21 solutions of 10 mg/l with 1 g of fly ash and MnO_2-coated fly ash for 90 min over a range of initial pH values from 2.5 to 11.5 at 298 K. About 0.1 N HCl or 0.1 N NaOH solutions were used for pH adjustment. The samples were shaken in 250 ml glass bottles having stoppers on a shaking incubator (Tempo Make) at a speed of 100 RPM. Blank solution containing 100 ml of only dye solution without any adsorbent was shaken simultaneously as control to determine the impact of pH change on solution color. The sample after 90 min of shaking were allowed to settle for 1 h followed by filtering the samples through 0.45-μm glass microfiber and subjected for the measurement of absorbance value.

14.2.3.2.2 Effect of Contact Time and Temperature

Studies were carried out to evaluate the effect of temperature on dye solution. Three temperatures were selected are 288, 298, and 308 K. Each dye solutions of 10 mg/l concentration were shaken in 250-ml glass bottle with constant dosage of both the adsorbent 1 g/100 ml of RB 21 on a shaking incubator (Tempo Make). One gram of fly ash and MnO_2-coated fly ash in 100 ml of the dye solution was shaken starting from 2 to 120 min, until equilibrium was reached and no further dye/color was removed upon continuous shaking. Blank runs with only the adsorbents in 100 ml of distilled water were conducted simultaneously at similar condition, to assess for any color leached by both the adsorbents. The samples were filtered through 0.45-μm glass microfiber prior to concentration measurements for equilibrium color removal.

14.2.3.2.3 Effect of Initial Concentration and Adsorbent Quantity

A series of 250-ml glass bottles containing 100-ml solution of RB 21 ranging from 2.5 to 50 mg/l with different quantities fly ash and MnO_2-coated fly ash varying from 0.5 to 2.0 g were shaken at 100 RPM in shaking incubator (Tempo) for 90 min at the temperature of 298 K. The samples after shaking were immediately filtered through 0.45-µm glass microfiber (Whatmann Make). The filtered solution was then measured for color in terms of concentration using calibration graph of RB 21 on a UV–visible spectrophotometer (Shimadzu). Blank were conducted simultaneously with varying amount of fly ash and MnO_2-coated fly ash in 100 ml of distilled water.

14.2.3.2.4 Adsorption Isotherm

Once pH, equilibrium time, temperature, optimum adsorbent dosage, and dye concentration were optimized for both the adsorbents—dye combinations, isotherm studies were conducted by varying the mass of adsorbent used.

Hundred milliliter of each dye solution was shaken with varying quantity of fly ash and MnO_2-coated fly ash, from 0.5 to 2.0 g on a shaking incubator (Tempo Make) at 100 RPM for optimum contact time period. The sample after shaking were allowed to settle for 1 h and then filtered through 0.45-µm glass microfiber. The filtered solutions were then measured for color in terms of concentration unit on a UV–visible spectrophotometer (Shimadzu). Blanks were conducted simultaneously with varying quantities of both the adsorbents in 100 ml of distilled water, the color leached by both the adsorbents were measured and subsequently used to adjust for the actual color values.

14.2.3.3 COLUMN STUDY

Column experiments were carried out to measure removal of RB 21 on both the adsorbents. Different combination of fly ash:sand (1:1, 1:2, 1:3) were tried in order to increase the filtration rate. The column experiments were conducted using a glass column (4.1 cm inner diameter, 10 cm in height). The column was packed with the different ratio of fly ash:sand between two supporting layers of glass wool and glass beads. The dye solution was fed through the fixed bed column in the down flow mode. Before starting the experiment, the bed was rinsed with distilled water and left overnight to ensure a closely packed arrangement of particles with no void or cracks. The dye solution was passed through the column using a peristaltic pump (Watson Marlow). The samples were collected at specific interval of time and measured for the dye concentration by UV–visible spectrophotometer. The flow to the column was continued until the effluent concentration approached nearly to influent concentration of the dye solution. The effect of both dye concentration at constant bed height and flow rate was examined.

14.3 RESULTS AND DISCUSSION

14.3.1 CHARACTERIZATIONS OF ADSORBENTS

14.3.1.1 BRUNAUER–EMMETT–TELLER

The specific surface area, pore size, and pore volume of fly ash obtained from the N_2 equilibrium adsorption isotherm (BET) were found to be 1.2530 ± 0.0112 m² g⁻¹, 70.9976 Å, and 0.002224 cm³ g⁻¹, respectively.

The specific surface area, pore size, and pore volume of MnO_2-coated fly ash are 69.5820 ± 0.1960 m² g⁻¹, 48.9785 Å, and 0.085200 cm³ g⁻¹, respectively, which causes drastic increase in BET surface area. Larger the specific surface area and the finer the particle size distribution of adsorbent, the greater its adsorption capacity and interaction with an adsorbates are.[28]

14.3.1.2 X-RAY FLUORESCENCE SPECTROMETRY

The chemical composition obtained using XRF shows that the major components of raw fly ash are SiO_2: 60.36%, Al_2O_3: 23.71% while for MnO_2-coated fly ash the major component are SiO_2: 44.65%, MnO_2: 19.75%. Other chemicals present in both the adsorbent as estimated are given in Table 14.2.

TABLE 14.2 Chemicals Present in Both the Adsorbent.

Sample	Al_2O_3	Fe_2O_3	SiO_2	MnO_2	MgO	CaO	LOI
Raw fly ash	23.71	6.60	60.36	0.06	0.57	0.78	0.22
MnO_2-coated fly ash	16.19	3.20	44.65	**19.75**	0.41	0.57	6.88

LOI, Loss on ignition. All values are in percentage.

14.3.1.3 SCANNING ELECTRON MICROSCOPY

Fine particles, that is, microstructure of fly ash showed an important role on the adsorption capacity. Fly ash particles are generally spherical in shape. The investigation revealed that most of the particles present in the fly ash are spherical in shape with a relatively smooth surface grain.

A marked change in surface morphology was observed, when the MnO_2-coated fly ash is compared with the raw fly ash. The raw fly ash comprises porous spherical particles (Fig. 14.2), whereas Figure 14.3 shows that MnO_2 accumulated on fly ash particles, covering it unevenly.

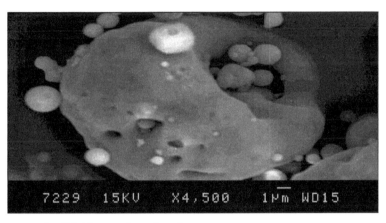

FIGURE 14.2 SEM image of raw fly ash.

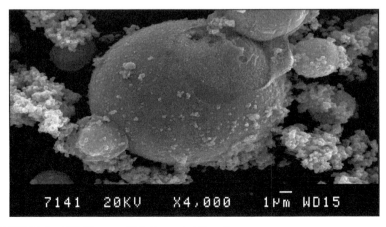

FIGURE 14.3 SEM image of MnO_2-coated fly ash.

14.3.2 BATCH ADSORPTION STUDY

14.3.2.1 EFFECT OF DYE SOLUTION PH

The effect of pH on the adsorption of RB 21 using raw fly ash and MnO_2-coated fly ash were studied by varying the pH of the solutions ranging from 2.5 to 11.5 using 0.1 N NaOH and HCl solutions at 298 K. The results (Fig. 14.4) indicate that both adsorbents showed removal was optimum at an initial pH of 2.5, ~71% for raw fly ash and ~100% for MnO_2-coated fly ash, whereas the percentage adsorption at pH 7 for raw fly ash was ~66% and for MnO_2-coated fly ash was ~79%, respectively. At alkaline pH (11.5), the removal efficiency decreased drastically (raw fly ash: ~7% and MnO_2-coated fly ash: ~28%). However, maintaining dye solution pH 2.5, the solution shows acidic behavior which may not be directly dispersed off into the environment.

Since the pH of an aqueous dye solution is about 7.0, the following experiments were made without the pH adjustment of the dye solutions.

14.3.2.2 EFFECT OF CONTACT TIME AND TEMPERATURE

Adsorbent have fixed capacity of adsorption and therefore adsorption is affected by the contact time. Temperature change will also change the free energy and entropy of adsorption. So, to evaluate the effect of contact time of the aqueous media between dye and adsorbent and temperature, the agitation time was varied between 2 and 160 min for both raw fly ash and MnO_2-coated fly ash. The results are shown in Figures 14.5 and 14.6.

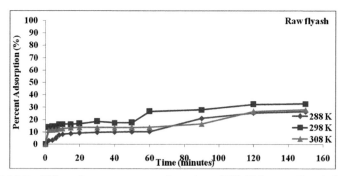

FIGURE 14.5 Effect of temperature on adsorption of RB 21 by raw fly ash (initial dye concentration: 10 mg/l, contact time: 2–160 min, adsorbent dose: 1.0 g/100 ml).

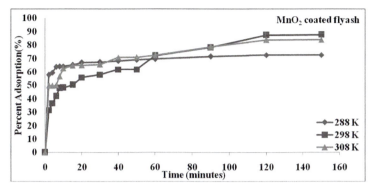

FIGURE 14.6 Effect of temperature on adsorption of RB 21 by MnO_2-coated fly ash (initial dye concentration: 10 mg/l, contact time: 2–160 min, adsorbent dose: 1.0 g/100 ml).

It was observed that optimum equilibrium adsorption time and temperature for raw fly ash was established within 90–120 min at 298 K, whereas for MnO_2-coated fly ash, the equilibrium time was same, that is, 90–120 min but optimum temperature was found to be 308 K. The removal of color in different time interval using both the adsorbent are shown in Figures 14.7 and 14.8.

FIGURE 14.7 Dye (RB 21) removal in different time interval using fly ash.

It was observed that adsorption of dyes followed the trend of initial rapid followed by gradual adsorption. The dye molecules first encounter the boundary layer effect and thereafter adsorb on the surface, and this is followed by diffusion into the porous structure of the adsorbent which takes a longer time.[12]

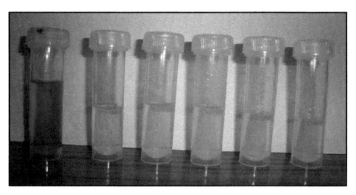

FIGURE 14.8 Dye (RB 21) removal in different time interval using MnO_2-coated fly ash.

14.3.2.3 EFFECT OF INITIAL DYE CONCENTRATION

The adsorption capacity of raw fly ash and MnO_2-coated fly ash for reactive dye RB 21 was determined at different initial dye concentrations (10–50 mg/l). It was observed that dye removal efficiency for RB 21 reached up to 33% and 88% (2.5 mg/l) at lower concentration, then decreasing to less than 6% and 77% at higher concentration (50 mg/l) for raw fly ash and MnO_2-coated fly ash, respectively. Higher dye adsorption achieved at low initial concentration was because of the availability of unoccupied binding sites on the adsorbents.[11] The results presented in Figure 14.9 show that the amount of dye adsorbed decreases with increase in the initial dye concentration

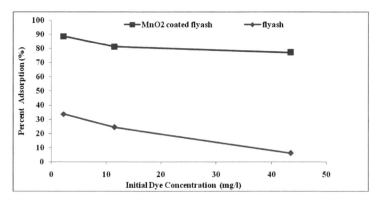

FIGURE 14.9 Effect of initial dye concentration on percent adsorption and dye adsorption capacity using raw fly ash and MnO_2-coated fly ash (initial dye concentration: 2.5–50 mg/l, contact time: 90 min, adsorbent dose: 1.0 g/100 ml, temperature: 298 K).

of solution from 2.5 to 50 mg/l at 298 K, indicating the inverse relation between adsorption and initial dye concentration confirming that the adsorption process to be highly dependent on the initial dye solution concentration.

14.3.2.4 ADSORPTION ISOTHERM

The calculation to determine the percent adsorption of the dyes was made according to the equation:

$$\% \text{ Removal} = \frac{C_0 - C_e}{C_0}. \tag{14.1}$$

The percentage removal of the dye and the amount adsorbed (mg g^{-1}) were calculated using the following relationship:

$$Q_e = \frac{(C_0 - C_e)}{m}. \tag{14.2}$$

Langmuir developed a theoretical equilibrium isotherm relating the amount of gas adsorbed on a surface to the pressure of the gas. The linear form of Langmuir isotherm equation is given as

$$\frac{1}{Q_e} = \frac{C_e}{Q_0} + \frac{1}{Q_0 K_L} \tag{14.3}$$

C_e/Q_e was plotted against C_e using linear regression analysis as shown in Figure 14.10. The constants Q_0 and K_L were determined from the intercept and slope of the linear plots, respectively. The Q_0 from Langmuir isotherm for raw fly ash were 0.536, 0.287, 0.255, and 0.189 mg g^{-1}, while the values of K_L were found to be 0.374, 0.420, 0.512, and 1.14 L mg^{-1}, respectively.

Freundlich developed an empirical model which can be applied to nonideal adsorption on heterogeneous surfaces[13] as well as multilayer adsorption. The Freundlich model[14] equation is expressed as

$$Q_e = K_F C_e^{1/n} \tag{14.4}$$

The linear logarithmic form of Freundlich is given by the following equation:

$$\log Q_e = \log K_F + \frac{1}{n} \log C_e \tag{14.5}$$

where K_F is defined as an adsorption or distribution coefficient and represents the amount of adsorbate adsorbed on an adsorbent for a unit equilibrium concentration. The slope $1/n$ is a measure of the adsorption intensity or surface heterogeneity.[11] The slope ranges between 0 and 1 and is a measure of adsorption intensity or surface heterogeneity becoming more heterogeneous as its value gets closer to zero. A value below unity implies chemisorptions process, where $1/n$ above one is an indicative of cooperative adsorption.[15]

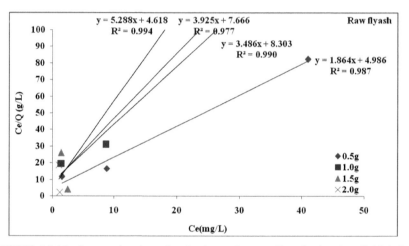

FIGURE 14.10 Langmuir adsorption isotherm for raw fly ash showing [initial dye concentration range: 2.5–50 mg/l, contact time: 90 min, temperature: 298 K].

Figure 14.11 presents the plot of ln Q_e against ln C_e. The Freundlich isotherm indicate that K_F values for MnO_2-coated fly ash were found as 0.238, 0.154, 0.121 and 0.095 mg g^{-1}, and the values of $1/n$ were 0.188, 0.339, 0.463 and 0.554, respectively.

Freundlich isotherms tend to fit the experimental data better at low concentrations, whereas Langmuir isotherms fit better at higher concentrations.[16] For both the isotherms, the accuracy balancing the isotherm model to experimental equilibrium data which was typically assessed based on the coefficient of correlation for determination for the linear regression, that is, the isotherm giving an R^2 value closest to unity was deemed to provide the best fit.

Equilibrium adsorption data of RB 21 onto raw fly ash and MnO_2-coated fly ash concludes that raw fly ash fitted to Langmuir adsorption isotherm indicating monolayer adsorption process was favored, while MnO$_2$-coated fly ash fitted to Freundlich isotherm confirming multilayer adsorption process occurring.

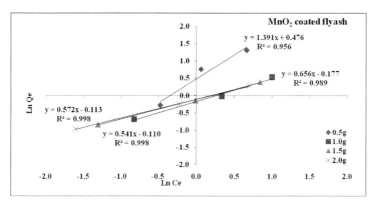

FIGURE 14.11 Freundlich adsorption isotherm for MnO$_2$-coated fly ash showing [initial dye concentration range: 2.5–50 mg/l, contact time: 90 min, temperature: 298 K].

14.4 COLUMN STUDY

The removal of dye by adsorption on column composed of (fly ash and sand) and (MnO$_2$-coated fly ash and sand) mixtures reach equilibrium condition when the adsorbent mixture stops adsorbing the solute dye. The adsorption equilibrium point is indicated by the concentration of the dye in the filtrate, the time beyond which no further color removal is observed. At this point, the concentration of the dye in the filtrate is nearly equal to the initial feed concentration.

Maximum removal was observed for both the adsorbent at low flow rate (1.2 ml/min). This is because exhaustion time was increased with decrease in flow rate. This may be due to the fact that at a low rate of influent the reactive dye had more time to be in contact with the adsorbent which resulted in a greater removal of all reactive dye molecules in column.[30] At higher flow rate, the adsorption capacity was lower due to sufficient in residence time of the solute in the column for diffusion of the solute into the pores of the adsorbent to take place, and therefore the solute left the column before equilibrium occurred.[31]

For column fed with 1:1 ratio of raw fly ash and sand, the removal efficiency was found to be 88–92% for RB 21 which slightly decreased to nearly 88% for 1:2 ratio and to below 80% for 1:3 ratio at low concentration as shown in Figure 14.12. In case of MnO_2-coated fly ash at low flow rate (1.2 and 1.5 ml/min), the removal efficiency for RB 21 was above 90%, whereas at 2.5 ml/min, the removal efficiency was decreased to 80–85% at low concentration, respectively, as shown in Figure 14.13.

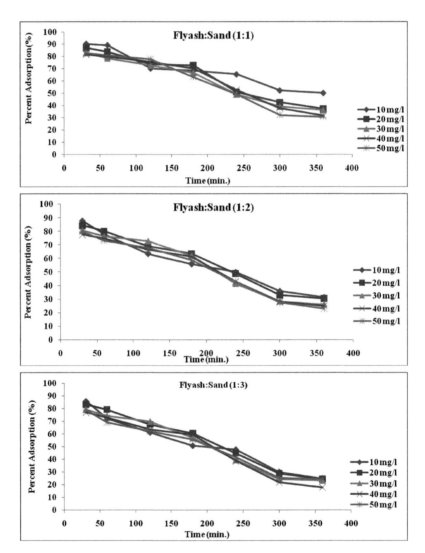

FIGURE 14.12 Percent adsorption of reactive blue 21 from aqueous solution of different concentrations in media mixture of fly ash and sand in different ratio.

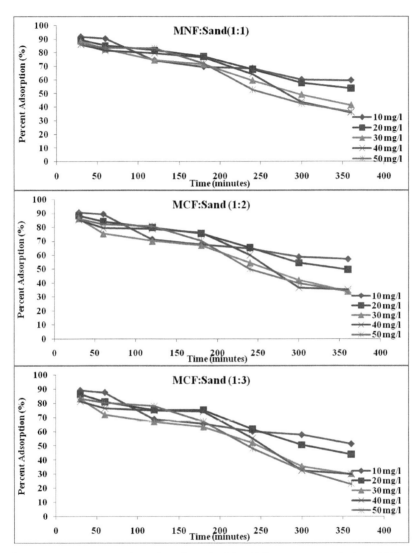

FIGURE 14.13 Percent adsorption of reactive blue 21 from aqueous solution of different concentrations in media mixture of MnO_2-coated fly ash (MCF) and sand in different ratio.

14.5 CONCLUSIONS

The chemical composition of raw fly ash and MnO_2-coated fly ash were observed to be the following:

Raw fly ash: SiO_2 (60.36%), Al_2O_3 (23.71%), Fe_2O_3 (6.60%), CaO (0.78%), MgO (0.57%), MnO_2 (0.06%), and loss on ignition (LOI) (0.22%).

MnO_2-coated fly ash: SiO_2 (44.65%), Al_2O_3 (16.19%), Fe_2O_3 (3.20%), CaO (0.57%), MgO (0.41%), MnO_2 (19.75%), and LOI (6.88%). The higher percentage of SiO_2 led to an increase in dye adsorption, whereas higher content of SiO_2 and MnO_2 helped in more efficient dye removal in aqueous solution.

The characteristics of raw fly ash and MnO_2-coated fly ash were as indicated in Table 14.3.

TABLE 14.3 The Characteristics of Raw Fly Ash and MnO_2-coated Fly Ash.

Parameters	Raw fly ash	MnO_2-coated fly ash
BET surface area ($m^2\ g^{-1}$)	1.2530 ± 0.0112	69.5820 ± 0.1960
Adsorption average pore width (Å)	70.9976	48.9785
Pore volume ($cm^3\ g^{-1}$)	0.002224	0.085200

The SEM images clearly show that raw fly ash particles are mainly spherical, smooth, and porous, whereas MnO_2-coated fly ash particles are irregular with rough surface and accumulated with cubic structure of MnO_2 unevenly. Dyes adsorption study for color removal conducted at different pH (2.5–11.5) and dosage 0.5–2.0 g/100 ml, varying reactive dye concentration 2.5–50 mg/l, and reaction time 2–120 min. Based on the study, the optimum dosage and operating conditions for dye adsorption were as shown in Table 14.4.

TABLE 14.4 The Optimum Dosage and Operating Conditions for Dye Adsorption.

Sr. No.	Parameter	Raw fly ash	MnO_2-coated fly ash
1	pH	7.0 (~66%)	7.0 (~79%)
2	Concentration	10 mg/l	10 mg/l
3	Time	90 min	90 min
4	Dose	1 g/100 ml	1 g/100 ml
5	Temperature	298 K	308 K

In column technique, at low flow rate (1.2 ml/min) both adsorbent shows good removal efficiency. Still, MnO_2-coated fly ash is best suitable and cheap adsorbent for treatability of the selected reactive dyes.

KEYWORDS

- adsorption
- reactive dye
- adsorption isotherm
- fly ash
- MnO_2-coated fly ash
- batch and continuous studies

REFERENCES

1. Panswad, T.; Luangdilok, W. Effect of Chemical Structures of Reactive Dyes on Color Removal by Anaerobic–Aerobic Process. *Water Res.* **2000,** 34.
2. Bhattacharya, G. K.; Sharma, A. Adsorption of Pb(II) from Aqueous Solution by *Azadirachta indica* (Neem) Leaf Powder. *J. Hazard. Mater.* **2004,** 113.
3. Tarley, C. R. T.; Arruda, M. A. Z. Biosorption of Heavy Metals Using Rice Milling By-products. Characterization and Application for Removal of Metals from Aqueous Effluents. *Chemosphere* **2004,** 54.
4. Ozacar, M.; Sengil, I. A. Adsorption of Acid Dyes from Aqueous Solution by Calcined Alunite and Granular Activated Carbon. *J. Hazard. Mater.* **2003,** 98.
5. Vaughan, T.; Seo, C. W.; Marshall, W. E. Removal of Selected Metal Ions from Aqueous Solution Using Modified Corncobs. *Biores. Technol.* **2001,** 78.
6. Gupta, V. K.; Suhas. Application of Low-cost Adsorbents for Dye Removal—A Review. *J. Environ. Manage.* **2009,** 90.
7. Lata, H.; Grag, V. K.; Gupta, R. K. Dye Removal from Aqueous Solution by Adsorption on Treated Sawdust. *Dyes Pigm.* **2007,** 74.
8. Hameed, B. H.; Din, A. T. M.; Ahmad, A. L. Equilibrium and Kinetic Studies on Basic Dye Adsorption by Oil Palm Fibre Activated Carbon. *J. Hazard. Mater.* **2007,** 141.
9. Hashemian, S.; Dadfarnia, S.; Nateghi, M. R.; Gafoori, F. Sorption of Acid Red 138 from Aqueous Solutions onto rice Bran. *Afr. J. Biotechnol.* **2008,** 17.
10. Panda, G. C.; Das, S. K.; Guha, A. K. Jute Stick Powder as a Potential Biomass for the Removal of Congo Red and Rhodamine B from their Aqueous Solution. *J. Hazard. Mater.* 2009, 164.

11. Khan, T. A. Ali, I.; Singh, V. V.; Sharma, S. Utilization of Flyash as Low-cost Adsorbent for the Removal of Methylene Blue, Malachite Green and Rhodamine b Dyes from Textile Wastewater. *J. Environ Protect. Sci.* **2009**, 3.
12. Malik, A.; Taneja, U. Utilizing Flyash for Color Removal of Dye Effluents. *American Dyestuff Reporter*, 1994.
13. Wang, S. B.; Li, H. Dye Adsorption on Unburned Carbon: Kinetics and Equilibrium. *J. Hazard. Mater.* **2005**, 126.
14. Gupta, V. K.; Mittal, A.; Krishnan, L.; Gajbe, V. Adsorption Kinetics and Column Operations for the Removal and Recovery of Malachite Green from Wastewater Using Bottom Ash. *Sep. Purif. Technol.* **2004**, 40.
15. Rachakornij, M.; Ruangchay, S.; Teachakulwing, S. *J. Sci. Technol.* **2004**, 26.
16. Dizge, N.; Aydiner, C.; Demirbas, E.; Kobya, M.; Kara, S. Adsorption of Reactive Dyes from Aqueous Solutions by Fly Ash: Kinetic and Equilibrium Studies. *J. Hazard. Mater.* 2008, 150.
17. Chatterjee, D.; Patnam, V. R.; Sikdar, A.; Moulik, S. K. Removal of Some Common Textile Dyes from Aqueous Solution Using Flyash. *J. Chem. Eng. Data* **2010**, 55.
18. Pengthamkeerati, P.; Satapanajaru, T.; Singchan, O. Sorption of Reactive Dye from Aqueous Solution on Biomass Fly Ash. *J. Hazard. Mater.* 2008, 1539.
19. Gupta, V. K.; Mohan, D.; Sharma, S.; Sharma, M. Removal of Basic Dyes (Rhodamine Band Methylene Blue) from Aqueous Solutions Using Bagasse Flyash. *Sep. Sci. Technol.* **2000**, 35.
20. Basar, C. A. Applicability of the Various Adsorption Models of Three Dyes Adsorption onto Activated Carbon Prepared Waste Apricot. *J. Hazard. Mater.* **2006**, 135.
21. Sivakumar, V. M.; Thirumarimurugan, M.; Xavier, A. M.; Sivalingam, A.; Kannadasan, T. *Int. J. Biosci. Biochem. Bioinform.* **2012**, 2.
22. Gidde, M. R.; Dutta, J.; Jadhav, S. Comparative Adsorption Studies on Activated Rice Husk and Rice Husk Ash by Using Methylene Blue as Dye. *Int. Congress Environ. Res.* **2004**.
23. Mittal, P. K. Dye Removal from Wastewater Using Activated Carbon Developed from Sawdust: Adsorption Equilibrium and Kinetics. *J. Hazard. Mater.* **2004**, 113.
24. Malik, P. K. Use of Activated Carbons Prepared from Saw Dust and Ricehusk for Adsorption of Acid Dyes: A Case Study. *Dyes Pigm.* **2003**, 56.
25. Chander, M.; Arora, D. S. Evaluation of Some White-rot Fungi for their Potential to Decolourise Industrial Dyes. *Dyes Pigm.* **2007**, 72.
26. Dabrowski. Adsorption—From Theory to Practice. *Adv. Colloid Interface Sci.* 2001, 93.
27. Ali, H. A.; Egzar, H. K.; Kamal, N. M.; Abdulsaheb, N.; Mashkour, M. S. Removal of Amaranth Dye from Aqueous Solution Using Pomegranate Peel. *Int. J. Basic Appl. Sci.* **2013**.
28. Koa, P. N.; Tzeng, J. H.; Huang, T. L. Removal of Chlorophenols from Aqueous Solution by Fly Ash. *J. Hazard. Mater.* **2000**, 76.
29. Bada, S. O.; Potgieter-Vermaak, S. Evaluation and Treatment of Coal Flyash for Adsorption Application. *Leonardo Electron. J. Practices Technol.* **2008**, 12.

30. Han, R.; Wang, Y.; Zhao, X.; Wang, Y. F.; Xie, F.; Cheng, J.; Tan, M. Adsorption of Methylene Blue by Phoenix Tree Leaf Powder in a Fixed-bed Column, Experiments and Prediction of Breakthrough Curves. *Desalination* **2009,** 243.
31. Tan, I. A. W.; Ahmad, A. L.; Hameed, B. H. Adsorption of Basic Dye Using Activated Carbon Prepared from Oil Palm Shell, Batch and Fixed Bed Studies. *Desalination* **2008,** 225.

CHAPTER 15

TURNING WASTE INTO ZEOLITE 4A RESIN AND DELINEATION OF THEIR ENVIRONMENTAL APPLICATIONS: A REVIEW

S. U. MESHRAM[1], B. R. GAWHANE[2], and P. B. SUHAGPURE[2]

[1]*Department of Applied Chemistry, Laxminarayan Institute of Technology, RTM Nagpur University, Amravati Road, Nagpur-33, Nagpur, India*

[2]*Department of Oil, Fats and Surfactants Technology, Laxminarayan Institute of Technology, RTM Nagpur University, Amravati Road, Nagpur-33, Nagpur, India*

*Corresponding author. E-mail: sidmesh2@gmail.com

CONTENTS

Abstract	292
15.1 Introduction	292
15.2 Framework Structure of Zeolite A	293
15.3 Nomenclature	294
15.4 Synthesis of Zeolite 4A	295
15.5 Commercial Method for Production of Zeolite 4A	296
15.6 Synthesis of Zeolite 4A Using Waste Sources	297
15.7 Environmental Applications of Zeolite 4A Resin	301
Keywords	305
References	306

ABSTRACT

Zeolites are hydrated, microporous crystalline alumina-silicate with three dimensional framework made up of T-O-T (T = Si, Al) bonds with enclosed cages/channels of uniform dimensions. This property enables them to avail as membrane to separate the molecules on the basis of their size; hence they are also known as molecular sieves. Zeolites are widely demanded inorganic resins in the field of water and waste water treatment, credited to its ion exchange, molecular sieving, adsorption, detergent builder, water softening, etc. properties. Many leading detergent manufactures use Zeolite 4A as builder material, which has proven to be a better alternative to phosphatic additives.

Although literature studies reveal the potential applicability of many types of Zeolites, it has been noticed that Zeolite 4A resin is manufactured commercially on a mega scale. In spite of its tremendous potential, the high cost of Zeolite 4A has limited its bulk utility. In this reference, many investigations have been reported on the eco-friendly production of Zeolite 4a resin using waste materials such as flash, rice husk ash, composite ash generated from coal and biomass co-firing, aluminum waste, ceramic and clay minerals, etc. The current paper accounts the efforts attempted by researchers towards the manufacture of Zeolite 4A with economic consideration along with delineation applications.

15.1 INTRODUCTION

Zeolite was first discovered by Swedish mineralogist A. F. Cronstedt in 1756. Zeolites are crystalline microporous aluminosilicates with very well-defined structures that consist of a framework formed by tetrahedral units of SiO_4 and AlO_4. The isomorphism substitution of Al^{3+} and Si^{4+} in the tetrahedra results in a negative charge on the Zeolite framework that can be balanced by exchangeable cations.[1] Interstitial cations in Zeolites can be exchanged to fine-tune the pore size of Zeolites. For example, the sodium form of Zeolite A has a pore opening of approximately 4 Å (4A molecular sieve). If Na^+ is exchanged with the K^+, the pore opening is reduced to approximately 3 Å; Ca^{2+} replaces $2Na^+$; thus, the pore opening increases to approximately 5 Å.[2] Zeolites are having the following formula: $Mx/n[(Al_2O_3) \times (SiO_2)y] \cdot zH_2O$, where M is an exchangeable cation located in the voids.[3] These voids define many special properties of Zeolites, such as the adsorption of molecules in the huge internal channels. The ability to adjust the pores precisely determines the uniform openings to allow for molecules smaller than its pore diameter to be adsorbed whilst excluding larger molecules, hence the name "molecular sieves." As a consequence of the peculiar structural properties of Zeolites, they have a wide range of industrial applications,[4] mainly based on

- *Ion exchange*: Exchange of inherent $Na^+/K^+/Ca^{2+}$ for other cations on the basis of ion selectivity.
- *Water adsorption*: Reversible adsorption of water without any desorption chemical or physical change in the Zeolite matrix.
- *Gas adsorption*: Selective absorption of specific gas molecules.

There is significant interest in the preparation of Zeolite 4A owing to its wide-spread industrial applications in separation processes as a sorbent, membrane, detergent builder, and catalyst, etc.

Initially, only natural Zeolites are used, but more recently, synthetic forms have been made on.

Industrial scale is giving rise to tailor made Zeolites that are highly reproducible.

15.2 FRAMEWORK STRUCTURE OF ZEOLITE A

The primary building unit for Zeolites is the tetrahedron and the secondary building units (SBUs) are the geometric arrangements of tetrahedra.[5] In most Zeolite structures, the primary structural units, the AlO_4 or SiO_4 tetrahedra, are assembled into SBUs which may be simple polyhedra, such as cubes, hexagonal prisms, or cubo-octahedra.[6] The structures can be formed by repeating SBUs and according to them, Zeolites can be classified into eight groups. Figure 15.1 is presenting the components of the Zeolite structure, respectively.

Zeolites have an open-structure framework which consist of many channels and/or interconnected voids of discrete size (in the range 0.3–20 Å) which are occupied by cations and water molecules. Each AlO_4 tetrahedron in the framework bears a net negative charge which is balanced by a cation. The cations can reversibly be exchanged for other ions possessing the same sign charge when aqueous salt solution is passed through channels and voids. This replacement results in the narrowing of the pore diameter of the Zeolite channels. The SBUs can be a simple arrangement of tetrahedra such as 4, 6, 8, 10, or more complicated rings. Other factors such as the location, size, and coordination of the extra-framework cations can also influence the pore size.

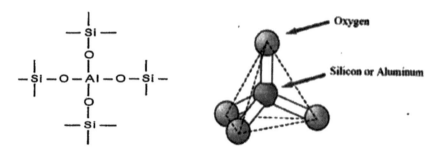

FIGURE 15.1 Chemical structure and primary building unit of Zeolite structure.

The Zeolites can be primarily categorized on the basis of (i) their silica/alumina ratio and (ii) their pore opening with specific dimensions. The pore diameter of Zeolites of type A is between 3 and 5 Å. Using these characteristic properties, the structures can be named accordingly. The nomenclature is represented in Table 15.1.

TABLE 15.1 Classification of Zeolite.

Sr. no.	Zeolite	SiO_2/Al_2O ratio	Pore openings (Å)
1	Zeolite A (Na$^+$)	1.0–1.2	4
2	Ca^{2+} Zeolite	1.0–1.5	5
3	K$^+$ Zeolite	1.0–1.5	3
4	Zeolite-X	1.5–2.0	7–8
5	Zeolite-Y	2.0–2.5	7–13
6	Siliceous-Y	2.5–4.0	7–13
7	Mordenite/ZSM-5	8.0–20	13–15

15.3 NOMENCLATURE

The wide variety of possible Zeolite structures is due to the large number of ways in which the SBUs can be linked to form various polyhedra. One such polyhedron is the truncated octahedron, better known as the sodalite cage. Each sodalite cage consists of 24 linked tetrahedral which are further linked to form different Zeolites with distinct framework topologies[7] as depicted in Figure 15.2. Each type of Zeolite has specific uniform pore size. According to database, Zeolite A is classified into three different grades 3A, 4A, and 5A, all of which possess the same general formula but have different cation type. On the basis of crystallographic reports, the Structure Commission of the International Zeolite Association determines and assigns a three letter code to Zeolite "topology."[8]

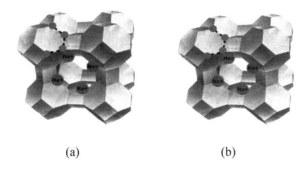

(a)   (b)

FIGURE 15.2 Frame work topologies of Zeolite 3A (a) and Zeolite 4A (b).

15.4 SYNTHESIS OF ZEOLITE 4A

With the awareness of society toward the depletion of energy resources, it is time to move on to develop other methods of fulfilling the requirement of energy. Biomass is an effective alternative to alleviate this problem. Today, synthetic Zeolites are used commercially more often than natural Zeolites. Greening the manufacture of Zeolites 4A for wide range of industrial applications and simultaneously lowering production cost is a pervasive challenge. The sources for early synthesized Zeolites were standard chemical reagents. The main advantages of synthetic Zeolites are that they can be engineered with a wide variety of chemical properties and pore sizes and that they have greater thermal stability. Conventional Zeolite synthesis involves the hydrothermal crystallization of aluminosilicate gels or solutions in a basic environment. The type of the Zeolite is affected by various factors such as composition of the reaction mixture, nature of reactants and their pretreatments, temperature of the process, pH of the reaction mixture (pH > 10), and other factors.[9,10]

At present, the main problem existing in Zeolite synthesis is the availability and cost of raw material specifically the silica sources. The preparation of synthetic Zeolites from silica and alumina chemical sources is expensive. Yet, cheaper raw materials such as clay minerals, natural Zeolites, coal ashes, municipal solid waste incineration ashes, and industrial slags are utilized as starting materials for Zeolite synthesis. The use of waste materials in Zeolite synthesis contributes to the mitigation of environmental problems.

15.5 COMMERCIAL METHOD FOR PRODUCTION OF ZEOLITE 4A

It is reported that Zeolite 4A can be commercially produced using the hydro-gel process. The typical raw materials include aqueous solution of sodium silicate (Na_4SiO_4)·H_2O, Al_2O_3·$3H_2O$, and sodium aluminate ($NaAlO_2$) with caustic soda to maintain high pH during reaction.[11] These materials are mixed with water in a tank, followed by gelation, aging, and hydrothermal crystallization, as shown in Figure 15.3. The reactions are presented as follows:

$2NaOH + Al_2O_3 \cdot 3H_2O \rightarrow 2NaAlO_2$ (sodium aluminate) $+ 4H_2O$

Zeolite 4A gel formation

Na_4SiO_4 (sodium silicate) $+ NaAlO_2$ (sodium aluminate) $+ H_2O \rightarrow 2Na_2O:Al_2O_3:1.75SiO_2 - 70H_2O$

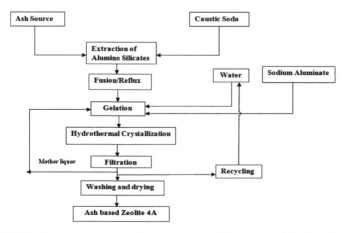

FIGURE 15.3 Commercial method for production of fly-ash-based Zeolite 4A.

15.6 SYNTHESIS OF ZEOLITE 4A USING WASTE SOURCES

15.6.1 FLY-ASH-BASED ZEOLITE 4A

In India, National Environmental Engineering Research Institute, Nagpur, has conducted pioneering work in the field of Zeolite synthesis. Various processes have been developed for synthesis of Zeolite from fly ash, which have been reported by several patents and research articles.[12,13] Currently, there is increasing interest in the synthesis of Zeolite from by-products such as coal fly ash (CFA), rice husk ash (RHA), and bagasse fly ash, etc.[14–19] CFA is the waste product of combustion of coal in a coal-fired thermal power station.[20] CFA represents major composition as oxides of silicon and aluminum with iron oxide up to certain extents are having the largest industrial solid-waste generation. It has been reported that approximately 750 million tons of CFA is produced globally each year; on average, only 25% of this is utilized and the rest is disposed of as a waste causing yet another environmental concern.[21] Shih and Chang investigated the effects of curing temperature and chemical composition on formulation of two types of Zeolites, MS 4A and MS 13A, using Class F coal ash at lower temperature (only 311 K), but the treatment time required for synthesizing Zeolite 4A coal more than 3 days was longer enough.[22]

A number of researchers have used hydrothermal process successfully by varying the temperature within the range of 333–573 K. Fusion of the alkali–fly ash mixture facilitates the formation of highly active Na-aluminate and silicates, which are readily soluble in water and enhance Zeolite formation. Addition of sodium aluminate to the fly ash before fusion brought the success in obtaining NaA Zeolite.[23] Chang observed that the addition of aluminum hydroxide to the fused coal ash solution followed by hydrothermal treatment at 60°C produced Zeolite 4A depending on the source of ash. The results confirm that the quantity of dissolved aluminum species is critical for the type of Zeolite formed from fused CFAs.[24] The present result is, therefore, very much useful in opening up a way to synthesize Zeolite at low cost with useful applications. Figure 15.4 is pointing various bituminous coal-based power plants in Chandrapur and Rajasthan, India.

FIGURE 15.4 Pure coal-based thermal power plants (TPP) in India.

15.6.2 RICE-HUSK-ASH-BASED ZEOLITE 4A

Globally, about 600 million tons of rice paddies are produced per annum.[25] This is an alternative silica source instead of the pure chemical sources earlier, because it is less selective and highly active.[26] The large amount of silica freely obtained from RHA provides an abundant and cheap alternative of silica sources for many industrial uses, including synthesis of Zeolite. Reports are available on process for production of Zeolite A from RHA generated from biomass-based power stations.

The RHA was obtained by combusting the rice husk at different temperature and durations, that is, 450–750°C for 2–6 h. The resulting black ash contains silica (SiO_2) which varies from 85% to 98% depending on the burning conditions, the furnace type, the rice variety, and the rice husk moisture content. NaA type of Zeolite was synthesized from RHA with high purity and absence of impurities via the hydrothermal condition. They produce more than 90% amorphous silica from RHA.[27]

Hadi Nur has reported the direct synthesis of Zeolite 4A from rice husk and carbonaceous RHA. The quality of carbonaceous RHA-based Zeolite 4A was higher than that of normal rice husk.[28] Younesi formulated nano-Zeolite NaA from rice husk at room temperature without organic additives by hydrothermal method.[29] Figure 15.5 is depicting the various rice husk-based mini power plants in Andhra Pradesh and Bihar, India.

FIGURE 15.5 Pure rice husk-based mini power plants in India.

15.6.3 BAGASSE FLY-ASH-BASED ZEOLITE 4A

Bagasse and sugarcane straw wastes are generated by the sugar and alcohol production processes. Sugarcane straw is the material that is removed before the cane is crushed and comprises the dried/fresh leaves and the top of the plant.[30,31] Combustion of sugarcane bagasse in boiler for steam and power generation produces a great amount of another solid waste which contains significant amount of carbon as well as silica, denominated sugarcane bagasse ash (SCBA).[32] Therefore, the development of new procedure for its productive reuse is relevant. By means of an alkali fusion extraction method, quartz particles can be dissolved and used as a silicon source for synthesizing silica materials such as Zeolites.[33] The SCBA can be successfully used as a raw material for the hydrothermal synthesis of Zeolite A. Some authors have reported the conversion of bagasse fly ash into Zeolites.[34] Zeolite was successfully developed from the low cost sugarcane straw ash in the absence of organic templates, without addition of aluminum solution and without fusion prior to hydrothermal treatment.[35] Figure 15.6 is pointing various biomass-based mini power plants in Maharashtra, Andhra Pradesh, and Karnataka, India.

FIGURE 15.6 Pure biomass (bagasse)-based mini-power plants in India.

15.6.4 COMPOSITE-ASH-BASED ZEOLITE 4A

Composite ash represents the physico-chemical composition of coal and biomass simultaneously. The conversion of this ash into a value-added product, Zeolite 4A, can help to reduce the environmental burden of composite ash, in addition to CFA, to a great extent.[36] Majority of the electrical energy in our country is produced from the combustion of coal and hence, the rate of consumption of coal from natural reservoirs is quite high. Searching for a suitable alternative of coal for energy generation is a need of hour. To resolve this issue, most of the mini-thermal power stations pertaining power generation capacity in the range of 5–20 MW have already started the process using rice husk as bio fuel. Since the calorific value of rice husk is almost two-third of coal, it is preferred to blend it with coal for cost-effective production of energy. During combustion, it reduces the CO_2 emission responsible for global warming up to certain extent, since rice husk is having low-carbon composition. Zeolite A resin was prepared using composite ash as a source material for extraction of sodium silicate which acts as the main ingredient in the composition.[37] The possibilities of synthesizing value-added product such as aluminosilicate resin commonly known as Zeolite 4A were explored as composite ash contains about 85% silica and 5% alumina. Zeolite A resin was prepared using composite ash as a source material for extraction of sodium silicate which acts as the main ingredient in the composition.[38] The other main component, sodium aluminate, was prepared using aluminum hydroxide gel as reported elsewhere.[39] Figure 15.7 is pointing composite (coal and biomass)-based power plants in Akaltara and Andhra Pradesh, India.

FIGURE 15.7 Rice husk and coal co-firing power plants in India.

15.7 ENVIRONMENTAL APPLICATIONS OF ZEOLITE 4A RESIN

Zeolite A has extensive applications in basic science, petrochemical science, energy conservation/storage, medicine, chemical sensor, air purification, environmentally benign composite structure, and waste remediation.

The specific objective of this study is to generate data toward synthesis of Zeolite 4A resin in a more economically viable fashion using waste materials such as coal ash, RHA, and composite ash and exploring their utility in major industrial applications.

15.7.1 PHOSPHATE-FREE DETERGENT BUILDER

Sodium tripolyphosphate and tetrasodium pyrophosphate are the most popular detergent builders and have been used extensively in detergents and soap industries. However, their use is environmentally hazardous due to the excessive deposition of phosphates and nitrates into water bodies that result into eutrophication as shown in Figure 15.9.[40] In 1978, "Procter and Gamble" introduced Zeolite NaA in its laundry detergents and nowadays, most of the commercial washing powders contain Zeolite, instead of harmful phosphates, banned in many parts of the world because of the risk of water eutrophication.[41] At present, more than 1000,000 tons of Zeolite 4A per annum is used in detergents and cleaning products worldwide (Fig. 15.8). The high exchange property helps to remove hardness-causing ions resulting in high sequestering power and exhibits phenomena like soil anti-redeposition and dye transfer. FAZ-A-based detergents can affect the optical brightness of fabrics to a certain extent, due to the presence of iron oxide. Recently, a

FIGURE 15.8 Zeolite-A-based detergents (Henkel Ltd.).

novel route to produce Zeolite A using composite ash and their applicability in detergents have been explored successfully and reported elsewhere.

FIGURE 15.9 Effect of eutrophication on fishes.

15.7.2 ION EXCHANGER/WATER SOFTENER

Hydrated cations within the Zeolite pores are bound loosely to the Zeolite framework and can readily exchange with other cations when in aqueous media. Hence, Zeolite 4A are used in water-softening process in boiler in many industries as these impurities can corrodes the internal machinery. Zeolites are effectively used as ion exchanger for removal of ammonium ions from water/wastewater. CFA-based Zeolite A is used in water decontamination and has good efficiency for metal uptake from wastewaters from electroplating baths.[42]

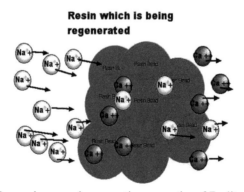

FIGURE 15.10 Ion-exchange and regeneration properties of Zeolite 4A.

15.7.3 SORPTION OF HEAVY METALS AND ANIONIC POLLUTANTS

The presence of the toxic metals generated by metal finishing or mineral processing industries in streams and lakes can result in major hazards to the environment and public health. Zeolites have been widely explored for heavy metal immobilization from natural or industrial water. Fly-ash-based Zeolite A with anionic characteristics have been developed for their applications for removal of arsenic from water using suitable surfactant HDTMABr. Unique properties and selectivity of Zeolites are being exploited for environmental remediation of soils contaminated with heavy metals and oxyanions.[43]

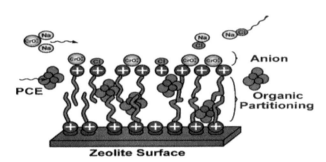

FIGURE 15.11 Schematic diagram of modification of a Zeolite surface by surfactants.

15.7.4 ADSORBENT FOR CO_2 CAPTURE AND VOCS

Zeolites are used to adsorb a variety of materials. This includes applications in drying, purification, and gas separation. The shape selective properties of Zeolites are also the basis for their use in molecular adsorption.[44] They can remove volatile organic chemicals from air streams, separate isomers, and mixtures of gases. Surface modified fly-ash-based Zeolites are used as versatile materials for VOCs monitoring in indoor environment. The FAZ material is routinely added to small air filters to adsorb harmful gases and reduce allergy problems. Attempts have also been made to immobilize microorganisms to impart microbial activities (*Azotobacter chroococcum*) in Zeolites. The in-situ incorporation of AMP (3-amino-2-methyl-1-propanol) resulted in an adsorbent with significantly improved characteristics to adsorb carbon dioxide at lower temperatures.[45]

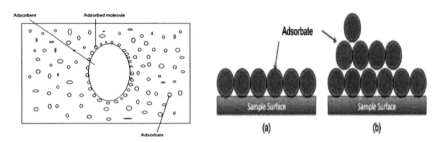

FIGURE 15.12 Surface modification of Zeolite A for removal of targeted pollutants,

15.7.5 MEMBRANE FOR SEPARATION OF MOLECULES

Zeolites are porous solid materials which exhibit the property of acting as sieves on a molecular scale. The unique properties of molecular sieve have led to their use in wide variety of processes for treating gas and liquid, namely, drying, sweetening (sulfur removal), removal of carbon dioxide, isomer separation, for designing thin films, coatings, membranes, and separation of gas mixture.[46–48] All these rely on the molecular sieve causing greater selectivity for one type of molecule as compared with others. However, selectivity mainly depends upon the shape and size of molecule and molecular polarity. This membrane was also used for other reverse osmosis and pervaporation applications.

FIGURE 15.13 Molecular sieve for separation of molecules.

15.7.6 CATALYST IN CHEMICAL/PETROCHEMICAL INDUSTRIES

Zeolites have been widely used as industrial catalysts because they are inexpensive and environmentally benign. The high surface area of Zeolite 4A (400–450 m²/g) provides large area for carrying out the catalytic reaction for reactant molecules. Zeolite A is used as catalysts in petroleum refining, particularly in the area of high octane gasoline, in the processing of heavy crudes and in the production of diesel.[49] Shape selectivity is a unique property of Zeolites. As a consequence, Zeolites are able to restrict or prevent the passage of organic molecules, based on size and stearic effects. An important application of reactant shape selectivity is in the cracking of linear alkanes by protecting the branched chain ones particularly in petrochemical industry.

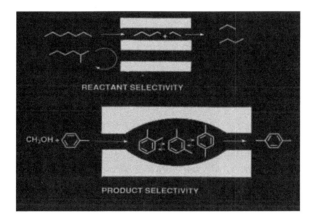

FIGURE 15.14 Shape selective catalysis in Zeolite 4A resin.

KEYWORDS

- Zeolite 4A
- waste ash
- synthesis
- environmental applications
- biomass

REFERENCES

1. Querol, X.; Moreno, N.; Umana, J.; Alastuey, A.; Hernandez, E.; Lopez-Soler, A.; Planan, F. Synthesis of Zeolites from Coal Fly Ash: An Overview. *Int. J. Coal Geol.* **2002**, *50*, 413–423.
2. http://www.asdn.net/asdn/chemistry/zeolites.html.
3. Fruijtier-Polloth, C. The Safety of Synthetic Zeolites Used in Detergents. *Arch. Toxicol.* **2009**, *83*, 23–35.
4. Breck, D. Chapter 7: Ion Exchange Reactions in Zeolites. In: *Zeolite Molecular Sieves, Structure, Chemistry, and Use*. Robert E. Krieger Publishing: Malabar, 1984, FL TIC: 245213.
5. Breck, D. *Zeolite Molecular Sieves: Structure, Chemistry and Use*; John Wiley and Sons: London, 1974; p 4.
6. Bekkum, V.; Flanigen, E.; Jacobs, P.; Jansen, J. *Introduction to Zeolite Science and Practice*, 2nd Revised ed., Elsevier: Amsterdam, 1991.
7. Georgiev, D.; Bogdanov, B.; Angelova, K.; Markovska, I.; Hristov, Y. Synthetic Zeolites Structure, Classification, Current Trends in Zeolite Synthesis Review. In: International Science Conference, Stara Zagora, Bulgaria Economics and Society Development on the Base of Knowledge, 2009.
8. Haag, W.; Lago, R.; Weisz, P. The Active Site of Acidic Alumino-silicate Catalysts. *Nature* **1984**, *309*, 589–591.
9. Ríos, C.; Williams, C.; Fullen, M. Nucleation and Growth History of Zeolite LTA Synthesized from Kaolinite by Two Different Methods. *Appl. Clay Sci.* **2009**, *42*, 446–454.
10. Baccouche, A.; Srasra, E.; Maaoui, M. Preparation of Na-P1 and Sodalite Octahydrate Zeolites from Interstratified Illite–Smectite. *Appl. Clay Sci.* **1998**, *13*, 255–273.
11. Farag, I.; Zhang, J. Simulation of Synthetic Zeolites-4A and 5A Manufacturing for Green Processing. *IRACST—Eng. Sci. Technol.: Int. J. (ESTIJ)* **2012**, *2*(2). ISSN 2250-3498.
12. Rayalu, S.; Bansiwal, A.; Meshram, S.; Labhsetwar, N.; Devotta, S. Fly Ash Based Zeolite Analogues: Versatile Materials for Energy and Environment Conservation. *Catal.: Surv. Asia* **2006**, *10*(2), 74.
13. Rayalu, S.; et al. Surface Modified Zeolite and Process for Synthesis Thereof for Sequestration of Anions. 2009, US Patent No. US 7,510,659 B2.
14. Machado, N.; Miotto, D. Synthesis of Na-A and Na-X Zeolites from Oil Shale Ash. *Fuel* **2008**, *84*, 2289–2294.
15. Purnomo, C.; Salim, C.; Hinode, H. Synthesis of Pure Na-X and Na-A Zeolite from Bagasse Fly Ash. *Microporous Mesoporous Mater.* **2012**, *162*, 6–13.
16. Tanaka, H.; Eguchi, H.; Fujimoto, S.; Hino, R. Two-step Process for Synthesis of a Single Phase Na-A Zeolite from Coal Fly Ash by Dialysis. *Fuel* **2006**, *85*, 1329–1334.
17. Tanaka, H.; Fujii, A. Effect of Stirring on the Dissolution of Coal Fly Ash and Synthesis of Pure-form Na-A and Na-X Zeolites by Two-step Process. *Adv. Powder Technol.* **2009**, *20*, 473–479.
18. Katsuki, H.; Komarneni, S. Synthesis of Na-A and/or Na-X Zeolite/Porous Carbon Composites from Carbonized Rice Husk. *J. Solid State Chem.* **2009**, *182*, 1749–1753.

19. Ghasemi, Z.; Younesi, H. Preparation and Characterization of Nano Zeolite NaA from Rice Husk at Room Temperature without Organic Additives. *J. Nanomater.* **2011,** 2011, 8 pp.
20. Chauhan, Y.; Talib, M. A Novel and Green Approach Synthesis and Characterization of Nanoadsorbents (Zeolite) from Coal Fly Ash: A Review. *Sci. Rev. Chem. Commun.* **2012,** *2*(1), 12–19.
21. Hollman, G.; Steenbruggen, G.; Janssen-Jurkovicova, M. A Two-step Process for the Synthesis of Zeolites from Coal Fly Ash. *Fuel* **1999,** *78*, 1225–1230.
22. Shih, W.; Chang, H. Conversion of Fly Ash into Zeolites for Ion-exchange. *Appl. Mater. Lett.* **1996,** *28*(4–6), 263–268.
23. Rayalu, S.; Sapre, J.; Meshram, S.; Munshi, K.; Singh, R. A Case Study for Synthesis of Zeolite A from Indian Flyash of Bituminous Origin. *Aust. J. Chem.* **2004,** *16*(2), 717–724.
24. Chang, H.; Shih, W. Synthesis of Zeolites A and X from Fly Ashes and their Ion-exchange Behavior with Cobalt Ions. *Ind. Eng. Chem. Res.* **2000,** *35*, 4185–4191.
25. Kumar, A.; Mohanta, K.; Kumar, D.; Parkash, O. Properties and Industrial Applications of Rice Husk: A Review. *Int. J. Emerg. Technol. Adv. Eng.* **2012,** *2*, 2250–2459.
26. Ramli, Z. *Rhenium-impregnated Zeolites: Synthesis, Characterization and Modification as Catalysts in the Metathesis of Alkanes*. University Technology Malaysia, 1995.
27. Thuadaij, P.; Nuntiya, A. Synthesis of Na-X Hydrate Zeolite from Fly Ash and Amorphous Silica from Rice Husk Ash by Fusion with Caustic Soda Prior to Incubation. In: *Proc. IPCBEE* **2011,** *10*, 69–74.
28. Nur, H. Direct Synthesis of Na Zeolite from Rice Husk & Carbonaceous Rice Husk Ash. *Indonesian J. Agric. Sci.* **2001,** *1*, 40–45.
29. Ghasemi, Z.; Younesi, H. Synthesized Nanozeolite NaA from Rice Husk at Room Temperature without Organic Additives by Hydrothermal Method. *J. Nanomater.* **2011,** 1–8.
30. Saad, M.; Oliveira, L.; Cândido, R.; Quintana, G.; Rocha, G.; Gonçalves, A. Preliminary Studies on Fungal Treatment of Sugarcane Straw for Organosolv Pulping. *Enzyme Microb. Technol.* **2008,** *43*, 220–225.
31. Moriya, R.; Gonçalves, A.; Duarte, M. Ethanol/Water Pulps from Sugar Cane Straw and their Biobleaching with Xylanase from *Bacillus pumilus*. *Appl. Biochem. Biotechnol.* **2007,** *137–140*, 501–513.
32. Balakrishnan, M.; Batra, V. Valorisationof Solid Waste in Sugar Factories with the Possible Applications in India: A Review. *J. Environ. Manage.* **2011,** *52*(11), 2886–2891.
33. Murilo, P.; Cleiser, T.; da Silva, T. P.; et al. Synthesis of Zeolite NaA from Sugarcane Bagasse Ash. *Mater. Lett.* **2013,** *108*, 243–246.
34. Villar-Cociña, E.; Valencia-Morales, E.; Gonzalez, R.; Hernandez-Ruiz, J. Kinetics of the Pozzolanic Reaction between Lime and Sugar Cane Straw Ash by Electrical Conductivity Measurement: A Kinetic–Diffusive Model. *Cem. Concr. Res.* **2003,** *33*, 517–524.
35. Fungaro, D.; Vitória, T.; da Reis, S. Use of Sugarcane Straw Ash for Zeolite Synthesis. *Int. J. Energy Environ.* **2014,** *5*(5), 559–566.
36. Hui, K.; Chao, C. Pure, Single Phase, High Crystalline, Chamfered-edge Zeolite 4A Synthesized from Coal fly Ash for use as a builder in detergents. *J. Hazard. Mater. B* **2006,** *137*, 401–409.

37. Meshram, S.; Khandekar, U.; Mane, S.; Mohan, A. Novel Route of Producing Zeolite A Resin for Quality-Improved Detergents. *J. Surfact. Deterg. AOCS* **2014**.
38. Chayakorn, B.; Pesak, R. Synthesis of Zeolite A membrane from rice husk ash. *J. Met. Mater. Min.* **2009**, *19*, 79–83.
39. Rayalu, S.; Labhsetwar, N.; Biniwale, R.; Meshram, S.; Udhoji, J.; Khanna, P. Producing Zeolites from Fly Ash. *Chem. Ind. Dig.* **1998**, *41A*, 1212.
40. Udhoji, J.; Bansiwal, A.; Meshram, S.; Rayalu, S. Improvement in Optical Brightness of Flyash Based Zeolite A for Use as a Detergent Builder. *J. Sci. Ind. Res.* **2005**, *64*, 367–371.
41. Yangxin, Y.; Zhao, J.; Bayly, A. Development of Surfactants and Builders in Detergent Formulations. *Chin. J. Chem. Eng.* **2008**, *16*(4), 517–527.
42. Hui, K.; Chao, C.; Kot, S. Removal of mixed heavy metal ions in wastewater by Zeolite 4A and Residual Products from Recycled Coal Fly Ash. *J. Hazard. Mater.* **2005**, *127*, 89–101.
43. Shaheen, S.; Derbalah, A.; Moghanm, F. Removal of Heavy Metals from Aqueous Solution by Zeolite in Competitive Sorption System. *Int. J. Environ. Sci. Dev.* **2012**, *3*(4), 362–367.
44. Mravec, D.; Hudec Janotka, J. Some Possibilities of Catalytic and Noncatalytic Utilization of Zeolites. *Chem. Pap. – Chemicke Zvesti.* **2005**, *59*(1), 62–69.
45. Kumar, V.; Labhsetwar, N.; Meshram, S.; Rayalu, S. Functionalized Fly Ash Based Alumino-Silicates for Capture of Carbon Dioxide. *AOCS, Energy Fuels* **2011**, *25*, 4854–4861.
46. Hui, K. S.; Hui, K.; Lee, S. A Novel and Green Approach to Produce Nano-porous Materials Zeolite A and MCM-41 from Coal Fly Ash and their Applications in Environmental Protection. *World Acad. Sci. Eng. Technol.* **2009**, *3*, 489–499.
47. Sanoap, T.; Yanagishitab, H.; Kiyozumib, Y.; Mizukamib, F.; Harayab, K. Separation of Ethanol/Water Mixture by Silicalite Membrane on Pervaporation. *J. Membr. Sci.* **1994**, *95*, 221–228.
48. Daramola, M.; Aransiola, E.; Ojumu, T. Potential Applications of Zeolite Membranes in Reaction Coupling Separation Processes. *Materials* **2012**, *5*, 2101–2136. ISSN 1996-1944.
49. Yilmaz, B.; Müller, U. Catalytic Applications of Zeolites in Chemical Industry. *Top. Catal.* **2009**, 888–895.

CHAPTER 16

STUDY OF ECO-FRIENDLY ADDITIVES FOR WOOD–PLASTICS COMPOSITES: A STEP TOWARD A BETTER ENVIRONMENT

S. A. PURANIK[1], DINESH DESAI[2,*], and KINTU JAIN[2]

[1]Atmiya Institute of Technology & Science, Rajkot, Gujarat, India

[2]Plastics Technology Department, Lalbhai Dalpatbhai College of Engineering, Ahmedabad, Gujarat, India

*Corresponding author. E-mail: dineshdesai2002@gmail.com

CONTENTS

Abstract		310
16.1	Introduction	310
16.2	Additives	311
16.3	Eco-Friendly Additives	312
16.4	Experimental Work	313
16.5	Innovation	317
16.6	Applications	317
16.7	Conclusion	318
Keywords		319
References		319

ABSTRACT

The additives are generally the polymer materials with dispersed matrix without affecting significantly the molecular structure of the polymer to attend certain desirable properties. There are number of advantages in adding these additives in the wood–plastics composites (WPCs). The major advantages are shown here. One of such additives, that is, palm kernel nut shell (PKNS) powder was prepared. This being an eco-friendly additive can replace other inorganic additives including the base material (up to certain percentage).

Addition of wood flour or such additives improves the mechanical properties of thermoplastics, but on the other hand, it increases the burning speed of the materials. Thus, to achieve good WPCs, additives are very critical and indispensible even though they are used at just small percentages, as they give WPCs sufficient stiffness, rigidity, and good stability against light and heat. Hence, a study is made to select such additives with right percentage, so as to provide the required properties.

WPCs can be made with a variety of plastics and wood additives, such as polyethylene, polypropylene, polyvinyl chloride, and polystyrene. What's more, WPCs are just one category of an emerging family of materials that can be termed "thermoplastic biocomposites." Besides wood, these biocomposites can also utilize other natural fibers such as rice hulls, palm fiber waste, or flax.

In this chapter, a study of physical and mechanical properties of polypropylene and untreated and treated PKNS powder composites is explained in detail. The result analysis and graph are prepared and discussed here.

16.1 INTRODUCTION

Wood–plastic composites (WPCs) are defined as composite materials that contain thermoplastics and wood in various forms. The average product contains about 50% wood in particulate form. The additives are generally the polymer materials with dispersed matrix without affecting significantly the molecular structure of the polymer to attend certain desirable properties. There are number of advantages in adding these additives in the wood plastics. The major advantages are to improve processing conditions, increase resin's stability to oxidation, obtain better impact resistance, increase or decrease hardness, control surface tension, facilitate extrusion molding, control blocking, reduce cost, increase flame resistance, and many more. The various types of additives are the fillers, antioxidants, heat stabilizers, UV stabilizers, colorants, antistatic, flame/fire retardants, cross-linking agents, blowing agents, lubricants, impact modifiers, processing aids, etc.[1]

Because the WPC industry is relatively young, the long-term 10–20-year durability of outdoor products like decking is still being proven in the field. To achieve good WPCs, additives are very critical and indispensible even though they are used at just small percentages, as they give WPCs sufficient stiffness, rigidity, and good stability against light and heat. A study is made to select these additives so as to provide these properties as required.

Till now, the wood–plastics composites (WPCs) were using 30–40% of wood or fillers. Remaining components were other chemicals including the base polymer material. But now the WPCs, with up to 90% wood fiber or wood flour content is gaining popularity. Wood filler, usually added in ratios of 40–60% of a given WPC formulation, adds stiffness, and decreases the tendency of plastic to creep. This, itself is an eco-friendly additive for the WPC. Thus, the demand for WPC is expected to rise in quest for minimal maintenance requirements, excellent weather ability, and high resistance to wear and tear in construction applications. Adding wood flour improves the mechanical properties of thermoplastics, but on the other hand, it increases the burning speed of the materials. This is one of the major disadvantages in WPCs.[2]

16.2 ADDITIVES

Plastic additives are specialty chemicals used to impart specific properties to plastics for making them more desirable for application in end-user segments. Plastic additives are available in different types based on the functions they perform. Some of the major types include plasticizers, heat and light stabilizers, flame retardants, antioxidants, antifog, slip, antiblock, lubricants, and master batch additives. The use of plastic additives provides certain properties to the end product such as durability, flexibility, antioxidant potential, and microbial resistance among others, which is driving the demand for plastic additives.[3]

The demand for plastic additives is growing in line with increasing demand for plastics, which is strongly correlated with economic growth of the country. Demand for plastics is growing as they offer a better alternative to traditional materials such as wood, metal, and glass in a wide range of applications. As per "Tech Sci Research" report in "India Plastic Additives Market Forecast & Opportunities, 2019," plastic additives market revenues are projected to grow at a CAGR of 14% during 2014–2019 due to increasing demand from end-user markets for plastics, such as packaging, construction, and automobiles.

Few of the important types of additives are fillers, antioxidants, heat stabilizers, UV stabilizers, colorants, antistatics, flame/fire retardants, cross-linking agents, blowing agents, lubricants, impact modifiers, and processing aids. They are gaining the popularity.

16.3 ECO-FRIENDLY ADDITIVES

Petroleum-based products such as resins in thermoset plastics are toxic and non-biodegradable. Recyclability and environmental safety are becoming increasingly important to the introduction of materials and products. Ecological concerns have resulted in renewed interest in natural materials. They are converted into the form of H_2O and CO_2. These H_2O and CO_2 are absorbed into the plant systems. The resins and fibers used in the green composites are biodegradable, when they dumped, decomposed by the action of microorganisms. This helps the green composite combine plant fibers with natural resins to create natural composite materials.[4]

A list of few green additives is given below:

- Soybean/starch and plant-based fibers soy protein, starch, etc. Polymer processing green composites trash collection and transport compost life cycle of "green" composites.
- Other polymers—lignin natural rubber, polyesters—poly-hydroxyl alkanoates (PHAs), proteins—collagen/gelatin casein, albumin, fibrogen, silks, elastins, protein from grains, polysaccharides—starch, cellulose, chitin, pullulan, levan, konjac, elsinan, natural—biodegradable polymers.
- Biodegradable polymers synthetic—(1) poly(amides); (2) poly(anhydrides); (3) poly(amide-enamines); (4) poly(vinyl alcohol); (5) poly(ethylene-*co*-vinyl alcohol); (6) poly(vinyl

acetate); (7) polyesters—poly(glycolic acid), poly(lactic acid), poly(caprolactone), poly(*ortho*-esters); (8) poly(ethylene oxide); (9) some poly(urethanes); (10) poly(phosphazines); (11) poly(imino carbomates); and (12) some poly(acrylates).
- The other fibers used are palm kernel nut shell (PKNS), coir (coconut), bamboo, pineapple, ramie, natural fibers such as kenaf, flax, jute, hemp, and sisal have attracted renewed interest, especially as a glass fiber substitute in the automotive industry. This includes nonwood fibers and wood fibers.

16.4 EXPERIMENTAL WORK

Wood–fiber/thermoplastic blends have already made a name for themselves in extruded decking and fencing boards. Now they are moving into injection molding. However, recent developments in the manufacture of WPC compounds have significantly improved the quality, consistency, and capability of this environmentally friendly material. In fact, the latest generation of WPCs can be run smoothly through traditional injection-molding equipment with minimal adjustments to process settings and no physical hardware modifications. However, these materials do have some processing characteristics that distinguish them from familiar molding resins.[5]

Out of above the additive selected here is PKNS. A fine powder was made of this shell of about 80 μm size. Two separate experimental set up were made. In one case, the shell powder was used as virgin material, whereas in other case, it was treated with NaOH solution of approximately 1–2 pH value. Various properties were tested for these two batches separately. These samples were casted in specific specimen for the set of tests. The tests conducted were density, melt flow index (MFI), and flexural strength.

The results obtained are discussed below:

16.4.1 DENSITY

The densities found for the both set are tabulated in Tables 16.1 and 16.2 separately. The values found are as under.

TABLE 16.1 Density of Composites at Different Loading of UPKSP.

Composition (wt%)		Density (g/cm³)
PP	UPKSP	
100	0	0.9097
90	10	0.9309
80	20	0.9398
70	30	0.9577

UPKSP, Untreated palm kernel nut shell powder; PP, polymer matrix.

TABLE 16.2 Density of Composites at Different Loading of TPKSP.

Composition (wt%)		Density (g/cm³)
PP	TPKSP	
100	0	0.9097
90	10	0.9319
80	20	0.9426
70	30	0.9595

TPKSP, treated palm kernel nut shell powder; PP, polymer matrix.

These two tables are graphically represented in Figure 16.1. Density of virgin PP is 0.9097 g/cm³. But the density increase with the increase in filler loading. For 30% untreated palm kernel nut shell powder (UPKSP) loading, the percentage increase in density of composites, compare to virgin PP is around 4.80%.

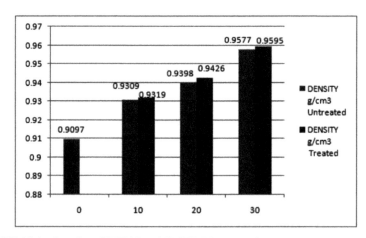

FIGURE 16.1 Density of UPKSP and TPKSP-PP composites.

16.4.2 MELT FLOW INDEX

The experimental values for the MFI for both set of samples are found and represented graphically as shown in Figure 16.2.

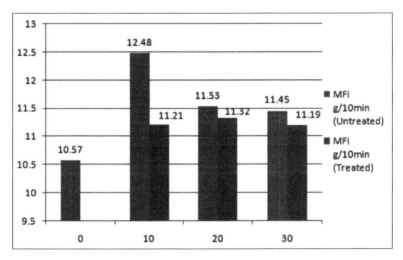

FIGURE 16.2 MFI of UPKSP and TPKSP-PP composites.

MFI of composites with different filler content are shown in Figure 16.2. MFI of virgin PP is 10.57. It is found that the MFI increases with filler contain. MFI of 30% UPKSP content composites is 11.45 and the percentage increase in MFI is 88%. MFI of 30% treated palm kernel nut shell powder (TPKSP) content composites is 11.19 and the percentage increase in MFI is 62%.

Results show that PKSP impart plasticity to the PP. The reason for this is high lignin (53.4%) contained in the PKSP. Lignin acts as processing aids and imparts flexibility to PP. After treating PKSP with NaOH, the treated PKSP-PP composite shows reduction in MFI due to increase in density with PPP compare to untreated PKSP-PP composite.

16.4.3 FLEXURAL STRENGTH

The experimental values for the flexural strength for both set of samples are found and represented graphically as shown in Figure 16.3.

It is found that the flexural strength of PP is 336 kg/cm^2, whereas the flexural strength of PKSP-filled PP composites is increased with increasing of filler loading. For example, the flexural strength at 30% UPKSP filled PP is 363 kg/cm^2. The percentage increase in flexural strength at 30% loading of UPKSP is around 8% compare to virgin PP. This is expected because the addition of filler increases the stiffness of the composites.

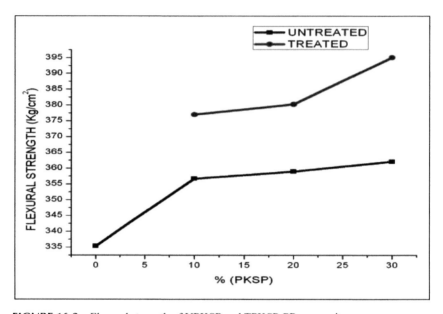

FIGURE 16.3 Flexural strength of UPKSP and TPKSP-PP composites.

After alkali treatment, flexural strength is increased substantially. Treated PKSP-PP composites show very high flexural strength compared to untreated PKSP-PP composites. Flexural strength at 30% TPKSP filled PP is 395 kg/cm^2. Percentage increased in Flexural strength at 30% loading of TPKSP is around 18% compared to Virgin PP. Treatment led to increase the flexural properties of composites by improving adhesion across the interface also overcome the dispersion problems.

Therefore, by addition of PKSP, it not only increases the biodegradation properties but also the physical and mechanical properties of polypropylene are increase.

For preparing articles made from such materials that can be used commercially, a study of biodegradable resin named PHAs was also made. This is a natural resin compound with translucent color having product description as given below. At present, this resin is used for injection-molded articles along with various polymers. It can be also be used for making the injection-molded WPC articles.

16.5 INNOVATION

Unlike traditional plastics, Ecomann PHA bioplastics are derived from agricultural waste; its finished products are completely biodegradable in soil, salt water, home-composting and industrial-composting environments. In accordance to ASTM D6400 and EN 13432 standards, Ecomann PHA bioplastics degrades up to 90% within 180 days in a composting environment, and it eventually turns completely into carbon dioxide and water. The basic properties of this compound made, are given below in Table 16.3.

TABLE 16.3 Properties and Value of Poly-hydroxyl Alkanoates.

Sr. No.	Property	Value
1	Density	1.29 g/cm^3
2	Melt flow rate	4 g/10 min
3	Tensile strength	25 MPa
4	Breaking extension	5%
5	Notched impact strength	3 kJ/m^2

16.6 APPLICATIONS

From this compound, the commercial articles were made successfully. The method applied to make these articles was injection-molding process. Various parts such as pen, soap container, tooth brush stand, etc. were made. A few of them are displayed in Figure 16.4.

FIGURE 16.4 PHA used for making molding parts such as pen, soap container, tooth brush, etc.

16.7 CONCLUSION

A new building material known as WPCs has emerged. WPCs are a combination of a thermoplastic matrix and a wood component, the former is usually recycled polyethylene or polypropylene, and the latter a wood-processing residual, for example, sawdust, natural fiber and wood shavings, etc.

Biobased materials made from renewable resources, such as wood or such additive, play an important role in the sustainable development of society. One main challenge of biobased building materials is their inherent moisture sensitivity, a major cause for fungal decay, mold growth, and dimensional instability, resulting in decreased service life as well as costly maintenance.

The disintegration of the modified wood components during processing also creates a more homogeneous compound of the WPCs, which may be beneficial from a mechanical performance perspective. This compound can be made suitable for processing by both "extrusion process" and "injection-molding process." Future studies are suggested to include analyses of the surface composition, in order to tailor new compatible wood–polymer combinations in WPCs and biocomposites.

KEYWORDS

- **wood–plastics composites (WPCs)**
- **untreated palm kernel nut shell powder (UPKSP)**
- **treated palm kernel nut shell powder (TPKSP)**
- **poly-hydroxyl alkanoates (PHAs)**
- **melt flow index (MFI)**

REFERENCES

1. Kord, B.; Hosseini, S. K. Hybrid Nanocomposites. *BioResources* **2011**, *6*(2), 1741–1751.
2. Shahi, P.; Behravesh, A. H.; Daryabari, S. Y.; Lotfi, M. *Polym. Composites* **2012**, *33*(5), pp 753–763.
3. Brydson, J. A. *Plastics Materials*, 7th ed., 1932; pp 124–155.
4. Manolis, S. L. Wood-filled Plastics—They Need the Right Additives for Strength, Good Looks & Long Life. *Plast. Technol.*, July **2004**.
5. Rozman, H. D.; Kumar, R.; Abusamah, A.; Saad, M. J. J. Rubber Wood–Polymer Composites Based on Glycidyl Methacrylate and Diallyl Phthalate. *J. Appl. Polym. Sci.* **1998**, *67*, 1221–1226.

CHAPTER 17

SYNTHESIS OF CRUDE OIL BY CATALYTIC PYROLYSIS OF WASTE PLASTICS

SIDDHARTH AVNESH MEHTA[*]

Department of Plastics and Polymer Engineering, Maharashtra Institute of Technology, Aurangabad, Maharashtra, India

[*]*E-mail: sidmehta1993@gmail.com*

CONTENTS

Abstract	322
17.1 Introduction	322
17.2 Pyrolysis	323
17.3 Laboratory Process	325
17.4 Results	327
17.5 Conclusion	328
Keywords	328
References	328

ABSTRACT

With the increase in consumption of plastics, we are also posing a threat to mankind. Since plastics take more than a century to degrade, we are left with a question, "How do we dispose plastics and where?" Since years, we have been dumping it either on grounds or in sea. Incineration has also not proved to be that effective. We have a solution. What if we could utilize this waste plastics and turn it back to raw material. As simple as water to ice and back to water? Yes, this chapter will guide you through the science of converting waste plastics to crude oil.

17.1 INTRODUCTION

Plastic products have become an integral part in our daily life as a basic need. It is produced on a mammoth scale worldwide and its production crosses 150 million tons per year globally. Plastics are a relatively new man-made material that provides vast material benefits throughout their useful lifespan. However, their end-of-life disposal currently leaves much to be desired. The US EPA estimates that 30 million tons of the municipal solid waste generated in the United States annually are in the form of plastics. Of this, only 7% is recovered for recycling, mostly in the form of polyethylene, and roughly 10% is combusted in waste-to-energy facilities to generate electricity. The remainder of plastic wastes is land filled, which is clearly a loss of nonrenewable, fossil-based resources. Also, plastics litter in some cases poses a threat to human health and also threatens other ecosystems. There is also an estimated 100 million tons of plastic litter in the ocean, with millions more tons added annually.[3]

In India, approximately 8 million tons plastic products were consumed every year (2008) which rose to 12 million tons by 2012. It is a fact that plastics will never degrade for several years. Further, the recycling of a virgin plastic material can be done 2–3 times only, because, after every recycling, the strength of plastic material is reduced due to thermal degradation. It is to mention that no authentic estimation is available on total generation of plastic waste in the country. However, considering 70% of total plastic consumption is discarded as waste, thus approximately 5.6 million tons per annum of plastic waste is generated in country, which is about 15,342 tons per day.[4] Conversion of waste plastics to oil is by far the only technique that can be used for plastics waste management as it solves two of the today's major problems, that is, disposal of plastics waste and generation of fuel at lower price.

17.2 PYROLYSIS

Pyrolysis process cracks/breaks down polymer chains into useful lower molecular weight compounds. The products of plastic pyrolysis process could be utilized as fuels or chemicals.

Three different pyrolysis processes are reported:

- hydro-pyrolysis,
- thermal pyrolysis, and
- catalytic pyrolysis.

Pyrolysis is the process wherein complex organic molecules such as heavy hydrocarbons are broken down into simpler molecules such as light hydrocarbons, by the breaking of carbon–carbon bonds. The rate of pyrolysis and the end products are heavily dependent on the temperature as well as the presence of catalysts. Basically, pyrolysis is the breakdown of a large alkane into smaller, more useful alkanes or alkenes. Simpler, pyrolysis is the process of breaking a long chain of hydrocarbons into short ones.

"Pyrolysis" is used to describe any type of splitting of molecules under the influence of heat, catalysts, and solvents, such as in processes of destructive distillation or cracking.

17.2.1 THERMAL PYROLYSIS

Thermal pyrolysis involves the degradation of the plastic materials by heating in absence of oxygen. The process is generally conducted at temperatures in the range of 350 and 900°C and results in the formation of a solid residue (mostly carbonized char) and a volatile fraction that may be separated into condensable hydrocarbon oil consisting of paraffins, isoparaffins, olefins, naphthenes, and aromatics, and a noncondensable high calorific value gas. The proportion of each fraction and their precise composition depends not only on the nature of the plastic waste but also on process conditions. The extent and the nature of these reactions depend both on the reaction temperature and also on the residence of the products in the reaction zone, an aspect that is primarily affected by the reactor design. However, the thermal degradation of high molecular weight plastics to low molecular weight materials requires high temperatures, and this has a major drawback. In that, a very broad product range is obtained. But, this problem can be addressed by using catalysts.

17.2.2 CATALYTIC PYROLYSIS

In this process, a suitable catalyst is used to carry out the pyrolysis reaction. The addition of catalyst lowers the reaction temperature and also reduces the cycle time. In addition, catalytic degradation yields a much narrower product distribution of carbon atom number with a peak at lighter hydrocarbons and occurs at comparatively lesser temperatures. From an economic perspective, reducing the cost even further will make this process cost-effective and an even more attractive option. This option can be optimized by reusing the catalysts and the use of more effective catalysts in lesser quantities. This method seemed to be the most promising to be developed into a cost-effective commercial plastics waste management system process to solve the acute environmental problem of plastic waste disposal.

17.2.3 MECHANISM OF THERMAL DEGRADATION

The plastic pyrolysis follows very complex routes that cannot be described by one or more chemical reactions, but still rather imperfectly by either empirical formulas which feature fractional stoichiometric coefficients or compressive systems of elementary reaction. A detailed study on the mechanism for the thermal decomposition of plastics is proposed by Cullis and Hirschler.[5]

The four different mechanisms proposed are as follows:

- end-chain scission or unzipping,
- random-chain scission/fragmentation,
- chain-stripping/elimination of side chain, and
- cross-linking.

The decomposition mode mainly depends on the type of plastic (the molecular structure).

17.3 LABORATORY PROCESS

17.3.1 EQUIPMENT

- Heating mantle,
- round-bottom flask,
- condenser,
- 90° bent,
- collection beaker, and
- thermostat.

17.3.2 PROCESS CONDITIONS

- Temperature: 350–400°C,
- condition: inert atmosphere,
- entire setup insulated with glass wool, and
- double insulation by asbestos.

17.3.3 PROCESS

Figure 17.1 will help to better understand the process. Waste plastic is stuffed inside the round bottom flask. All the connections are done, that is, bents, condenser, and beaker. The heating mantle is switched on. Temperature is to be maintained between 450 and 500°C. Adjust the regulator accordingly. Insulation is provided all around the equipment. Glass wool is the best option that we could use. To make it further effective, we covered the glass wool with asbestos belt. This provided double insulation. At the same time, it decreased the heating time drastically as the heat loss was minimized. The plastic material inside the round bottom flask starts to melt. As we know that generally plastics burn directly into vapors on heating. But, one must know that what we see is direct heating of plastics. Whereas in this case, the plastics are inside the flask, that is, we heat the flask and the plastic directly. This is called indirect heating of plastics. Thus on indirect heating of plastics, the plastics start to melt. By this time, one should start the water circulation through the condenser. On further heating, the plastics tend to boil like water. On extensive heating, that is, above 300–350°C, the plastics start to evaporate. They start converting to vapors.

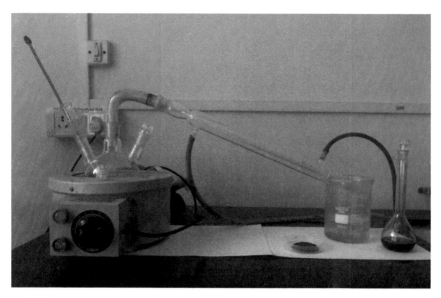

FIGURE 17.1 Experimental setup.

These vapors then are made to pass through the condenser. Here, the temperature is much lower because of the circulation of cool water. Thus, the vapors condense.

Chemically, when the plastics are heated to such high temperatures, they break the bonds between them. The chain breaks and reduces in size. And then, while we condense it, these shorter chains condense to a fluid. This fluid is nothing but precious crude oil. This is chemically proved since oils have lesser chain lengths than plastics. Thus, breaking long chains of plastics yields us to shorter chains of oils. This is then passed through another 90° bent. A beaker is placed below this bent which is half-filled with water. The bent is kept dipped inside the water in the beaker. This is an additional safety measure used to avoid any possible leakage of vapors. There are two types of vapors. One is condensable vapors and the other is noncondensable vapors. The condensable vapors condense on passing through the condenser, thus converting to oil and further collected in the beaker. The noncondensable vapors do not condense at all and remain in vapor phase. These vapors are collected separately through the opening provided separately. These vapors are then collected together to be enclosed in a container. These are used as fuel gas. They are combustible vapors. Major application is in cooking gas. The condensed gas, that is, oil is collected into the beaker. It is then separated from water using separating funnel. The separated oil is then stored and sent to the refinery. There it is further distilled into various products like petrol, kerosene, diesel, etc. through a fractional distillation tower according to its various boiling points. These then further are used for several applications like automobiles, engines, generators, ships, etc.

17.3.4 EXPERIMENTAL WORK

17.3.4.1 1ST TRIAL

- Operating time—1 h
- Raw material—40.19 g
- Oil obtained—55 ml
- Residue—4 g

17.3.4.2 2ND TRIAL

- Operating time—4 h
- Raw material—300 g
- Oil obtained—420 ml
- Residue—2 g

17.4 RESULTS

17.4.1 TESTS RESULTS

Test	ASTM	Sample	Petrol
Flash point	D93	47°C	43°C
Kinematic viscosity	D445	4.8345 mm^2/s at 35°C	2–4.5 mm^2/s at 40°C
Cloud point	D2500	15°C	About 10°C
Pour point	D97	8°C	3°C
Centrifuge	D4007	No phase separation	No phase separation

17.5 CONCLUSION

In this way, we can conclude that the waste plastics that we generally consider waste are not waste but an invaluable resource from which crude oil can be obtained. This technique solves world's two biggest problems for which it is not able to find a solution, that is, disposal of waste plastics and generation of crude oil. This will not only help the world to get rid of plastics waste but also get it converted to precious oil. This technology if triggered properly could change the present scenario of the world. It will also play an important role in changing the Indian economy as well.

KEYWORDS

- **pyrolysis**
- **waste plastics**
- **oil**
- **fuel**
- **energy**

REFERENCES

1. Zadgoankar, A. Eco-friendly Plastic Fuel. Conversion of Waste Plastic into Liquid Hydrocarbons/Energy—A Major Breakthrough in the Arena of Non-conventional Sources of Energy. Information Brochure and Technical Write-Up.
2. Panda, A. K.; Singh, R. K.; Mishra, D. K. Thermolysis of Waste Plastics to Liquid Fuel. *Renew. Sustain. Energy Rev.* **2009**, *743*, 1–16.
3. Bhatti, J. A. Current State and Potential for Increasing Plastics Recycling in US. College Report, Columbia University, 2010.
4. Central Pollution Control Board. An Overview of Plastic Waste Management. *J. Parivesh* **2013**, 139–144.
5. Garforth, A.; Lin, Y. H.; Sharratt, P. N.; Dwyer, J. Production of Hydrocarbons by Catalytic Degradation of High Density Polyethylene in a Laboratory Fluidized Bed Reactor. *Appl. Catal. A: Gen.* **1998**, *169*(2), 331–342.

6. Singh, B.; Sharma, N. Mechanistic Implications of Plastic Degradation. *Polym. Degrad. Stab.* **2008,** *93,* 561–584.
7. Pyrolysis of Rejects of a Waste Packaging Separation and Classification Plant. *J. Anal. Appl. Pyrol.* **2009,** 85, 384–391.
8. Microwave-induced Pyrolysis of Plastics Waste. *Ind. Eng. Chem. Res.* **2001,** 40, 4749–4756.

PART V
Modification of Inorganic Materials

CHAPTER 18

GROWTH OF KNbO$_3$ CRYSTALS AND THEIR APPEARANCE

NARESH M. PATIL[1*], VIVEK B. KORDE[1], and SANJAY H. SHAMKUWAR[2]

[1]*Department of Applied Physics, Laxminarayan Institute of Technology, RTM Nagpur, Nagpur 440033, India*

[2]*Arts, Commerce & Science College, Kiran Nagar, Amravati 444606, India*

[*]Corresponding author. E-mail: nmpatil70@gmail.com

CONTENTS

Abstract 334
18.1 Introduction 334
18.2 Experimental Studies 335
18.3 Physical Appearance 335
18.4 Conclusion 336
Keywords 337
References 337

ABSTRACT

The doped $KNbO_3$ single crystals have been synthesized using flux method by taking K_2CO_3 and Nb_2O_5 in the molar ratio of 1:2:1 with an impurity of Al_2O_3 (25 and 50 mg). Slow heating and cooling techniques were adopted to get good quality single crystals. The good quality single crystals were obtained by this method. The size of the crystal is about 0.4–0.6 cm in length with very small thickness (i.e., 0.1–0.2 cm). The crystals are partially transparent which are suitable for optical properties. The color of the grown crystals is silver black. The crystals are irregular in shape and hard to cut. It has smooth cleavage. The details of growth technique will be given separately in another chapter. The study of characterization and their dielectric properties is under process.

18.1 INTRODUCTION

Recently, much attention has been given to the research on new lead free ferroelectric material, because of the concern regarding the detrimental effect of lead on environment.[1] Such material are $BaTiO_3$, $KNbO_3$ (KN), $Ba_5Ti_2O_7Cl_4$,[2] etc. potassium niobate $KNbO_3$ are considered to be promising candidate for lead free ferroelectric material,[3] as they have strong room temperature ferroelectricity and high Curie temperature.[4] The KN has an orthorhombic symmetry at room temperature and undergoes phase transition at −10, 225, and 425°C from rhombohedral → orthorhombic → tetragonal → cubic, respectively.[5] The lead free complex perovskite, KN single crystals are promising materials for technological applications as high permittivity capacitors, ferroelectric devices, electro-optic devices, actuator, sensors, transducers.[6,7] Therefore, it is of special interest due to very large electro-optic properties and good dielectric properties.[8]

In this chapter, we study about the growth technique of single crystals, and their interesting properties which appear due to addition of external impurity (Al_2O_3), further detailed study of synthesis and physical appearance is given in this chapter.

18.2 EXPERIMENTAL STUDIES

18.2.1 PREPARATION OF SAMPLE

Doped $KNbO_3$ single crystals were prepared by flux method. The appropriate amount of the composition of AR grade potassium carbonate (K_2CO_3), niobium pentaoxide (Nb_2O_5), and aluminum oxide (Al_2O_3) are taken in the molar ratio of 1.2:1 with an impurity of Al_2O_3 (25 and 50 mg) separately. Then the mixture was ground together in a mortar for 5–6 h, and in last stage, the mixture were taken in two separate crucible, that is, Al_2O_3 (25 mg) and Al_2O_3 (50 mg) separately. The total weight of the mixture was 10 and 20 g, respectively.

18.2.2 GROWTH MECHANISM

In growth mechanism, the mixture was placed in to a 50-ml flat-bottomed platinum crucible covered with a platinum lid and introduced in a programmable furnace for crystal growth (made by Electronic Equipment Company, Mumbai) at 1080°C. The mixture was heated in furnace up to 1100°C. The mixture was soaked at this temperature for 24 h and cooled slowly at the rate of 20°C/h up to 900°C. It was reheated till 1000°C and kept at this temperature for 18 h. Finally, the crucible was cooled at the rate of 20°C/h till room temperature and the large sized single crystals were obtained. The reheating mechanism is important in order to get large-sized single crystals. Conventional method supports stray nucleation due to which small crystallites develops. The reheating is expected to redissolve the number of small crystals that might be nucleated initially as a result of stray nucleation. In the reheating process, the smaller crystallites dissolve rapidly, while the larger crystallites are also attacked and get reduced in size. Gradually, while during re-soaking and recooling process the crystallites in the solution acts as a crystal growth centers or as seeds for crystal growth and large crystal growth take place.

18.3 PHYSICAL APPEARANCE

The larger single crystals were obtained by this method. The size of crystal is about 0.4–0.6 cm in height, having thickness 0.1–0.2 cm. The crystals are semitransparent with smooth cleavage. The KN crystals doped with Al_2O_3 (25 mg) and Al_2O_3 (50 mg), are labeled as A and B, respectively, for the discussion. The color of sample labeled as A is silver black color and for the sample labeled as B is white with blackish tints which are prominent throughout the sample. The coloration of crystals A and B are presented in Table 18.1. Technological conditions of crystal growth are also presented in Table 18.1. Both the samples are brittle in strength. In fact, the crystal breaks even by putting a nominal pressure on it.

TABLE 18.1 Comparison of Crystals.

Crystal	Coloration	Remarks
A	Silver black color	Semitransparent with smooth cleavage
B	White blackish color	Semitransparent with smooth cleavage

The physical properties to cover a broad range of technological important dielectric, piezoelectric, ferroelectric, and electrical properties, carrying out further systematic studies, with varying composition and preparative conditions, appropriate materials for different industrial and technological application can be developed out of these systems.

18.4 CONCLUSION

In this chapter, we have studied growth technique of Al-doped $KNbO_3$ single crystals. The properties of these crystals are strongly affected by the technique of crystal growth as well as by adding external impurities, that is, dopants. The doping may cause coloration of crystals, generation of both point defects and extended defects. It also studied about physical appearance of the grown crystals. Grown crystals are semitransparent in nature, so it has good optical property.

KEYWORDS

- KNbO$_3$ single crystal
- physical appearance
- doping effect
- growth technique
- memory devices.

REFERENCES

1. Chen, C.; Jiang, X.; Li, Y.; Wang, F.; Zhang, Q.; Lao, H. *J. Appl. Phys.* **2010**, *108*, 124106.
2. Ingle, S. G.; Pati, N. M. *Jpn. J. Appl. Phys.* **2000**, *39*, 2670–2674.
3. Wongasaenmai, S.; Laosiritaworn, Y.; Ananta, S.; Yimnirun, R. *Mater. Sci. Eng. B* **2006**, *128*(B), 83–88.
4. Masuda, I.; Kakimoto, K.-I.; Ohsata, H. *J. Electroceram.* **2004**, *13*, 555–559.
5. Kimura, H.; Zhao, H.; Tanahashi, R. *Crystals* **2014**, *4*, 190–208.
6. Ko, J.-H.; Kim, D. H.; Kajima, S.; Chen, W.; Ye, Z.-G. *J. Appl. Phys.* **2006**, *100*, 066105.
7. Laishram, R.; Thakur, O. P.; Bhattacharya, D. K. *Mater. Sci. Eng.* **2010**, *172*(B), 172–176.
8. Kumar, S.; Varma, K. B. R. *Mater. Sci. Eng.* **2010**, *172*(B), 177–182.

INDEX

A

Additives, 311
  eco-friendly, 312–313
  PHA used for making molding parts, 318
  poly-hydroxyl alkanoates
    properties and value of, 317
  results from experimental work
    applications, 317–318
    density, 313–314
    flexural strength, 315–317
    innovation, 317
    melt flow index, 315
  TPKSP
    density of composites at different loading, 314
    flexural strength of, 316
    MFI of, 315
  types of, 312
  UPKSP
    density of composites at different loading of, 314
    flexural strength of, 316
    MFI of, 315
Adsorption of reactive blue 21 (RB 21) on fly ash, 270
  adsorbents, characterizations of
    Brunauer–Emmett–Teller (BET), 276
    SEM image of MnO_2-coated fly ash, 277
    SEM image of raw fly, 277
    X-ray fluorescence spectrometry, 276
  details of
    chemical used, 272
    formula weight, 272
    molecular formula, 272
    structure and color CAS, 272
  materials and methods, 271–272
    adsorbent, characterization of, 273
    adsorption of, 273–276
    treatment of, 272–273
  optimum dosage and operating conditions for, 286
  results and discussion
    adsorbents, characterizations of, 276–277
Al-doped SnO_2/PANI, 124–125
  materials and methods
    preparation of, 126–127
  results and discussion
    hydrogen gas sensing, 131–132
    resistance for hydrogen gas, mechanism of change in, 132–134
    SEM, 128
    UV-vis spectroscopy, 128–129
    XRD, 129–130
Aloe vera, 238
  acemannan, 239
  properties of, 240
  treating osteoarthritis
    components, beneficial in, 239–242
    therapeutic role in, 255–257
Aromatic polybenzimidazoles (PBIs), 56
  DMFC uses, 67–69
    paper published from 1995 to 2015, 67
  H_2-PEMFCs and DMFCs, energy density of batteries, 57
  membranes, formation of, casting from
    alkaline solution, , 61
    dimethyl acetamide, 60–61
    trifluoroacetic acid and phosphoric acid, 61

properties of
    acid doping, 62–64
    conductivity, 63, 65
    mechanical strength, 64–65
    membrane degradation, 66
    methanol crossover, 66–67
    thermal and chemical stability, 65–66
    water uptake and conductivity, 61–64
synthesis, 58
    catalytic polymerization, 60
    solution polymerization, 60
    thermally rearranged polymerization, 58–59

B

Batch adsorption study, 273
  effect of
    contact time and temperature, 278–280
    dye solution pH, 278
    initial dye concentration, 280–281
  isotherm, 281–283
  operating conditions and effect
    contact time and temperature, 274
    dye solution pH, 274
    initial concentration and adsorbent quantity, 274–275
    isotherm, 275
Bone morphogenetic protein-2 (BMP-2), 253

C

Carbon nanotube (CNT), 184
  conducting paints with, 197–198
  dispersion of, 190–191
    characterization, 194–195
    criteria for, 191
    mechanical, 191–193
    organic solvents in, 194
    surfactants use, 193
    water-soluble, 193–194
  MWCNTs, 185
  SWCNTs, 185
  synthesis of
    CVD, 185–187
    functionalization, 187–190
Carbonaceous nanoparticles, 179
  activated carbon, 179–180
  graphene, 180
Chemical vapor deposition (CVD) method, 185
  catalytic, 185–186
  non-catalytic, 186–187
Chitosan-based modified bio-adsorbents
  advantages, 215
  limitations, 215
Chronic cadmium toxicity, 4
  harmful effects of Cd, 4–5
  ion-exchange process, 5

Color-change fibers
  SPs, applications of, 84–85
Column study, 275–276, 283
  adsorption of reactive blue 21
    concentrations in media mixture of fly ash and, 284
    concentrations in media mixture of MCF, 285
Conductive paints, 184
  spraying, 200–201
Core-shell nanoparticles (CSNPs), 96

D

Denture base acrylic resin, 44
Denture stomatitis acrylic resin, 42
  Candida albicans, 42, 43, 44
  QAMS, 46
Direct methanol fuel cells (DMFCs), 57
  protonic electrolyte-based, 57
Doped $KNbO_3$, 334
  crystals, comparison of, 336
  growth mechanism, 335
  physical appearance, 335–336
  sample preparation, 335
Dyeing process, 270

Index

F

Flame retardancy, nanocomposite, 149
Food packaging, 138
  biopolymer nanocomposites, role of, 159–160
  intelligent/smart packaging polymer nanocomposite, 156–157
    indicators, 158–159
    sensor, 157–158
  polymer nanocomposites, 151
    active, testing of, 152–156
    improved, 152–153
    scope for research in, 161
Functionalization
  chemical, 187–188
  physical
    endohedral method, 189–190
    polymer wrapping, 188
    surfactant adsorption, 188–189

H

Heat-cured systems
  experimental process, 47–48
    color retention of denture, 48
    decolorization of denture, 48
    microbial analysis, 49–50
    QAS dose optimization studies, 49
  materials and methodology, 47
  results and discussion
    incorporation of QAS, 50–52
Heat deflection temperature (HDT), 148
Heavy metals, 208–209
  adsorption on hydrogels, 216–219
  MCL standards for, 210–211
  modified bio-adsorbent
    chemical reaction, prepared by, 213–215
    starch-based, 213
  modified biopolymers, adsorption on, 211
    chitosan-based modified bio-adsorbents, 212–213
  sources in industrial wastewater, 209–210

I

Indicators
  freshness, 158
    time and temperature, 159
Insulin-like growth factor-1 (IGF-1), 254
Ion-exchange process, 5
  analysis of metal ion uptake
    at different electrolytes, 7
    at different pH, 8
    rate of, 8
  effect of metal ion uptake in different
    electrolytes with variation in concentrations, 11–15
    electrolytes with variation in rate, 17–19
    pH, 15–17

L

Lead free ferroelectric material, 334
Liquid detergent
  preparation of
    analysis of, 232
    compositions of, 231
    physico-chemical properties of novel biopolymer, 232
    stain removing properties, 233
Liquid paint
  electrical resistivity, test methods for, 198–199
    method A, 199
    method B, 200

M

Matrix metalloproteinases (MMPs), 239
Medical textiles
  SPs, in the field of
    applications of, 85
MnO_2-coated fly ash (MCF), 270
  characteristics of, 286
Multiwall carbon nanotubes (MWCNTs), 185

N

Nanocomposite, 140

applications of, 150
characterization of, 145–146
food packaging
  correlation between, 151
  layered silicate, 141–142
    modification of layered silicate, 143
    structure of, 142
  polymer matrix, 141
  preparation of, 140, 144
    intercalation of polymer from solution, 145
    melt intercalation, 144–145
    in situ intercalative polymerization, 144
    in situ template synthesis, 144
  properties of
    barrier, 147
    flame retardancy, 149
    mechanical, 146–147
    optical, 149
    rheological, 148–149
    thermal, 147–148
  thermal properties
    coefficient of thermal expansion, 148
    HDT, 148
    thermal stability, 147–148
Nanofibers
  electrospinning apparatus, used for fabrication, 127
Nanofillers like carbon nanotubes (CNTs), 96
Nanoparticles
  carbonaceous, 179
    activated carbon, 179–180
    graphene, 180
  metal-containing
    oxide nanoparticles, 177–178
    size and shape, 177
  zeolite, 178
Nanotechnolgy and nanomaterials, 94, 175
  dendrimer polymers
    silica-based PAMAM, 177
  dendritic polymers, 176–177

nanoparticles, 176
wastewater treatment, 176
Novel biopolymer
  composition of, 226
  result and discussion, 234–235
Novel copolymer
  chronic cadmium toxicity, 4
    harmful effects of Cd, 4–5
    ion-exchange process, 5
  copolymer composites, 5, 6
  copolymer resin, 5
Novel copolymer and composite
  analysis of metal ion uptake
    at different pH, 8
  analysis of rate of metal ion uptake, 8
  p-HBSF-II copolymer
    effect of metal ion uptake by, 12
    effect of $NaClO_4$ electrolyte on metal ion uptake by, 14
    effect of $NaNO_3$ electrolyte on metal ion uptake by, 14
  experimental
    materials, 6
    p-HBSF-II copolymer, synthetic route of, 7
    synthesis of, 6
  experimental analysis of metal ion uptake
    at different electrolytes, 7
  metal ion uptake at different pH by
    copolymer and composite, 16
    p-HBSF-II composite and copolymer, 17
  rate of metal ion uptake by
    p-HBSF-II composite, 18–19
    p-HBSF-II copolymer, 18–19
  results and discussion
    elemental analysis, 8
    FTIR spectral analysis, 9
    ion-exchange studies, 11–19
    NMR spectral analysis, 10
    surface, 10–11
    thermal degradation, kinetics of, 21–25
    thermal studies, 19–21

Index 343

O

Osteoarthritis
  Aloe vera
    pathway of action in osteoarthritic cartilage repair, 257
    therapeutic role in treating osteoarthritis, 255–257
  cartilage repair, factors contributing to, 250–251
    growth factors, 251–254
    macrophages, 251
    Wnt-signaling pathway, 255
  chondrocyte cells, 243–244
  collagen, 244
  disease, progression of, 245–246
    bradykininf, 250
    complement cascade, pathway of, 248
    complement system, 247
    inflammatory response, 248–249
    MMPs, effect of, 250
    normal and osteoarthritic joint, difference between, 246
    structural changes, 246–247
  proteoglycan, 244–245
  structure of cartilage, 242–243
Osteogenic protein-1/bone morphogenetic protein-7 (OP-1)/BMP-7, 253–254

P

Poly(aniline-*co*-*N*-phenylaniline) copolymers (PANI-*co*-PNPANI), 30
  characterization, 32
  synthesis of, 31–32
Poly(methyl methacrylate) (PMMA), 42, 43
Poly(*N*-phenylaniline) (PNPANI)
  experimental synthesis of, 31
  results and discussion, 32
    copolymers, repeat unit for, 33
    FTIR spectra of, 35
    FTIR spectral data of, 35
    homopolymers, repeat unit for, 33

    temperature dependence of, 37
    transport properties of, 37
    UV-Vis absorption bands of, 33
Paint preparation
  binder, 195–196
  HLB number, 196
  pigments, 196
  solvent serves, 196
Palm kernel nut shell (PKNS), 310
Plastic, 322
  experimental work trials, 327
  laboratory process
    equipment, 325
    process, 325–327
    process conditions, 325
  products, 322
  pyrolysis
    catalytic, 324
    thermal, 323–324
    thermal degradation, mechanism of, 324
  tests results, 327
Platelet-derived growth factor (PDGF), 252
Polyaniline (PANI), 30
  experimental synthesis of, 31
  issue with, 31
  results and discussion
    copolymers, repeat unit for, 33
    FTIR spectral data of homopolymers and copolymers, 35
    homopolymers, repeat unit for, 33
    temperature dependence of, 37
    transport properties of, 37
Polymer electrolyte membrane fuel cells (PEMFCs), 57
Polymer nanocomposites (PNCs), 94, 139
  applications, 112–114
    automobile sector, 115
    barrier packaging and sports goods, 114
    coatings, 115
    energy storage devices and sensors, 114

low-flammable products, 115
transparent materials, 114–115
preparation methods
    in-situ polymerization, 100, 102
    melt mixing and compounding, 98–99
    melt-extrusion processing, 97–98
    microemulsion method, by, 101
    microemulsion polymerization, 100
    solution blending, 99
    solution casting, 103
    ultrasound-assisted polymerization, 102
properties of
    electrical, 111–112
    flame-retardant, 108
    mechanical, 103–104
    physical, 110
    rheological, 110–111
    surface morphological, 104–107
    swelling index (SI), 109–110
    thermal, 108–109
types
    bio-material, 96–97
    conducting material, 96
    core-shell material based, 97
    elastomeric nanocomposites, 95
    inorganic nanomaterial-filled, 95
    natural filler-based polymer composites, 95
    polymer nanoblends, 96
Polymeric surfactants, 224
    biopolymer, preparation of
        composition of, 226
        -COOH groups, calculation for, 227
        -OH groups, calculation for, 227
        reactor programing, 228–229
        stiochometrical, calculation for individual ingredients in, 226
    physicochemical analysis, methods of
        acid value, 229
        hydrophilic lipophilic balance, 231
        molecular weight viscosity-average method, 231
        saponification value, 230
        solid content, 230
        volatile content, 230
    raw material analysis
        coconut oil, 225
        maleic anhydride and phthalic anhydride, 225
        rosin, 225
        sorbitol, 225
Pristine polymer materials, 138
Protective paint systems, 196
Pyrolysis, 323
    catalytic, 324
    thermal, 323–324
    thermal degradation, mechanism of, 324

Q

Quaternary ammonium compounds (QACs), 43
Quaternary ammonium silane-functionalized methacrylate (QAMS)
    hexadecyl trimethyl ammonium bromide, 46–47

R

Raw fly ash, characteristics, 286
Reactive dyes, 270
    pollutants in wastewater, separation of, 271

S

Sensors
    gas, 157–158
        biosensors, 158
        florescence-based oxygen sensor, 158
Single-wall carbon nanotubes (SWCNTs), 185
Skincare textile
    SPs, in the field of
        applications of, 85–86
Smart polymers (SPs), 76
    applications of
        biomaterial, 77–80
        optical data storage, 86–88

Index 345

textile applications, 82–86
tissue engineering, 80–82
as biomaterial, 77
  drug delivery systems, 78
  glucose sensors, 78–79
  pH-sensitive polymers, 78
  temperature-responsive polymers, 79–80
classification, 77
Solar paints, 184
Starch-based modified bio-adsorbents, 216
Synthetic acrylic resins, 42
  PMMA, 43
  QACs, 43
Synthetic polymers, 76

T

Tin oxide/polyaniline (SnO_2/PANI), 124–125
  materials and methods
    preparation of, 126–127
  results and discussion
    hydrogen gas sensing, 131–132
    resistance for hydrogen gas, mechanism of change in, 132–134
    SEM, 128
    UV-vis spectroscopy, 128–129
    XRD, 129–130
Thermal degradation
  kinetics of, 21
    Freeman–Carroll plot (n), 24–25
    parameters of p-HBSF-II and composite, 22
    Sharp–Wentworth plot (E_a), 23
Thermogravimetric analysis, 19
  p-HBSF-II copolymer composite
    TGA data of, 21
    thermograms of, 20
Transforming growth factor β (TGF-β), 252–253
Treated palm kernel nut shell powder (TPKSP), 314

U

Untreated palm kernel nut shell powder (UPKSP), 314

W

Water pollution, 174
  nanotechnolgy and nanomaterials, 175–177
Wood-plastics composites (WPCs)
  advantages, 310–311
Wound dressing textile
  SPs, in the field of
    applications of, 86

Z

Zeolites, 292
  commercial method for production of, 296
  environmental applications of
    chemical/petrochemical industries, catalyst in, 305
    eutrophication on fish, effect of, 302
    ion exchanger/water softener, 302
    ion-exchange and regeneration properties of, 302
    phosphate-free detergent builder, 301–302
    separation of molecules, membrane for, 304
    sorption of CO_2 capture and VOCS, 303
    sorption of heavy metals and anionic pollutants, 303
    zeolite-A-based detergents, 301
  framework structure of, 293
    chemical structure, 294
    classification of, 294
    primary building unit of, 294
  industrial applications, based on, 293
  nanoparticles, 178
  nomenclature, 294–295
    topologies of, 295

surface modification of, 304
synthesis of, 295–296
synthesis, using waste sources
    bagasse fly-ash-based zeolite 4A, 299
    fly-ash-based zeolite 4A, 297–298
    rice-husk-ash-based zeolite 4A, 298–299

PGSTL 01/25/2018